Reports of China's Basic Research

Editor-in-Chief

Wei Yang, National Natural Science Foundation of China, Beijing, China
Zhejiang University, Hangzhou, Zhejiang, China

The National Natural Science Foundation of China (NSFC) was established on February 14, 1986. Upon its establishment, NSFC was an institution directly under the jurisdiction of the State Council, tasked with the administration of the National Natural Science Fund from the Central Government. In 2018, it became managed by the Ministry of Science and Technology (MOST) but kept its due independence in operation. Since its establishment, NSFC has comprehensively introduced and implemented a rigorous and objective merit-review system to fulfill its mission of supporting basic research, fostering talented researchers, developing international cooperation and promoting socioeconomic development.

Featuring science, basics, and advances, the series of *Reports of China's Basic Research* is organized by the NSFC to present the overall level and pattern of China's basic research, share innovative achievements, and illustrate excellent breakthroughs in key fields. It covers various disciplines including but not limited to, computer science, materials science, life sciences, engineering, environmental sciences, mathematics, and physics. The series will show the core contents of the final reports of the Major Programs and the Major Research Plans funded by NSFC, and will closely follow the frontiers of basic research developments in China.

If you are interested in publishing your book in the series, please contact Qian Xu (Email: violetta_xuqian@zju.edu.cn) and Mengchu Huang (Email: mengchu.huang@cn.springer.com).

Weicheng Fan
Editor

Unconventional Emergency Management Research

ZHEJIANG UNIVERSITY PRESS
浙江大学出版社

Springer

Editor
Weicheng Fan
Hefei Public Safety Research Institute
Tsinghua University
Hefei, Anhui, China

ISSN 2731-8907 ISSN 2731-8915 (electronic)
Reports of China's Basic Research
ISBN 978-981-99-6797-1 ISBN 978-981-99-6798-8 (eBook)
https://doi.org/10.1007/978-981-99-6798-8

Jointly published with Zhejiang University Press
The print edition is not for sale in China (Mainland). Customers from China (Mainland) please order the print book from: Zhejiang University Press.

Translation from the Chinese language edition: "非常规突发事件应急管理研究" by Weicheng Fan, © Zhejiang University Press 2018. Published by Zhejiang University Press. All Rights Reserved.

This Springer imprint is published by the registered company Springer Nature Singapore Pte Ltd.
The registered company address is: 152 Beach Road, #21-01/04 Gateway East, Singapore 189721, Singapore

Editorial Board

Preface to the Series

As Lao Tzu said, "A huge tree grows from a tiny seedling; a nine-storied tower rises from a heap of earth." Basic research is the fundamental approach to fostering innovation-driven development, and its level becomes an important yardstick for measuring the overall scientific and national strength of a country. Since the beginning of the twenty-first century, China's overall strength in basic research has been consistently increasing. With respect to input and output, China's input in basic research increased by 14.8 times from 5.22 billion yuan in 2001 to 82.29 billion yuan in 2016, with an average annual increase of 20.2%. In the same period, the number of China's scientific papers included in the Science Citation Index (SCI) increased from lower than 40,000 to 324,000; China rose from the 6th to 2nd place in global ranking in terms of the number of published papers. In regard to the quality of output, in 2016, China ranked No. 2 in the world in terms of citations in 9 disciplines, among which materials science ranked No. 1; as of October 2017, China ranked No. 3 in the world in the number of both Highly Cited Papers (top 1%) and Hot Papers (top 0.1%), with the latter accounting for 25.1% of the global total. In talent cultivation, in 2006, China had 175 scientists (136 of whom from the Chinese mainland) included in Thomson Reuters' list of Highly Cited Researchers, ranking 4th globally and 1st in Asia.

Meanwhile, we should also be keenly aware that China's basic research is still facing great challenges. First, funding for basic research in China is still far less than that in developed countries—only about 5% of the R&D funds in China are used for basic research, a much lower percentage than 15%–20% in developed countries. Second, competence for original innovation in China is insufficient. Major original scientific achievements that have a global impact are still rare. Most of the scientific research projects are just a follow-up or imitation of existing research, rather than groundbreaking research. Third, the development of disciplines is not balanced, and China's research level in some disciplines is noticeably lower than the international level—China's Field-Weighted Citation Impact (FWCI) in disciplines just reached 0.94 in 2016, lower than the world average of 1.0.

The Chinese government attaches great importance to basic research. In the 13th Five-Year Plan (2016–2020), China has established scientific and technological innovation as a priority in all-round innovation and has made strategic arrangements to strengthen basic research. General Secretary Xi Jinping put forward a grand blueprint of making China into a world-leading power in science and technology in his speech delivered at the National Conference on Scientific and Technological Innovation in 2016, and emphasized that "we should aim for the frontiers of science and technology, strengthen basic research, and make major breakthroughs in pioneering basic research and groundbreaking and original innovations" at the 19th CPC National Congress on October 18, 2017. With more than 30 years of unremitting exploration, the National Natural Science Foundation of China (NSFC), one of the main channels for supporting basic research in China, has gradually shaped a funding pattern covering research, talent, tools and convergence, and has taken action to vigorously promote basic frontier research and the growth of scientific research talent, reinforce the building of innovative research teams, deepen regional cooperation and exchanges, and push forward multidisciplinary convergence. As of 2016, nearly 70% of China's published scientific papers were funded by the NSFC, accounting for 1/9 of the total number of published papers all over the world. Facing the new strategic target of building China into a strong country in science and technology, the NSFC will conscientiously reinforce forward-looking planning and enhance the efficiency of evaluation, so as to achieve the strategic goal of making China progressively share the same level with major innovative countries in research total volume, contribution and groundbreaking researchers by 2050.

The series *Advances in China's Basic Research* and the series *Reports of China's Basic Research* proposed and planned by the NSFC emerge against such a background. Featuring science, basics and advances, the two series are aimed at sharing innovative achievements, diffusing performances of basic research and leading breakthroughs in key fields. They closely follow the frontiers of basic research developments in China and publish excellent innovation achievements funded by the NSFC. The series *Advances in China's Basic Research* mainly presents the important original achievements of the programs funded by the NSFC and demonstrates the breakthroughs and forward guidance in key research fields; the series *Reports of China's Basic Research* shows the core contents of the final reports of Major Programs and Major Research Plans funded by the NSFC to make a systematic summarization and give a strategic outlook on the achievements in the funding priorities of the NSFC. We hope not only to comprehensively and systematically introduce backgrounds, scientific significance, discipline layouts, frontier breakthroughs of the programs and a strategic outlook for the subsequent research, but also to summarize innovative ideas, enhance multidisciplinary convergence, foster the continuous development of research in concerned fields and promote original discoveries.

As Hsun Tzu remarked, "When earth piles up into a mountain, wind and rain will originate thereof. When waters accumulate into a deep pool, dragons will come to live in it." The series *Advances in China's Basic Research* and *Reports of China's*

Basic Research are expected to become the "historical records" of China's basic research. They will provide researchers with abundant scientific research material and vitality of innovation, and will certainly play an active role in making China's basic research prosper and building China's strength in science and technology.

Wei Yang
Academician of the Chinese Academy of Sciences
Beijing, China

Preface

China's public safety and emergency management have received more and more attention since the outbreak of SARS in 2003. The General Office of the State Council has set up a working group on emergency response plans, focusing on promoting the work of "One Plan and Three Sub-systems" (an emergency management plan with a response system, mechanism and legal system). In 2006, the State Council issued the National Emergency Response Plan for Public Incidents. In 2007, the Emergency Response Law of the People's Republic of China was promulgated and implemented. Besides, the State Council has issued a series of policy documents on emergency management in various fields. All of the reports on the 16th, 17th, 18th and 19th National Congress of the Communist Party of China (CPC) have emphasized the demand to improve the public safety and emergency management system. In the 23rd collective study of the Political Bureau of the CPC Central Committee, General Secretary Xi Jinping emphasized the demand to build an all-round and three-dimensional public safety net.

In view of the urgent need for national public safety and emergency management and the need for cutting-edge and basic science research, the National Natural Science Foundation of China (NSFC) launched the Major Research Plan of "Unconventional Emergency Management Research" (hereinafter referred to as the Plan) in 2009, following the principle of "definite objective, stable support, integration and promotion, and leap-forward development" and carrying out innovative research around major strategic fields and directions on emergency management. It focuses on summarizing scientific objectives, proactively promoting interdisciplinary communication and cultivating innovative talent through top-level design. A total of 121 projects were funded by the Plan, including 92 fostering projects, 25 key projects and four integrated projects, with a total funding of 120 million yuan. All the funded projects have been successfully completed by the end of 2017. In the past eight years, with the joint efforts of scientists from many disciplines, the Plan has focused on three key scientific issues: information processing and evolution modeling of unconventional emergencies, decision-making theory on unconventional emergencies, and psychological and behavioral response laws of individuals and groups in emergencies. It systematically applies the theoretical methods of management, informatics, psychology and other relevant disciplines. Through observation, experiment and theoretical innovation and comprehensive integration of related disciplines, a series of innovative research achievements have been made, which has helped the leapfrog development of China's unconventional emergency management and provided a solid scientific basis for national emergency management decision-making. At the same time, the Plan contributes to the establishment and construction of the discipline of Safety Science and Engineering and the China Association for Public Safety. It has also enhanced the reputation of China's public safety and emergency management in international academic circles.

In 2014, China formally established the National Safety Commission of the CPC. In July 2015, the 15th plenary of the Standing Committee of the 12th National People's Congress passed the National Safety Law of the People's Republic of China, which indicates that China attaches great importance to safety issues and has raised it to a new strategic height. At present, China is still at a stage where emergencies are prone to happen frequently. The task of maintaining safety is important and arduous, and scientific research on public safety and emergency management still has a long way to go.

In order to share the research achievements of the Plan with researchers, emergency management workers and the public who are concerned about public safety and emergency management, this book is published with the support of the Management Science Department of the NSFC, hoping to provide powerful support for the research and exploration in public safety and emergency management.

Weicheng Fan
Academician of the Chinese Academy
of Engineering
Beijing, China

Contents

Contributors

Weicheng Fan Tsinghua University, Beijing, China

Tiemin Liu China Academy of Safety Science and Technology, Beijing, China

Yi Liu Institute of Public Safety Research, Tsinghua University, Beijing, China

Xiaofeng Meng Renmin University of China, Beijing, China

Chunchang Shan Counsellor's Office of the State Council, Beijing, China

Feiyue Wang Institute of Automation, Chinese Academy of Sciences, Beijing, China

Lei Wang Peking University, Beijing, China

Shouyang Wang Academy of Mathematics and Systems Science, Chinese Academy of Sciences, Beijing, China

Wenguo Weng Institute of Public Safety Research, Tsinghua University, Beijing, China

Xiaolin Zhou Peking University, Beijing, China

Project Overview

Weicheng Fan, Chunchang Shan, Shouyang Wang, Tiemin Liu, Feiyue Wang, Xiaofeng Meng, Lei Wang, Xiaolin Zhou, Wenguo Weng, and Yi Liu

1 Project Introduction

The major research program of "Unconventional Emergency Management Research" (hereinafter referred to as the major research program) is the second batch of major research programs initiated by the National Natural Science Foundation of China during the Eleventh Five-Year Plan period. Since its official launch in February 2009, the major research program has funded 121 projects, including 92 cultivation projects, 25 key supported projects and 4 integration projects, with a total funding of 120 million RMB. All the funded projects have been successfully completed by the end of 2017.

W. Fan (✉)
Tsinghua University, Beijing, China
e-mail: wfan@tsinghua.edu.cn

C. Shan
Counsellor's Office of the State Council, Beijing, China

S. Wang
Academy of Mathematics and Systems Science, Chinese Academy of Sciences, Beijing, China

T. Liu
China Academy of Safety Science and Technology, Beijing, China

F. Wang
Institute of Automation, Chinese Academy of Sciences, Beijing, China

X. Meng
Renmin University of China, Beijing, China

L. Wang · X. Zhou
Peking University, Beijing, China

W. Weng · Y. Liu
Institute of Public Safety Research, Tsinghua University, Beijing, China

W. Fan (ed.), *Unconventional Emergency Management Research*, Reports of China's Basic Research, https://doi.org/10.1007/978-981-99-6798-8_1

Emergency refers to an incident that suddenly occurs and has a seriously negative impact on the life and property of the society and the public. It can be manifested as natural disasters, accidents, public health incidents and social security incidents. Unconventional emergencies refer to emergencies without enough precursors but with obvious complexity characteristics such as rarity, urgency of time, severity of consequences, etc., which are not easy to cope with by conventional management methods. Modern civilized society is actually a coupling system composed of human beings and nature, showing the typical characteristics of complex giant system, and the more complex the system is, the more fragile the system will be. In recent years, unconventional emergencies are happening with high frequency and in multi-fields. Under the current development goal of our country, how to improve the understanding of the characteristics and scientific laws of disasters and accidents under modern conditions, so as to enhance the ability to prevent and deal with emergencies and the ability to prevent and reduce disasters, is an important problem to be solved. It is to ensure the normal operation of state management and healthy development of society.

Since the 16th National Congress of the Communist Party of China, China's government has comprehensively enhanced emergency management. In January 2006, the State Council issued the "National Emergency Plan for Public Emergencies," and the national emergency plan system was basically built up; In August 2007, the 29th plenary of the Standing Committee of the Tenth National People's Congress passed the Law of the People's Republic of China on Emergency Response, and the legal system framework of emergency management with this law as the core was gradually formed; In 2007, the report of the 17th National Congress of the Communist Party of China requested that "adhere to safe development" and "improve the emergency management mechanism for emergencies"; The Government Work Report of the State Council in 2008 pointed out that it is necessary to "strengthen the construction of emergency response system and mechanism and improve the ability to prevent and deal with emergencies."

Before the establishment of this major research plan, the relevant research was mainly deployed at the technical and engineering levels, and less involved in the scientific issues of emergency management. There are many researches on "prediction-response" problems in academic circles, but they don't attach great importance to "scenario-response" problems. There is a lack of scientific and systematic research on emergency management to support response decision-making, and the research on interdisciplinary content is relatively weak. This major research plan takes emergency decision-making as the core, organizes interdisciplinary research, takes "national characteristics" as the basic parameter of unconventional emergency management research, and considers special event constraints-real-time, extreme environment, resource shortage, information shortage or overload, psychological pressure, conflict of interest and complex system structure, and integrates multidisciplinary research results into unconventional emergency management system, aiming at providing a solid scientific basis for national emergency management decision-making.

1.1 Project Deployment

Focusing on major projects, major scientific problems and large simulation platforms, this major research program realizes "integration, sublimation and leapfrogging," follows the basic guiding idea of "limited goal, stable support, integrated sublimation and leapfrog development," and embodies the purpose of "relying on experts, scientific management, relaxing environment and favorable condition to innovation," and constructs a sustainable platform and integrated system.

During 8 years of implementing this major research plan, the major research plan follows the established goals and propels the research plan in stages. (i) Project approval and research stage (2009–2011): the innovative idea of integrating while studying was implemented, and the integrated projects, cultivation projects and key supported projects were funded at the same time. (ii) Interim evaluation (2011): According to the suggestions of the interim evaluation, the adjustment plan for the research plan was put forward, which laid the foundation for the later project research. (iii) After the interim evaluation (2011–2017): The major research plan selected the research direction that played a decisive role in realizing the overall goal of the research plan. We also set up a general integration project, and provided financial support in the form of key supported projects for the current hot issues.

According to the research objectives and scientific problems, there are 121 projects funded by this major research plan, including 92 cultivation projects (including 53 general cultivation projects, 23 small-scale integrated platform projects, 6 newly added cultivation projects, 2 Yutian earthquake emergency projects, 1 series publication project and 7 final evaluation projects), 25 key supported projects and 4 integrated projects. Among them, cultivation projects and key supported projects are divided according to three core scientific reference indexes: information processing and evolution simulation of unconventional emergencies, emergency decision-making theory of unconventional emergencies, and psychological and behavioral reaction laws of individuals and groups in emergency situations.

1.2 Comprehensive Integration

This major research plan encourages different projects to carry out substantive interdisciplinary communication, set up small integration projects based on the research fruits of each project and combine the deployment of new projects with the ongoing research projects. It has set up four integrated projects: dynamic simulation of unconventional emergencies and integration of computing experimental systems; study on the integration of emergency plan and emergency preparation; integrated sublimation research on basic scientific issues of national emergency platform and system; research on sublimation of general integration of unconventional emergency management. Their layout and implementation ideologies are as follows.

(1) **Research on integration of dynamic simulation and computational experiment system of unconventional emergency**

On the basis of the research fruits of cultivation projects and key supported projects in the early stage of the major research plan, this integration project focuses on integration research, breaking through the scientific and technical problems related to dynamic simulation and calculation of unconventional emergencies, constructing an artificial social model, optimizing algorithms. And it is to carry out simulation deduction and make great progress in dynamic simulation and calculation experiments of unconventional emergencies. Specific research contents include the following aspects. (i) Stimulation method and technology on emergency management: In view of unconventional emergencies such as epidemic, public opinion incidents, nuclear, biochemical events, etc., multi-paradigm simulation method for emergency management is used to simulate. (ii) Artificial society model construction method and technology: To achieve this, we need to adopt the idea of model driven architecture, MDA), put forward formalizing and explicit design based on model framework, and realize models in relevant fields. (iii) Large-scale artificial society's computing and experiment method and technology: To achieve this, we need to create multi-client cloud service technology for parallel emergency management computing experiment, cloud simulation resource management and scheduling method for artificial society model, and heterogeneous computing acceleration method of central processing unit (CPU)/graphics processing unit (GPU) for large-scale artificial society model simulation have been proposed. (iv) Parallel system technology for emergency management: To achieve this, we need to create a series of parallel system technologies for emergency management, such as real-time sensing and processing of big data information, deep integration of heterogeneous data, social network analysis and association mining, and data mining in artificial society model's computing experiments. It is to realize information interaction between artificial society model and real society. (v) Unconventional emergency dynamic simulation and calculation experiment: To achieve this, we need to take calculation experiment, provide demonstration of evolution process of specific scenarios for decision-making of the general integrated platform. It is to realize scenario evolution, improve platform's function and performance and integrate models and data.

(2) **Research on integration of emergency plan and emergency preparedness**

In view of the lack of pertinence, practicability and operability in emergency response and preparation in China and the lack of emergency system's construction (risk assessment, infrastructure, emergency drills, etc.) of governments at all levels, this integration project is of great help. It is based on the general description of unconventional incidents obtained by simulation system's integration project, supplemented by the research of emergency decision-making theory. We focus on breaking through the relevant theoretical and scientific problems of emergency response plan system and emergency preparedness system, providing emergency response plan preparation and revision guidelines, and completing the verification of relevant decision theory, system and practice to realize great progress in emergency response plan and emergency preparedness. Specific research contents include the following aspects. (i) Theory and method of assessment on emergency risk and vulnerability: The system vulnerability analysis and security defense system design model are proposed, the multi-level network input-output model and emergency strategy are established, and the protection strategy and method of vulnerable system are put forward. (ii) Theory and method of scenario construction of serious and extraordinary emergencies: The theory and method of scenario construction of serious and extraordinary emergencies are put forward, and the evaluation index and evaluation method of emergency response plan are formed. (iii) The theory and method of optimizing and perfecting the emergency plan system: the framework and core elements of the emergency plan system, the evolution model of the emergency plan system, the digital management strategy of emergency plan, and the theory and method of protecting the city's critical infrastructure are put forward. (iv) Theory, method and model for effective operation of emergency plan: The evaluation method and model of emergency drill based on "scenario-task-ability" are built up. The dynamic allocation model of emergency resources under uncertain condition is put forward. The critical conditions, destabilization process and evolution mechanism of regional crowd system are put forward, and the prototype system of plan integration system platform with "scenario-ability-plan-deduction" as the main line is designed and developed.

This integration project integrates the research fruits of various aspects into the "Emergency Plan Integrated Operation and Display Platform" in the form of case base, knowledge base, database and model base, and it integrates, promotes and displays the research fruits. It also selects typical unconventional emergency scenarios to comprehensively display the research fruits of emergency preparedness and emergency plan system to show the possible effects of the research fruits on emergency management practice.

(3) **Research on integration and sublimation of basic science problems of national emergency platform system**

This integration project focuses on the major national demand for the construction of unconventional emergency platform, solving the basic discipline problems of the national emergency platform system. It closely focuses on core scientific issues, model deduction, data integration, case deduction and psychological behavior evolution. In order to build up "scenario-response" emergency decision-making theory and method. From the perspective of comprehensive integration sublimation research, it is built around network integration, computing integration and application system integration to build "scenario-response" emergency decision-making theory and method. The research focuses on three main fields: Data Integration, Decision-making Model and Psychological Behavior (i) Data Integration: it mainly integrates the emergency information management model of multi-channel (news, Weibo, social network) information monitoring and multi-stage early warning (identification of potential incidents and false information, incident evolution and early warning at critical time), the real-time mining model of network's visual data stream, the topic recognition and tracking, public emotion tendency analysis and public opinion early warning model, and the temporal text flow's emergency detection model, etc. (ii) Decision-making model: it mainly integrates the technical prototype system of the national emergency platform system, the research and application platform for comprehensive integration of water disaster emergency management, the inquiry and collective decision-making model of unconventional emergency response, the auxiliary decision-making support system for subway construction emergency, the reconstruction model and system for the evolution process of bioterrorism incidents, the prototype simulation system for flood emergency decision-making (taking the flood in the Three Gorges area as the background), a special epidemic emergency prototype system combining artificial society model with computational experiments and its corresponding model base, case base and knowledge base, a deduction model of emergency dynamic evolution and emergency handling, a digital emergency plan system for natural disaster emergency rescue, a situation assessment model and a prediction and evaluation system for social impact and situation development of unconventional emergencies, a large-scale crowd evacuation network model, a model of prediction, early warning and emergency decision-making based on incident chain and plan chain and a multi-party online consultation mode. It also includes large-scale regional evacuation model and simulation system, emergency decision-making model and system for deliberately-caused disasters, dynamic model of large-scale infectious disease transmission, scenario-oriented emergency decision-making method and model, dynamic prediction and decision-making method and model of unconventional emergencies based on case deduction, integrated emergency resource support model and emergency resource allocation simulation platform and emergency resource allocation optimization model, etc. (iii) Psychological behavior: It mainly focuses on stress psychological and relevant physiological response characteristics and stress-handling model, models of rumor propagation dynamics and group behavior's characteristics in complex network environment, individual risk perception, group risk perception and handling characteristics in sudden epidemic situation, individual resilience and organizational resilience models of rescue workers, etc.

(4) **Sublimation research on general integration of unconventional emergency management**

On the basis of the fruits of the above three integration projects, a general integration platform is built up to connect with the relevant platforms of the national emergency response system (such as the national emergency response platform and system, etc.), to make relevant systems continue to develop, integrate and verify the latest basic research fruits and directly serve the national emergency response system, so as to provide scientific basis, technical support and decision-making reference for the country to respond to unconventional emergencies scientifically, orderly and efficiently.

The project "Research on Integration of Unconventional Emergency Dynamic Simulation and Computational Experiment System" will provide the general integrated platform with typical scenario inversion and computational experiment results of unconventional emergencies, the project "Research on Integration of Emergency Plan and Emergency Preparedness" will provide the general integrated platform with a theoretical basis for response plans in decision-making, and the project "Research on Integration Sublimation of Basic Science Problems of National Emergency Platform System" will provide the general integrated platform with tools and implementation platforms for technical aspects in decision-making. Therefore, it is necessary to enhance the demand communication and content coordination among the above three integration projects. And it is necessary to focus on the universality and extensibility, standardization, sharing and interactivity of the integration platform and how to sublimate after integration, so as to achieve more effective coordination and comprehensive integration. Specific research contents include the following two aspects. (i) Comprehensive risk judgment of "Four in One." It is necessary to considering the basic theory of "incident chain-public opinion spread-psychological behavior" coupling feedback and the dynamic demonstration of the whole process of all risks, the whole process, all aspects and the whole society, the evolution process of physical risk and social risk, so as to confirm the key nodes and paths of risk occurrence, dissemination and diffusion. Through pre-intervention, in-process disposal and risk warning mechanism, a risk management mode covering all risks and the whole process, is formed. The risk information's transmission, physical transmission, policy transmission mechanism and risk evolution theory of unconventional emergencies are revealed, and the theory and method of comprehensive judgment of unconventional emergencies are established. (ii) Panoramic security management and collaborative decision-making paradigm. It is necessary to establish a method of multi-agent, multi-objective and multi-task collaborative decision-making and scenario deduction, reveal the cognitive-behavioral theory of the government, the public and other participants in the public security governance system, and establish an emergency management and collaborative decision-making method that integrates cyberspace and real space.

1.3 Interdisciplinary Situation

The object of unconventional emergency management is an open and complex giant system, which contains rich and complex scientific problems. This major research project needs the intersection and integration of management science, information science and psychological behavior science. Its theoretical basis is management science, especially emergency management, operational research, decision-making theory and method, as well as decision-making support system and technology. The basic theory of data integration is information science, including cloud computing, data and knowledge management, computational intelligence, data mining and management system. Psychobehavioral science embodies the combination of consciousness psychology and neuroscience, and pays attention to individual behavior from the perspective of psychology, pays attention to the factors affecting brain's function and mechanism from the perspective of neuroscience, and tries to collect relevant biological indications.

According to incomplete statistics, the disciplines involved in this major research plan include: public safety and crisis management; public management and public policy; management of information resources; decision-making theory and method; information system and management; evaluation theory and method; prediction theory and method; psychology and behavior management system and engineering; risk management techniques and methods; organizational behavior and organizational culture; information theory and information system; information processing methods and technologies; computer software; computer architecture; computer application technology; information security; computer network; system science and system engineering; epidemiological methods and health statistics; psychology; neurobiology; cognitive science; epidemiology of infectious diseases; probability theory and stochastic analysis; mathematical statistics; operations research; applying mathematical methods; fluid mechanics; basic physics; heat and mass transfer; combustion science; environmental engineering; traffic engineering; disaster prevention engineering; hydrology and water resources; geographic information system; sociology; jurisprudence; research on ethnic issues; linguistics; journalism and communication.

Therefore, this major interdisciplinary research plan has three characteristics: (i) There are many types of disciplines with great span; (ii) the application and development of some basic disciplines, such as the basic theories and methods of statistical physics; (iii) Some cutting-edge theories and methods are applied, such as social computing, artificial society model, complex network. Each project team of this major research plan consists of researchers with different professional backgrounds. These researchers give full play to their respective advantages and complement each other in project research. In this way, the multi-disciplinary and multi-sectoral cooperation provides favorable research conditions and broad research horizons for this major research project to carry out high-level research.

This major research plan is funded by the Management Science Department of NSFC (National Natural Science Foundation of China), and has close contact and intersection with the Life Science Department and the Information Science Department. This major research plan gathers the relevant research fruits of different disciplines in China, has initially established a new interdisciplinary direction: Emergency Management Interdiscipline (second-class discipline), and has propelled the establishment of the first-class discipline of "Security Science and Engineering" (Fig. 1), and has completed the first step of the research on emergency management from scattered research to systematic research.

1.4 Guidance Expert Group and Management Work Group

Guidance expert group and management work group are set up in this major research plan, as shown in Tables 1 and 2.

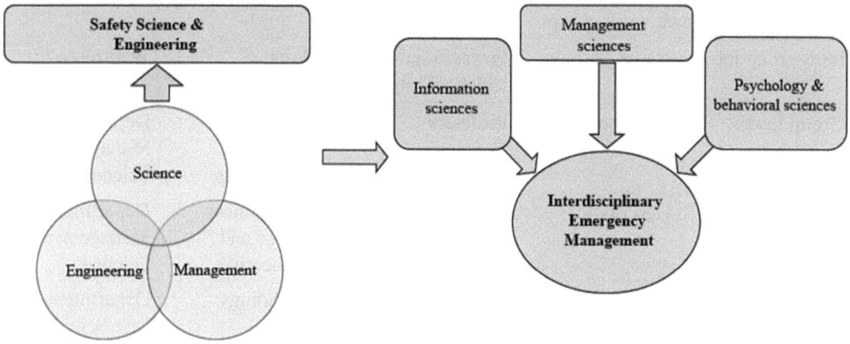

Fig. 1 Interdiscipline diagram

Table 1 Guidance expert group

Intra-group job	Name	Professional and technical job	Major	Workplace
Group leader	Fan Weicheng	Academician	Public security	Tsinghua University
Deputy group leader	Shan Chunchang	Counsellor	Emergency management	Counsellor's Office of the State Council
Member	Wang Shouyang	Researcher	Management and decision-making	Institute of Mathematics and Systems of Chinese Academy of Sciences
Member	Liu Tiemin	Researcher	Security science	China Academy of Safety Science and Technology
Member	Wang Feiyue	Researcher	Automatization	Institute of Automation, Chinese Academy of Sciences
Member	Meng Xiaofeng	Professor	Data process	Renmin University of China
Member	Wang Lei	Professor	Psychological behavior	Peking University
Member	Zhou Xiaolin	Professor	Psychological behavior	Peking University

Table 2 Management work group

Intra-group job	Name	Professional and technical job	Major	Workplace
Group leader	Li Yijun	Professor	Management science and engineering	Department of Management Science
Member	Gao Ziyou	Professor	Management science and engineering	Department of Management Science
Member	Cao Heqi	Associate researcher	Psychology	Department of Life Sciences
Member	Liu Ke	Professor	Data process	Department of Life Sciences
Member	Yang Liexun	Associate researcher	Management engineering	Department of Management Sciences
Member	Wang Qidong	Researcher	Environmental chemistry	Planning Bureau

2 Research Review

2.1 Overall Research Goal

The overall scientific objectives of this major research plan are as follows: (i) Integrating multidisciplinary observation, experiment and theoretical fruits, forming a profound scientific understanding of the objective laws of "monitoring, early warning and response command," which is the core link in emergency management, and providing scientific methods; (ii) Making breakthrough in the way of dealing with unconventional emergencies, building a theoretical system of "scenario-response"-oriented unconventional emergency management, and enhancing the independent innovation ability of emergency management science and technology; (iii) Improving the scientificalness of the national emergency management system (including emergency platform/pre-plan system), and providing decision-making reference for the country to respond to unconventional emergencies in an orderly and efficient manner; (iv) Building an interdisciplinary subject of emergency management to help it occupy an important position in the international emergency management science field and help cultivate creative talents for emergency management in China.

2.2 Core Research Issues

This major research plan is guided by the development situation and demand of national public security, closely combined with the major strategic planning of

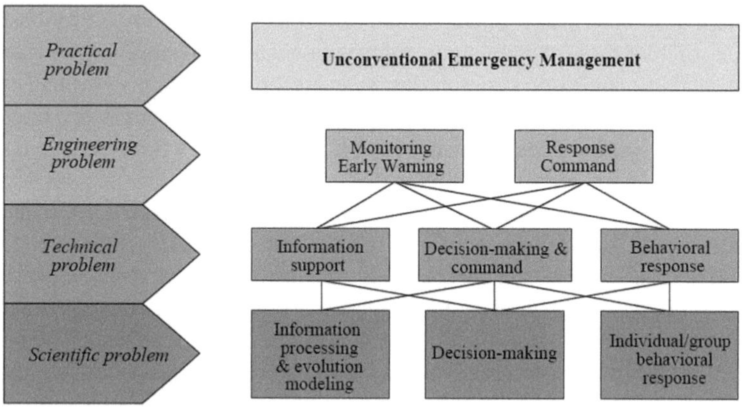

Fig. 2 Core scientific issues' framework

national security and public security, and it focuses on solving the common and key scientific problems of unconventional emergencies. It focuses on: the practical problems of emergency management of unconventional emergencies, overcoming the scientific difficulties at the technical and engineering levels, designing the core scientific problem framework of unconventional emergencies, injecting scientific connotation into the "shell resources" of the national emergency platform and plan system, and directly assisting China's emergency management practice. This major research plan, through observation, experiment and theoretical innovation and comprehensive integration of related disciplines, also focuses on three core scientific issues (the framework of it is shown in Fig. 2): information processing and evolution modelling of unconventional emergencies, emergency decision-making theory of unconventional emergencies, and psychological and behavioral response laws of individuals and groups in emergency situation.

(1) **Information processing and evolution modelling of unconventional emergencies**

The core scientific issue focuses on the possible precursors of the unconventional emergencies and the massive, heterogeneous and real-time data in the event evolution process. It studies the relevant scientific issues of collecting, acquiring, analyzing, disseminating, visualizing and sharing the information, and studies the non-traditional (such as data-driven and computational experiment-based) complex model construction theory and method for the evolution law of the unconventional emergencies. This core scientific issue can be divided into the following two aspects.

(i) Information processing of unconventional emergencies. It includes: theory and method for constructing a geospatial/social sensor network; uncertain and semi/unstructured information construction and mining methods for emergency management; theory of data cleaning, mining, dissemination, visualization and sharing in emergency management; new theory and methods for fast analysis of real-time and dynamic information and compensation of data loss; social data acquisition system's design and information monitoring and analysis method; theory and method of multi-platform heterogeneous data's integration management: incident reconstruction method based on multi-source dynamic information and relevant evolution law.

(ii) The modelling of evolution law of unconventional emergencies. It includes: early acquisition of symptom information and new theory of analysis; theory of relationship between abnormal data and unconventional emergencies' evolution: theory and method of incomplete information modeling for unconventional emergencies; identification, evaluation and action mechanism of influence factors during the process of unconventional emergencies; nonlinear dynamic laws of derivation, secondary emergency and coupling between unconventional emergencies; decoupling theory of complex systems formed by unconventional emergencies; complexity and vulnerability theory of urban lifeline system.

(2) **Decision-making theory of unconventional emergencies**

This core scientific issue studies the theoretical method of dynamic evaluation and decision-making contained in the on-site decision-making for unconventional emergency response; it studies the organization design, operation and evaluation theory and method of emergency decision-making command system, rescue/execution system and resource mobilization system; it also studies the technology platform of integrated decision-making support system for multi-event coupling and scenario construction. This core scientific issue can be divided into the following three aspects.

(i) Theories and methods of emergency assessment, judgment and decision-making. It includes: the interaction law of emergency decision-making and the evolution of unconventional emergencies; the whole process evaluation theory and method of unconventional emergencies; comprehensive judgment theory of unconventional emergencies; conflicting multi-objective, multi-stage, complex and dynamic emergency decision-making model; resource allocation model for unconventional emergencies; dynamic collaborative planning and real-time adjustment method based on complex tasks; information control and public communication and decision-making mechanism in unconventional emergency response.

(ii) Architecture design and operation optimization of emergency command system. It includes: topology expression of emergency command system and its completeness and process redesign; the effectiveness' evaluation theory and method of emergency decision system; theory and method of generation, adjustment and evaluation of emergency plan; design theory of flexible organization and coordination operation mechanism of multi-agent and multi-level rescue system: the multi-party emergency financing mechanism of fiscal fund-finance-donation and the dynamic and optimized raising mode for emergency funds; the rapid formation mechanism of the emergency supply chain and the methods and mechanisms for the disaster relief resources' collection and storage, coordination and supply, expropriation compensation and supervision over the use of the rescue resources of the whole society.

(iii) Theory and method of comprehensive support for emergency decision-making. It includes: theories of unconventional emergencies' scenario construction and decision-making simulation, models and simulation technology platform; New theories and technical methods of emergency decision-making support system based on unconventional emergencies; the design theory and implementation method of emergency decision-making drill's technology system; emergency decision-making's information system and heterogeneous public data platform's design basis; technical prototype and design method of emergency decision-making support system: software and hardware's integration theory and method of emergency decision-making support platform; design theory and implementation method of mobile emergency platform.

(3) **Psychological and behavioral responses of individuals and groups in emergency situations**

This core scientific issue studies the psychological and behavioral responses of some types of main participants, such as managers, rescuers and the public. They are observed as individuals and groups in emergency. This core scientific issue can be divided into the following two aspects.

(i) The action mechanism of pressure environment on individual psychology. It specifically includes: individual risk cognition, decision-making characteristics under unconventional emergencies and influencing factors on individual psychology; the influence of cultural differences on individual risk cognition and decision-making behavior: Individual's psychological and cultural characteristics of physical and emotional response to unconventional emergencies; the formation, evolution law and intervention of individual attitudes towards unconventional emergencies and related official handling; dynamic assessment and management strategies for the behavior and needs of disaster victims; the influence law of group identity on individual behavior under conflict condition: psychological quality/competence of managers, who need to deal with unconventional emergencies.

(ii) Group behavior in emergencies. It specifically includes: group structure characteristics and its evolution law; self-organization phenomenon and intervention mechanism in unconventional emergencies; group's social cognition and psychological behavior in unconventional emergencies: social psychological dynamic characteristics caused by unstable factors in social change; dynamic and optimal evacuation for ultra-large crowd under complex conditions; the process of strategy transformation and interactive construction of group decision; the influence of emergency decision-making's defect on group psychology and social stability.

3 Major Fruits of the Research

On the basis of the research fruits of the three core scientific issues, this major research plan's fruits have been integrated and sublimated and we have established a "risky government" model covering all risks and including the whole process, revealed the risks' evolution mechanism of unconventional emergencies, and formed short-term and medium- and long-term monitoring, forecasting and early warning methods and mechanisms; This paper reveals the evolution law of security incidents, predicts the situation and has formed a paradigm of panoramic security management and collaborative decision-making; It also reveals the cognitive-behavioral mechanism of the government, the public and other participants in the public security governance system, and it puts forward the "China Plan" in emergency management. This major research plan has propelled the leap-forward development in unconventional emergency management, which is mainly reflected in four aspects: unconventional emergency response, emergency response mode, management mechanism and influence scope (Fig. 3).

The basic research of unconventional emergencies provides a theoretical basis for the research and development of critical emergency handling technologies. The national key R&D plan of the 13th Five-Year Plan has deployed new emergency platform, national security platform and other technology research and development projects, and the national public health emergency platform system 2.0. The "China-WHO" public health emergency's action plan on science and technology innovation have also been put on the agenda.

The comparison of field development trends after the completion of the projects funded by this major research plan is shown in Table 3.

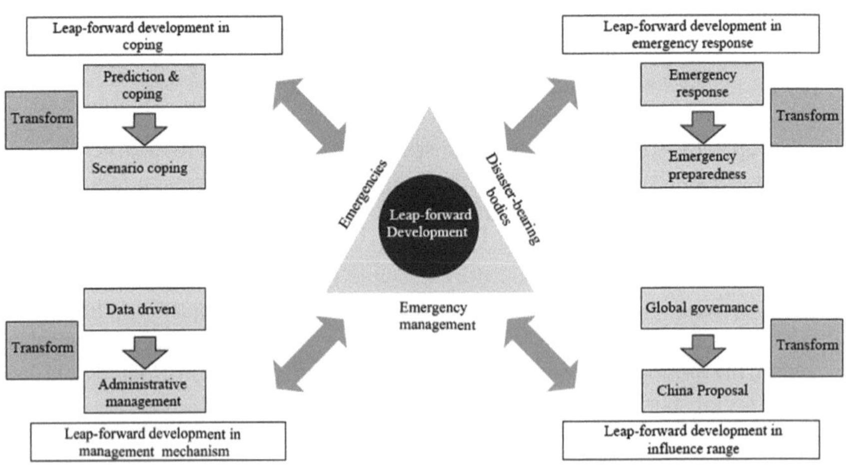

Fig. 3 The leapfrog development in unconventional emergency management

Table 3 Comparison of field development trends after the completion of the project

Core science issues	Subproblem	When the plan started	The domestic research status when the plan ended	The international research status when the plan ended	Advantages and gaps compared with international research status
Information processing & evolutionary modeling	Information processing of unconventional emergencies	The problems of data diversity and technical complexity in emergency management can't be solved	It develops and improves the theory and method of unconventional emergency's information processing and evolutionary modeling, formed a large-scale data cloud service system for unconventional emergency management, and integrated technology and methods of Internet of Things to build a big data storage and analysis system in the sensation of Internet of Things for emergency command	Foreign countries' researches on unconventional emergency information processing are more abundant than that of China, which provides technical support for big data processing of emergency management; Scientific issues such as data analysis, data dissemination, data visualization and data sharing in emergency management have received particular attention	The algorithm and software of the experimental system developed by China rank the top two in the world's top-level meetings. The developed platform can be applied to food safety's information monitoring and analysis. Its economic and financial public opinion's index analysis has reached the international advanced level as a whole

(continued)

Table 3 (continued)

Core science issues	Subproblem	When the plan started	The domestic research status when the plan ended	The international research status when the plan ended	Advantages and gaps compared with international research status
	Modeling of evolution law of unconventional emergencies	Unconventional emergencies have the characteristics of explosiveness, particularity, environmental complexity, and uncertainty in evolution. Therefore, it is unable to conduct a relatively reliable dynamic modeling of incidents and accurately reflect the evolution process of incidents	It puts forward the decision-making support theory, model, method and algorithm of unconventional emergency situation judgment based on multi-source crisis information's combination, which breaks through the method of topological structure calculation modeling and evolution law analysis of social network	Foreign researches focus on system analysis and modeling, integrated data management, sensor networks, robotics, data acquisition systems and monitoring biometrics, human and organizational behavior, with special attention to the coordination and integration of technology research and development projects	It has made systematic contributions to the scientific issues, basic theories and critical technologies of online perception of unconventional emergencies, and has made significant findings on the methodology of researching emergency and emergency response, and has made breakthroughs in basic theories and key algorithms, reaching the international advanced level as a whole

(continued)

Table 3 (continued)

Core science issues	Subproblem	When the plan started	The domestic research status when the plan ended	The international research status when the plan ended	Advantages and gaps compared with international research status
Emergency response	Theory and method of emergencies' evaluation, judgment and decision-making	It is unable to complete the information integration and expression of multiple disaster carriers, and it is difficult to predict the time variation law of disasters. It also lacks corresponding dynamic risk assessment methods and disaster response's decision-making and pre-assessment methods	Theoretical fruits are obtained in the aspects of temporal and spatial evolution law of emergency disasters, dynamic risk assessment, and pre-assessment of emergency plan, which are applied in China's national emergency platform system	There are richer theories and methods of emergency assessment and judgment in foreign countries than that in China, which can support decision-making to a large extent	A pre-evaluation system for disaster response and decision-making has been constructed, and a comprehensive prototype system for emergency's collaborative decision-making simulation has been built up. It has made some progress, but still lags behind compared with foreign countries
	Organization design and operation optimization of emergency command system	The formation process of emergency management system is a gradual progress, short-term and needs accumulation, and it is difficult to adapt to the new situation of frequent emergencies, compound and destructive increase	It has fundamentally propelled the transformation of China's current emergency management system to an open, collaborative and flexible standardized system, and has provided a theoretical support for comprehensively improving the ability and efficiency of China's emergency management	Foreign countries have richer experience in the research on the organization design and operation optimization of the emergency command system, which is related to national system of government and so on	It has constructed the emergency management system model, and in improving the operation efficiency of the emergency management system, it has reached the international leading level

(continued)

Table 3 (continued)

Core science issues	Subproblem	When the plan started	The domestic research status when the plan ended	The international research status when the plan ended	Advantages and gaps compared with international research status
Individual and group behavior and response	Action principle of pressure environment on individual psychology	The relevant research mainly comes from the case that individual breaks through the physiological limit to create the miracle of life in the disaster. The research on the characteristics and performance of physical resources and psychological system of individual in stress and depletion state and the research on the psychological process and law of individual and group in crisis situation are quite limited	It puts forward and verifies the theoretical model of the interaction efficiency of body and mind, which provides a new perspective for understanding the psychology and behavior of individuals and groups under crisis situation, and also provides new theoretical ideas and solid experimental evidence for crisis intervention	There are abundant and plentiful studies on stress response laws of human's psychological and physiological systems under limiting conditions in foreign countries. Thus, it provides methods for studying the influence of psychological characteristics on individual cognition	It reveals the rules of individual's physiological and psychological stress responses in emergency situations and makes certain progress in the research on the influence law of psychological characteristics in emergency situations on individual cognition and risk cognition. Compared with foreign countries, it still lags behind

(continued)

Table 3 (continued)

Core science issues	Subproblem	When the plan started	The domestic research status when the plan ended	The international research status when the plan ended	Advantages and gaps compared with international research status
	Group behavior in emergencies	Large-scale crowd's security and emergency management is relatively limited and research on the theory and method of large-scale crowd's evacuation risk, behavior monitoring and personnel tracking is limited	A large-scale crowd's evacuation behavior model is constructed from the macroscopic view to predict the large-scale crowd's evacuation risk under coupled disasters. We have developed an algorithm based on large-scale crowd's movement and behavior to detect abnormal behavior of large-scale crowd, which provides important data support for large-scale crowd security emergency management	Foreign countries have rich experience in dealing with large-scale crowd's gathering activities and there are more types of research fruits, which provide important technical support for risk prediction and emergency management	China's urbanization process started later than that in the world, and there are fewer studies on the response to large-scale group activities. It is in a state of parallel running with that of foreign countries

Domestic and International Research Situation

Weicheng Fan, Chunchang Shan, Shouyang Wang, Tiemin Liu, Feiyue Wang, Xiaofeng Meng, Lei Wang, Xiaolin Zhou, Wenguo Weng, and Yi Liu

Since the 16th century, there have been five scientific and technological revolutions in human history. On the one hand, the scientific and technological revolution has brought new tools and new methods, which have greatly changed our production and lifestyle, and gradually enhanced the ability of public security management. On the other hand, the revolutions have also brought new problems and challenges to public security. Emergency management starts with the government's setting up a special management organization or clarifying the emergency management responsibilities of the original related organizations. According to this standard, the history of emergency management can be divided into three stages: pre-emergency management period, emergency management's standardization period and emergency management's expansion period, as shown in Fig. 1.

W. Fan (✉)
Tsinghua University, Beijing, China
e-mail: wfan@tsinghua.edu.cn

C. Shan
Counsellor's Office of the State Council, Beijing, China

S. Wang
Academy of Mathematics and Systems Science, Chinese Academy of Sciences, Beijing, China

T. Liu
China Academy of Safety Science and Technology, Beijing, China

F. Wang
Institute of Automation, Chinese Academy of Sciences, Beijing, China

X. Meng
Renmin University of China, Beijing, China

L. Wang · X. Zhou
Peking University, Beijing, China

W. Weng · Y. Liu
Institute of Public Safety Research, Tsinghua University, Beijing, China

W. Fan (ed.), *Unconventional Emergency Management Research*, Reports of China's Basic Research, https://doi.org/10.1007/978-981-99-6798-8_2

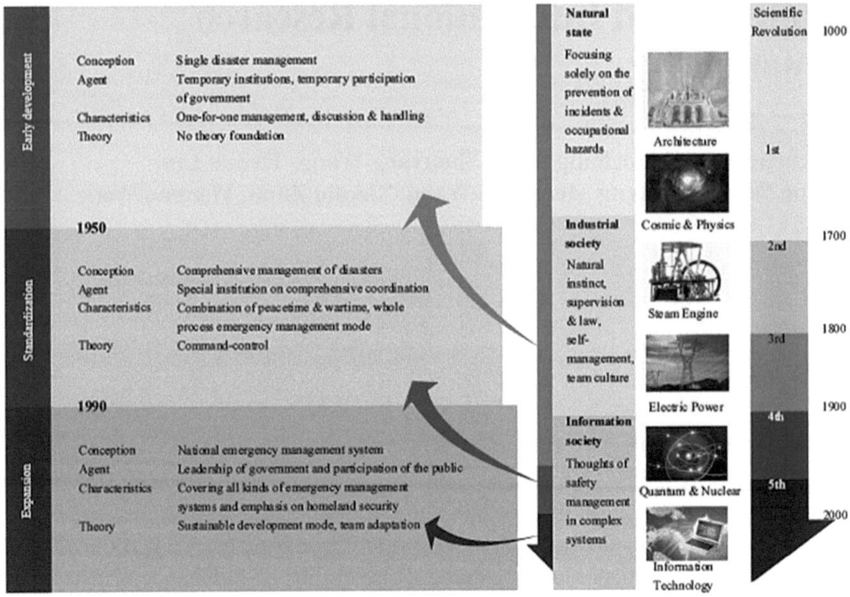

Fig. 1 The development history of emergency management

1 International Development History of Public Security and Emergency Management

Security is the premise and foundation of sound economic and social development and normal operation of state management in all countries of the world. Under the great impact of "9/11" incident in 2001, public security has been attached great importance by countries all over the world and has been raised to the height of national strategy. The United States, Britain, Japan, Germany and other countries have established emergency management and response systems, and formulated relevant bills and various emergency plans to ensure fast and prompt response to major emergencies.

1.1 The Development History of American Emergency Management System

American emergency management system is a typical representative of "comprehensive emergency model," and its emergency management system is both stable and flexible. After plentiful major disasters, the United States has gradually formed a relatively perfect emergency management system, which show high handling ability and efficiency in emergencies.

After the "9/11" incident, the American emergency management system has undergone three major adjustments (Table 1) and completed the transformation from "emergency response" to "emergency preparation." The motivation for each adjustment is that the emergency response capability is found to be insufficient in the major emergencies in the country or other countries.

On March 11, 2011, the Great East Japan Earthquake caused a strong shock to American society, and the US Senate soon held a hearing to consider how to deal with the risk of similar serious emergencies. On March 30, 2011, PPD-8 issued by US President Barack Obama replaced HSPD-8, which became a programmatic

Table 1 Three major adjustments of American emergency management system in recent years

Time	Motivation	Main contents	Result	Representative documents
2001	"9/11" Incident	Adjustment of national security strategy and emphasizing on national territorial security	On February 28, 2003, the No. 5 Homeland Security Presidential Directive 5 (HSPD-5): Management of Domestic Incidents was issued In March 2003, the United States Department of Homeland Security (DHS) was established. On December 17, 2003, the No. 8 Presidential Decree on Homeland Security (HSPD-8): National Preparedness was issued	National Incident Management System, NIMS; National Response Plan (NRP)
2008	Hurricane Katrina	Emphasizing the unified command in emergency management	In October, 2006, the Post-Katrina Hurricane Emergency Management Reform Act was passed, which brought great influence to the reform of the emergency organization structure, resource system, plan, command and other links of the US government (including federal, state, county and city) and even communities and various organizations	Post-Katrina Hurricane Emergency Management Reform Act; National Response Framework, (NRF)
2011	Great East Japan Earthquake	Strengthening national preparation and enhancing the construction of core capacity	Presidential Policy Directive No. 8 was issued on March 30, 2011 (PPD-8), the logic, pertinence and controllability of the emergency management system have been significantly enhanced; In November 2011, the United States launched the Strategic National Risk Assessment (SNRA), which carried out comprehensive, systematic and quantitative analysis and measurement of various risks in the national security field	PPD-8; National Preparedness Goal (NPG)

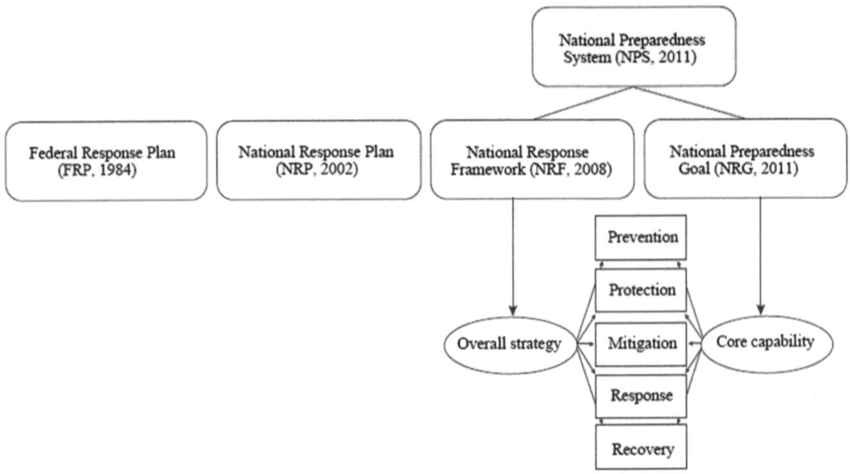

Fig. 2 Evolution of American emergency management's policy framework

document for developing and perfecting the national preparation system, and opened the prelude of building the "national preparation system."

The release of PPD-8 marks a new round of reconstruction of the emergency management system of the U.S. government, and formally puts forward the emergency management concept of "whole of community." Meanwhile, it takes National Preparedness as a basic strategy and Core Capability construction as the basic direction of emergency management system's construction. The establishment of emergency management framework, guided by the construction of core competence of emergency management, makes the logical system of American emergency management system clearer, more practical and rational.

The United States established the National Preparedness System (NPS) after 2011, and took the National Response Framework (NRF) as a part of the National Preparedness System. The National Response Framework and National Preparedness Goal, NPG) define the overall strategy and core competence of national preparedness according to five stages: prevention, protection, mitigation, response and recovery. The evolution of the US emergency management's policy framework and system is shown in Fig. 2. It can be seen that the focus of emergency management has shifted from emergency response to emergency preparation.

1.2 Development History of EU Emergency Preparedness and Response

The emergency preparedness and response of EU is mainly based on civil protection mechanism. When any type of natural disasters or man-made disasters (including environmental disasters, marine pollution, health incidents, etc.) occur within the EU, the EU can better protect human beings, environment, property and cultural heritage by pooling the civil defense forces of its member States. After the "9/11" incident, the Council of the European Union passed the "Council Decision of 23 October 2001, 2001/792/EC, and established a community mechanism to facilitate reinforced cooperation in civil protection assistance interventions), which marked the formal establishment of the civil defense mechanism of the European Union. Under the guidance of a series of EU resolutions, the EU civil protection mechanism has developed rapidly and is playing an important role in major international emergencies. On March 5, 2007, the EU adopted the decision, Council Decision of 5 March 2007, 2007/162/EC, Establishing a Civil Protection Financial Instrument. On November 8th of the same year, the Council Decision of 8 November 2007, 2007/779/EC Establishing a Community Civil Protection Mechanism (Recast) was adopted. In the development of the civil protection mechanism, the implementing regulations, 2001/792/EC [C(2003)5185] and 2007/779/EC [C(2010)5090], were also formulated as supplements to the legal documents. Considering the rapid increase in the number and severity of natural and man-made disasters, future disasters may become more extreme and complex, and may have long-term and wide-spread impacts and consequences, so it is very important to adopt a comprehensive approach in disaster management. Therefore, focusing on improving the systems' effectiveness of prevention, preparedness and response, the EU adopted the decision, 1313/2013/EU on a Union Civil Protection Mechanism, on December 17, 2013.

The EU civil protection mechanism covers the main links of disaster management—prevention, preparation and response. The aim is to mobilize the resources and strength of all EU member states and to bring all relevant government services, professionals, networks and systems under the coordination and command of EU institutions. According to the requirements, each EU member state must designate an official to be responsible for its own civil protection work and coordination. The European Commission appoints a coordinator to be solely responsible for coordinating civil protection-related measures. The civil protection mechanism mainly includes the following contents.

(i) Monitoring and Information Centre (MIC), was replaced and upgraded by Emergency Response Coordination Centre (ERCC) in 2012. In addition to carrying out an all-weather duty system and efficiently coordinating the response actions of EU member states, the center can also quickly collect and analyze real-time risk information to ensure real-time monitoring and immediate response to various emergencies. At the same time, the center is equipped with various types of risk management experts, further enhancing the abilities of risk monitoring and analysis, and greatly improving the work efficiency of dispatching intervention teams of EU experts.
(ii) Common Emergency Communication and Information System (CECIS) is a platform designed to ensure emergency communication among Member States and between MIC and Member States. Through the platform, early warning information can be sent and received, assistance requests can be made and the

situation of emergencies can be observed in an online log. The MIC, CECIS and the member states' contact points administered by the European Commission form the organization structure of EU Civil Protection Organization.

(iii) Experts team for evaluation and coordination, includes technical experts, evaluation experts, coordination group members and coordination supervisor. Selection of experts is subject to uniform criteria; MIC must clarify the mandate and dispatch procedures of the expert group; Expert information is provided by participating countries and regularly updated, and the expert database is compiled and consulted in CECIS; Experts are required to follow training programs.

(iv) Training program's purpose is to coordinate and ensure the compatibility and complementarity among rescue teams, improve the evaluation ability of rescue experts, improve the response ability and shorten the response time. Training projects included joint courses and exercises, exchange systems among member states, response procedures, common languages and sharing of lessons and experience.

(v) Civil protection modules, have confirmed 17 general requirements for civil protection modules such as forest fire, flood, rescue, and medical treatment, and it is conducive to improving the rapid response ability of civil protection.

In 2000, the EU constructed a unified public security management's technology support system, e-Risk. Based on satellite's broadband transmission technology, the e-Risk system assists member states in carrying out monitoring and information analysis on various risks, including floods, tsunamis, earthquakes, fires, nuclear leaks and terrorist incidents, and it provides support for member states to efficiently and timely handle emergencies and natural disasters across national borders, across disciplines and across police services. In 2010, the EU formulated the "Guidelines on National Risk Assessment and Risk Mapping," and it takes into account the existing EU laws and regulations, including the European Flood Risk Directive, the European Directive on the protection of critical infrastructure, the Directive on the control of major disasters and a series of European regulations.

1.3 The Development History of Emergency Management System in Britain

The British government emphasizes the coordination among organizations and all resources' integration, and puts forward the goal of comprehensive integration from four aspects: horizontal direction, vertical direction, concept and system. Horizontal integration refers to the establishment of cooperation mechanism among government departments that are at the same level. At the central and national level, the integration is mainly realized through the Civil Contingencies Committee (CCC) and the Civil Contingencies Secretariat (CCS). At the regional and district level, it mainly relies on regional or Local Resilience Forum to achieve integration. This forum is equivalent to China's joint conference system, and is the leading organization of regional and local emergency management. There are several working committees under the forum, such as the Health Committee, Transportation Committee, Public Utilities Committee and Volunteer Organization Committee under the London Emergency Management Seminar. The Committee also has several working groups, such as evacuation working group, risk assessment group, site clean-up group and drill group. In order to ensure the coordination of various departments in the process of emergency response. Britain has established a three-level handling mechanism of "gold, silver and copper." The "gold" level is the strategic decision-making level, which mainly solves the problem of "what to do," the "silver" level is the tactical decision-making level, which mainly solves the problem of "how to do it," and the "copper" level is the operational and executive level and includes the first-line disposers, such as fire brigade and police.

In 2004, according to the new situation of emergency management, the British government promulgated the Civil Contingencies Act 2004, which emphasized that accident prevention is the key to emergency management and required the government to combine emergency management with normal management to minimize the risk of disasters. At the same time, the law clearly stipulates the responsibilities of the central government and local governments for risk identification and assessment, emergency planning, emergency response, recovery and reconstruction. After that, legal documents such as the Draft Regulations for the Implementation of the Domestic Emergency Law in 2005, Emergency Preparedness and Emergency Handling and Recovery were issued one after another as a supplement to the basic emergency bill.

Carrying out business continuity management (BCM) is a distinctive feature of emergency management in Britain. It means that after an emergency, it is necessary to ensure the normal operation of social functions of various organizations, so as to improve the anti-risk ability of the whole society. In order to accurately understand the development level of national anti-risk ability, the British government published a report, "Risk: Improving Government's Capability to Handle Risk and Uncertainty," in 2002. The National Capability Survey (NCS) is also conducted nationwide every two years, and the risk checking and registration work was started in 2005 to assess the major disasters and threats that Britain may face in the next five years. According to the survey results, the government will start to study the shortcomings and solve problems from the perspective of public policy.

2 Development History of China's Public Security Emergency Management

Since the founding of People's Republic of China, China's emergency management system has experienced three stages: monotonization, diversification and structuring, as shown in Fig. 3. From 1949 to 2003, although there were many cases of emergency management and much relevant experience, there was no special emergency management system whose response mechanism was a typical "impact-response" model, and the focus of emergency management was disaster relief. The early preparation, prevention and early warning of unexpected incidents, and the afterward recovery and reconstruction were not paid enough attention to. The whole process of emergency management system's construction in a modern sense is in a blank state. The severe acute respiratory syndrome (SARS) occurred in China in 2003 greatly promoted the development of China's emergency management system.

On July 28, 2003, the Central Committee of CPC and the State Council put unusual management on the agenda for the first time, which was the first milestone in the construction of China's emergency response system. Later, China proposed to speed up the construction of public emergency response mechanism of major issues, the general office of the State Council set up a special "public emergency response plan working group," which aims to promote public emergency work, that is, "One Case, Three Systems" (It refers to the emergency plan and emergency system, emergency mechanism, emergency legal system). By the end of 2005, the emergency preplan system's framework for public emergencies in China has been basically completed, and it is considered as the second milestone in China's construction of emergency management system. In May 2006, the Emergency Management Office

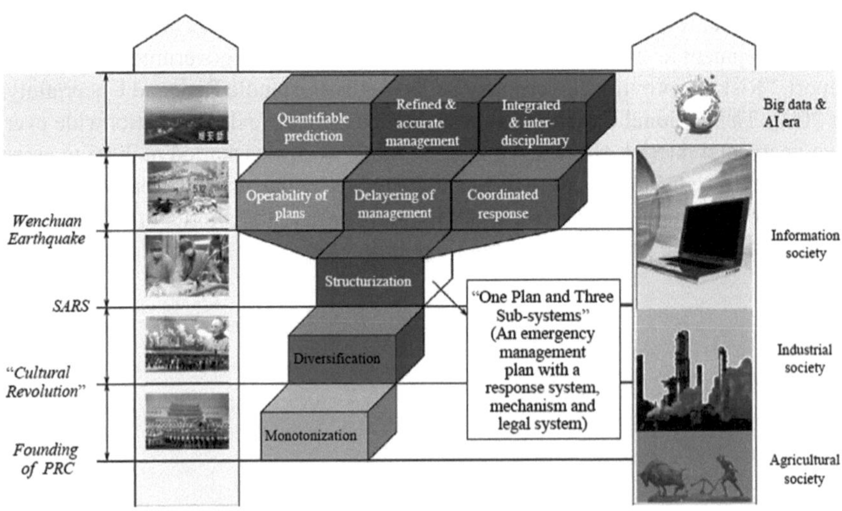

Fig. 3 Development history of China's emergency management system

of the State Council was established under the general office of the State Council. It is directly responsible to the Premier of the State Council and performs duties of on-duty emergency response, information collection and comprehensive coordination, thus playing a pivotal role in the operation. This initiative is considered as a sign of "normality and specialization" in emergency management. The National Public Emergency Plan and Emergency Response Law of the People's Republic of China were promulgated in 2006 and have been implemented since 2007, marking the basic completion of China's legal system of emergency management, and it is considered as the third milestone in the construction of China's emergency management system.

Since 2003, when public security was included in the national economic and social development plan and the national science and technology plan as an important field, the scientific and technological level and guarantee ability of public security in China have been rapidly improved with remarkable achievements. The national emergency platform has been basically built up, the capacity of handling and managing emergency has been greatly improved, the construction of emergency management ability and the ability to respond and manage emergencies have been significantly enhanced; the timeliness and accuracy of natural disaster monitoring, forecasting and early warning have been significantly improved, the social security's risk prevention and control network have basically taken shape, the ability of rapid response and on-site handling have been significantly enhanced, and the trend of integration and socialization of comprehensive public security has become increasingly obvious; the complete set of technical equipment system has been exported to foreign countries, and the international technical competitiveness has obviously improved. For example, ECU911 technical system has played a great role in the rescue and post-earthquake reconstruction of Ecuador's M 7.8 earthquake.

3 International Trend of Public Security Emergency Management

In recent years, with the development of new technologies and the promotion of globalization, it has become an international consensus to attach great importance to public security, and the innovation of public security science and technology has become an international trend. In 2015, the United Nations adopted Sendai Framework for Disaster Risk Reduction 2015–2030, which proposed to greatly reduce the impact of disasters on global population, economy, important infrastructure and services. The United States issued "Crisis Response and Disaster Resilience 2030: Take Strategic Action in an Uncertain Era," which focused on the change of individual role, the protection of key infrastructure and the application of new technologies. The European Union issued Horizon 2020 Plan, which specifically put forward "Secure Societies: Protecting Freedom and Security of Europe and its Citizens," with protecting citizens' security, combating crime and terrorism, and protecting people from natural disasters and man-made incidents as the main research directions. In the Basic Plan of Science and Technology (2016–2020), Japan has confirmed 13 key directions of scientific and technological innovation, among which four directions, such as national security, are directly related to public security. China's "National Medium-and Long-Term Science and Technology Development Plan (2006–2020)" has systematically studied relevant issues and deployed the development blueprint of public security science and technology.

3.1 Sendai Framework for Disaster Risk Reduction 2015–2030

On March 18th, 2015, the Third World Conference on Disaster Reduction closed in Sendai, Japan. On the basis of evaluating the implementation of the 2005–2015 Action Plan: Strengthening the Resilience of Nations and Communities to Disasters (Hyogo Action Plan for short), the conference exchanged the experiences of disaster reduction work in countries and regions around the world, such as post-disaster reconstruction and disaster reduction's decision-making by using science and technology. Meanwhile, it analyzed the economic impact of disaster reduction, and adopted the Sendai Framework for Disaster Risk Reduction 2015–2030. It predicted the fruits and goals of global disaster reduction work in the next 15 years, and 7 specific goals, 13 principles and 4 priority actions have been confirmed. The roles and responsibilities of various stakeholders and the specific implementation methods of international cooperation were clarified. The conference also adopted Sendai Declaration and Voluntary Commitment of Stakeholders.

(1) **The overall goal**

The Sendai Framework for Disaster Risk Reduction 2015–2030 is based on the Hyogo Action Plan, and strives to achieve the following fruits in the next 15 years: greatly reducing the disaster risks and losses of life, livelihood and health, and greatly reducing the disaster risks and losses of economic, social, cultural and environmental assets from the aspects of individuals, enterprises, communities and countries. In order to achieve the expectations, we must try to realize the following goals: adopting comprehensive and inclusive measures in economic, structural, legal, social, health, cultural, environmental, technical and political aspects to prevent new disaster risks, reduce existing disaster risks, prevent and reduce exposure in risky environment and vulnerability after disasters, and lastly, strengthening preparedness for disaster relief and recovery, so as to improve ability of disaster resistance.

In order to effectively evaluate the global progress of the goals and results of the framework, The Sendai Framework for Disaster Risk Reduction 2015–2030 puts forward the following seven global specific goals:

(i) By 2030, the global disaster mortality rate should be greatly reduced, and the average global disaster mortality rate per 100,000 people from 2020 to 2030 should be lower than that from 2005 to 2015.

(ii) By 2030, the number of people affected by disasters in the world should be greatly reduced. To achieve this specific goal, the average number of people affected by disasters per 100,000 people in 2020–2030 must be lower than that in 2005–2015.
(iii) By 2030, the direct economic losses' proportion in global GDP caused by disasters should be reduced.
(iv) By 2030, the damage to important infrastructure (including health and education facilities) and the interruption of basic services caused by disasters should be greatly reduced by enhancing the resilience of important infrastructure and basic services.
(v) By 2020, the number of countries formulating national and local reduction strategies of disaster risk should increase significantly.
(vi) By 2030, the participation of developing countries in international cooperation on disaster reduction should be greatly increased, and adequate and sustainable support should be provided for developing countries which implement this framework to complete their national actions;
(vii) By 2030, the availability of Multi-hazard Early Warning Systems and the opportunity to obtain disaster risks' information and assessment results should be greatly increased.

Among them, the first three targets give quantitative indicators for casualties, the number of people affected by disasters and direct economic losses, while the last four targets specify comparable targets for infrastructure security, implementation of risk reduction strategies, action support of developing countries and utilization of multi-hazard early warning systems.

(2) **Priority actions**

Sendai Framework for Disaster Reduction has formulated the following four priority actions, and given specific action plans for different districts, countries and regions.

(i) Understand the disaster risk. The policy and practice of disaster risk management should be based on a full understanding of all dimensions of disaster risk, including vulnerability, ability, exposure of personnel and property, hazard characteristics and environment, etc. Such knowledge is helpful for pre-disaster risk assessment, disaster prevention and mitigation, and the formulation and implementation of appropriate and effective disaster preparedness and response plans.
(ii) Strengthening disaster risk management to manage disaster risk. The second priority action highlights disaster prevention, mitigation, preparedness, relief, recovery and reconstruction, and propels close cooperation among all parties in the fields of comprehensive management, disaster reduction and relief and sustainable development. It is necessary to formulate clear ideas, plans, terms of reference, guidelines and coordination methods within and between departments, and it also needs the participation of stakeholders.

(iii) Increasing investment in disaster risk reduction and improving the ability of disaster resistance. The third priority is to advocate all-round investment in disaster prevention, mitigation and relief. The public and private sectors can invest in disaster risk prevention and reduction through structural and non-structural measures, and strengthen the resilience of individuals, communities, countries and their property in economic, social, health and cultural aspects.
(iv) Strengthening disaster preparedness to respond effectively and ensure a better reconstruction and rehabilitation. Sendai Framework for Disaster Reduction holds that disaster risk is increasing and the exposure of people and property in disaster is increasing. It advocates strengthening disaster preparedness as a priority, taking actions based on event prediction, integrating disaster risk reduction into emergency preparedness, and ensuring the ability to carry out effective disaster relief and reconstruction at all levels. The key is to give power to women and people with disabilities, and take the lead in adopting a gender-equitable and universally applicable approach to disaster relief, recovery, rehabilitation and reconstruction. Post-disaster reconstruction and rehabilitation also need to be prepared before the disaster. This stage is also an important opportunity to incorporate disaster risk reduction into various development measures, so that the country and community have better resilience to disasters.

(3) **The function of stakeholders**

In terms of the roles of various stakeholders, Sendai Framework for Disaster Reduction clarifies the responsibilities of civil society, volunteers and their organizations and community organizations: providing expertise and practical guidance in formulating and implementing normative frameworks, standards and plans for disaster risk reduction; participating in the implementation of local, national, regional and global plans and strategies; assisting and supporting enhancing public understanding, cultivating prevention culture and carrying out disaster risk education; advocating the establishment of resilient communities as appropriate, carrying out inclusive disaster risk management for the whole society, and enhancing synergy among groups. Special attention should be paid to the functions of women, children and adolescents, the disabled, the elderly, indigenous residents, immigrants, academic institutions and organizations, enterprises, professional associations, financial institutions in the private sector, and the media.

(4) **International cooperation and global partnership**

In terms of international cooperation, Sendai Framework for Disaster Reduction clarifies the general factors that should be considered in international cooperation for disaster risk reduction, such as differences in development levels, and special attention and assistance should be given to developing countries which have frequent disasters; making use of existing mechanisms and platforms to integrate information sharing and technology transfer in reduction of disaster risk into multilateral, regional and bilateral development assistance programs; and clarifying fields and main contents supported by international organizations. Strategic coordination has been emphasized, and it is necessary to support the application of the existing platforms, support the role of the United Nations International Strategy for Disaster

Reduction in supporting implementation and review, and call on international financial organizations to provide financial support and provide more loans to developing countries, so as to enhance the overall ability of the United Nations system to assist developing countries in disaster risk reduction field.

3.2 The Development Trend of EU's Emergency Management

(1) **Horizon 2020 Plan**

In 2010, the European Union launched a new ten-year economic development plan—"Europe 2020 Strategy" at the end of the Lisbon Strategy. The 7th Framework Program, FP7) ended at the end of 2013 as the main operational tool for implementing the EU development strategy, and the new research and innovation framework program 2020 (Horizon 2020 was officially launched in 2014 for a period of seven years (2014–2020).

The Horizon 2020 has a total budget of 77 billion euros, and its implementation period is from 2014 to 2020. Its three strategic priority fields are excellent scientific research, industrial leadership and social challenges. In the strategic priority field, "social challenges," the plan puts forward "Safe Society-Guaranteeing the Freedom and Security of Europe and Its Citizens." The research and innovation activities of this section involve protecting citizens, society, economy, infrastructure and basic services, economic prosperity, political stability and social welfare. The main objectives of building a "safe society" include: (i) improving the resilience of human society against natural and man-made disasters, developing new crisis management tools and interactive modes of communication, and providing new solutions to protect key infrastructure; (ii) Combating crime and terrorism, using new legal tools to prevent explosions and other incidents; (iii) Improving border security, enhancing the protection of maritime border, ensuring the security of island chain security, and supporting EU's foreign security policies, including conflict prevention and the construction of peace; (iv) Enhancing cyber security, including adopting safe information' sharing mode and new forms of information protection.

Disaster prevention is one of the core elements of social operation, and almost all social sectors are threatened by disasters and problems in resilience and security to a certain extent. Combating crime and terrorism requires new technologies and capabilities, including the ability to combat and prevent crimes (including cybercrime), illegal trafficking and terrorism (including cyber terrorism), as well as the ability to understand and respond to the thoughts and beliefs of terrorists. European border protection requires research on systems, equipment, tools, processes and rapid identification methods, including ensuring supply chain's security under EU tariff policy, and carrying out research on rescue schemes to support EU external security strategy and the application in civilian fields. The challenges in digital security focus on improving the security of existing applications, services and infrastructure. It also focuses on supporting the establishment of market orientation and market incentive mechanism in Europe by integrating the most advanced solutions or processes of national security. The driving force of this challenge should be end users, including law enforcement agencies, first responders, key infrastructure operators, ICT providers, ICT manufacturers, market operators and citizens. In addition, the security and interest of all stakeholders should be considered comprehensively, and the participation of end users is extraordinarily important.

Generally speaking, the funded projects of Horizon 2020 mainly focus on the new generation of emergency services, monitoring, response and communication systems, and have carried out research on important infrastructure's protection, resilience, security culture, security devices, law enforcement agencies and emergency rescue planning in the fields of traffic security, environmental security, oil and gas security, urban security, combating crime and terrorism, border security and digital security. Carrying out research on social security will help implement the policy objectives of "Europe 2020 Strategy," as well as the security industry policy, internal security strategy and network security strategy.

(2) **Action plan of Sendai disaster reduction framework of EU**

Corresponding to the four priority actions of Sendai Framework for Disaster Reduction, the EU proposed the following four key fields and relevant implementation priorities.

(i) Improving the knowledge base of risk in all EU policies (corresponding to "It is necessary to understanding disaster risk"). It focuses on: it is necessary to strengthen the collection and sharing of basic databases on loss and damage; to use prediction, scenarios and risk assessment to better respond to existing and new risks as well as risks' new forms; to further work with research groups to better address the lack of knowledge and foundation in disaster risk's management, and to enhance the science and policies' interaction in decision-making processes.

(ii) The whole society's participation in disaster risk management (corresponding to "It is necessary to strengthen the governance of disaster risk and managing disaster risk"). It focuses on: It is necessary to study the potential of educational measures in disaster risk's reduction; to promote practice and exchange and improve disaster management policies through mutual learning and expert review; to work with stakeholders, including local governments, civil society and communities, to develop strategies for risk perception; to cooperate with the private sector to encourage business-driven innovation in all areas of disaster risk management; to strengthen the links among disaster risk management, climate change adaptation and biodiversity strategies; to strengthen the links between disaster risk management, climate change adaptation and urban policy and planning; to support the formulation of inclusive local and national disaster risk's reduction strategies and mobilize the local officials, communities and civil society to actively participate; to assist regional organizations to support national institutions in implementing the Sendai Framework for Disaster Reduction and to build up national and regional platforms for disaster reduction.

(iii) Enhancing investment in risk informing within the EU (corresponding to "It is necessary to increase investment in disaster risk reduction and improving disaster's resistance ability"). It focuses on: it is necessary to implement multilateral and bilateral development aid in all external financial instruments of the EU to promote investment in risk informing; to track the investment trend of disaster risk reduction in all humanitarian projects and development aid projects; to promote investment in disaster prevention within the EU; to strengthen the use of disaster risks' financing mechanism, risk transfer and insurance mechanism, risk sharing and risk retention mechanism and to encourage and implement disaster reduction methods based on ecosystem.

(iv) Support the development of methods of overall disaster risk management (corresponding to "It is necessary to strengthen disaster preparedness to respond effectively and ensure better rehabilitation and reconstruction"). It focuses on: in the national disaster risk reduction strategy formulated by EU member States, it is necessary to establish good practices for integrating cultural heritage; to improve the ability to cope with and prevent disasters that will affect health, and enhance cooperation with the health institutions and other stakeholders; to strengthen the capacity construction of national government, community and other parties in disaster risk management; to develop and integrate transnational monitoring and early warning and warning systems to better carry out disaster prevention and response actions; to integrate the goal of "rebuilding a better future" into the assessment methods, projects and standards for managing disaster risks and improving resilience.

(3) **Main experience of EU emergency management system's construction**

The EU mainly emphasizes the following aspects in the construction of emergency management system.

(i) Reducing the risk of multiple disasters. It includes various disaster risks caused by natural disaster factors (such as seismic geology, meteorological hydrology, marine, biological and ecological environment disaster risks), technical and environmental disaster risks and related man-made disaster risks.

(ii) Multiscale reduction of disaster risk. From local districts, enterprises, country and regions to the whole world, a complete emergency management system has been formed.

(iii) Disaster risk reduction in various fields. It includes economic, political and social fields, as well as cultural, health and environmental fields.

(iv) Reducing disaster risk from multi-level organizations. It includes individuals, enterprises, communities and the whole country, as well as various stakeholders, regional and international organizations.

(v) Application of big data technology. The flow of people and materials is transformed into various forms of big data. Through the analysis of these data sets, the emergency resources are optimally allocated according to the time and space law of disasters, and the route of emergency materials' transportation under the crisis situation is optimally designed.

(vi) Application of virtual reality technology. For public security incidents, simulation and real scene reconstruction are carried out to form a set of reconfigurable event scenes. It can be applied to emergency training such as first responder training, interactive team cooperation training, preparation of plans, and allocation and storage of equipment, materials and teams.

3.3 Major Research Program on Resilient Infrastructure in the United States

Foresight can help emergency management departments or institutions make decisions in uncertain situations. In order to improve the foresight ability in emergency management, the Federal Emergency Management Agency (FEMA) has implemented the Strategic Foresight Initiative. In January 2012, FEMA released the progress report "Crisis Response and Disaster Resilience 2030: Strategic Action for Creating an Uncertain Era," focusing on the change of individual roles, the protection of key infrastructure, the prompt application of information technology, and the dependence on technological change, emphasizing the cross-departmental preparation and the consensus of grassroots individuals. Through the overall planning of the state, taking 15 national-level catastrophes as the starting point, an integrated comprehensive support capability was built up.

Interdependent Critical Infrastructure (ICIS) is composed of interconnected network-physical-social systems. Resilient infrastructure and emergency resources are the main means to mitigate disasters. In March 2015, the National Science Foundation (NSF) has set up a major research program called "Critical Resilient Interdependent Infrastructure Systems and Processes" (CRISP), which studies the design and performance of important and interdependent infrastructure systems, and aims to improve the service resilience of different scales to resist the hazards caused by natural disasters, terrorism, network attacks or unexpected software errors. In 2015, 20 projects were launched, each of which will last for 3 to 4 years. These projects were funded a total of 20 million US dollars; In 2017, 15 to 20 projects were launched, each of which will last for 2 to 4 years, with a total funding of 22.9 million US dollars.

The purpose of this major research plan is: (i) creating new knowledge, methods and engineering solutions to improve the resilience, performance and handling ability of ICIs; (ii) creating the theoretical framework and multidisciplinary model of important infrastructure systems, processes and services, and improving the analysis and prediction ability of complex behaviors to cope with the changes of systems and policies; (iii) studying the correlation framework among ICIs's physical, network, social, behavioral and economic elements, including but not limited to ICIs's physical design and layout optimization, the use of new materials, ICIs's software and computing system's integration framework, ICIs's modeling, simulation, monitoring and control system's frameworks, and innovative software engineering methods; (iv) studying the organizational, social, psychological, legal and economic issues and countermeasures of ICIs improvement.

It summarizes the funded projects of CRISP plan, and the research's focuses are as follows: through the interdisciplinary intersection of engineering, socio-economic science, behavioral cognitive science, information science and engineering, and aiming at important infrastructure systems (including physical infrastructure, resources, energy, transportation, finance, supply chain, physical and social networks and their coupling systems), the plan carries out research on new material use, layout optimization, sustainability and resilience design, data-driven and real-time simulation, adaptive control, behavior perception's resumption, comprehensive decision-making framework/model, comprehensive network's resilience analysis, and resilient city/community's computing framework, etc.

4 The Situation of China's Social Changes and Technology Development

Since the reform and opening-up, China has achieved great success. The report of the 16th National Congress of the CPC (Communist Party of China) pointed out that "China has entered a critical period of reform and development. The economic system, the social structure, the structure of interests and the ideological concept have undergone profound changes." The report of the 17th National Congress of CPC pointed out that "we should have a comprehensive understanding of the new situation and new tasks of industrialization, informatization, urbanization, marketization and internationalization." As a large developing country, China will, for a long time, be in the process of industrialization, informatization, urbanization, marketization and internationalization in the future. This is a problem that a majority of developed and developing countries have never encountered.

Among the four trends of industrialization, marketization, urbanization and internationalization, urbanization is the most core and complex proposition. The main reason is that urbanization is the carrier of industrialization, market-oriented platform and international stage. When talking about China's future trend, Liu He pointed out that faced with plentiful challenges of domestic urbanization, it is necessary to learn a lesson from the "big city disease" in developed countries and some developing countries, handle various practical problems prudently and responsibly, and find a new path that is in line with China's national conditions in constructing urban functional areas, accepting a large number of transferred labor forces, controlling noise, air and water pollution, traffic congestion and solving social problems brought by big cities.Urbanization makes the security risk and its governance show the characteristics of complexity and comprehensiveness, which requires emergency management to transfer from passive relief after disasters to proactive prevention before disasters, from responding to single disaster to complex disasters, and from reducing disaster losses to mitigating disaster risks. The Opinions of the Central Committee of the Communist Party of China and the State Council on Promoting the Reform of the System and Mechanism of Disaster Prevention, Mitigation and Relief issued in December 2016 clarified five basic principles for promoting the reform of the system and mechanism of disaster prevention, mitigation and relief: We should adhere to the people-oriented principle and effectively protect the lives and property of the people; we should adhere to the combination of prevention, resisting and rescue; we should adhere to comprehensive disaster reduction and make overall plans to resist various natural disasters; we should adhere to differential responsibility and territorial management; and we should adhere to the leadership of the Party Committee, government's arrangement, and extensive participation of social forces and market mechanisms.

Security is the life of the city, the most important element of a modern city, and the red line and the bottom line that must be adhered to in the construction of new urban areas. With the rapid advancement of the "Five Modernizations," various changes and adjustments have been made with unprecedented speed, scope and influence. Public security is facing some prominent contradictions and problems. For the present, public security incidents is prone to happen in China. The task of maintaining public security is important and arduous. In recent years, international organizations and developed countries have begun to widely use the concept of resilience in the security field. The United States, Britain, Japan and other countries have proactively propelled the construction of safe and resilient cities. Safe and resilient cities can be defined as cities that alleviate future shocks and pressures on their social, economic, technological systems and infrastructure, while maintaining the cities' basic functions, structures, systems and identities. In order to quickly respond to some major management issues arising in the process of economic, technological and social development and provide scientific analysis and policy suggestions for the high-level decision-making of the Party and the government, in 2017, the fourth emergency management project of the Management Science Department of National Natural Science Foundation of China, that is, "Research on Theoretical Methods and Strategies for the Security and Resilience Construction of Xiong'an New Area" explored the theoretical methods and countermeasures for the security and resilience construction of Xiong'an New Area from eight aspects respectively: (i) research on the top-level design for the construction of security and resilience construction of Xiong'an New Area; (ii) research on strategies for improving the disaster prevention capability of Xiong'an in natural disaster response; (iii) research on strategies for improving production's risk management and comprehensive supervision capabilities in Xiong'an New Area; (iv) analysis on countermeasures of improving public health emergency's response ability in Xiong'an New Area; (v) research on new situation of social security and new model of social governance in Xiong'an New Area; (vi) research on risk assessment of public security and resilience improvement strategy of urban grass-roots community in Xiong'an New Area; (vii) research on ecological security's protection mechanism of Xiong'an New Area; (viii) research on the guarantee of water safety and its governing mechanism in the process of planning and construction of Xiong'an New Area. These studies are of great significance for building a solid new urban security system in Xiong'an New Area.

In 2017, Ministry of Science and Technology of the People's Republic of China issued the "13th Five-year" Special Plan for Scientific and Technological Innovation of Public Security, which pointed out that there were still some weak links and deep-seated problems in scientific and technological innovation of public security in China, mainly manifested as: insufficient research on basic theories and lack of independent innovative achievements; there is still a gap in the overall technical level compared with foreign leading countries, and some key security and emergency technical equipment rely on imports; the state's key laboratory and other scientific research bases and talent team's construction are still weak. At the same time, the above-mentioned plan has clarified the development strategy, development goals, main tasks, policies and measures in the field of public security and technology during the "13th Five-Year Plan," which covers public security and technology fields such as social security, production security, comprehensive protection and emergency response. It is pointed out that China's public security capacity should be improved from the following aspects: common basic scientific problems of public security, comprehensive public security's guarantee technology, social security monitoring, early warning and control technology, guarantee of production security and major incidents' prevention and control technology, national major infrastructure safety guarantee and intelligent management technology, guarantee technology of urban public security, key technology of security and emergency industry, application of technology achievements in public security, and security and emergency industry. All these measures are to increase China's emergency management ability in public security.

Under the background of information age, the development of China's society and cities have new changing trends such as networking, virtualization and delayering. Management is featured by delayering, information is ubiquitous, and individual's role is prominent. With the rapid development of mobile Internet, Internet of Things, big data, cloud computing and other technologies, on the one hand, disaster-bearing bodies have developed into two societies: physical society and society of network information; On the other hand, it is an inevitable trend for individuals and communities to participate in emergency management (Fig. 4).

Facing the great development needs of integration of "risk-prediction-disposal-guarantee," a complex giant system on public security and needs of intelligent and resilient management in the future, we should continue to improve the technical level of risk assessment, monitoring and early warning, disposal and rescue, and comprehensive support, and realize the comprehensive perception, efficient prediction, intelligent decision-making and active support of the emergency management system; we should also build up an all-round public security net, and build up all-round guarantee measures for public security across different fields, levels, times and regions.

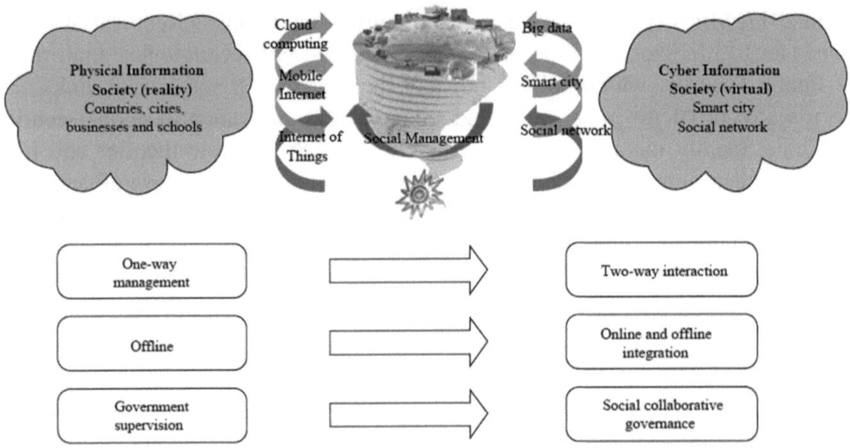

Fig. 4 Governance model of physical information society and Internet information society

Major Research Fruits

Weicheng Fan, Chunchang Shan, Shouyang Wang, Tiemin Liu, Feiyue Wang, Xiaofeng Meng, Lei Wang, Xiaolin Zhou, Wenguo Weng, and Yi Liu

This major research plan is oriented to the major strategic needs of national public security and emergency management. It focuses on three practical issues: information support, decision-making and behavior response of individuals and groups, emphasizes on three core scientific issues: information processing and evolution's modeling, emergency decision-making, behavior response of individuals and groups, takes root in China's problems, reality and practice, puts forward China's plan, and systematically applies the theoretical methods of management, information, psychology and other related disciplines from the following angles to take multi-disciplinary research. It has effectively promoted the leapfrog development of national emergency management ability: (i) It puts forward the theory and method of scenario construction, which injects scientific connotation into the construction of public security network

W. Fan (✉)
Tsinghua University, Beijing, China
e-mail: wfan@tsinghua.edu.cn

C. Shan
Counsellor's Office of the State Council, Beijing, China

S. Wang
Academy of Mathematics and Systems Science, Chinese Academy of Sciences, Beijing, China

T. Liu
China Academy of Safety Science and Technology, Beijing, China

F. Wang
Institute of Automation, Chinese Academy of Sciences, Beijing, China

X. Meng
Renmin University of China, Beijing, China

L. Wang · X. Zhou
Peking University, Beijing, China

W. Weng · Y. Liu
Institute of Public Safety Research, Tsinghua University, Beijing, China

W. Fan (ed.), *Unconventional Emergency Management Research*, Reports of China's Basic Research, https://doi.org/10.1007/978-981-99-6798-8_3

based on bottom-line thinking. (ii) It puts forward the theory and method of emergency preparedness, realizes the transition from focusing on disaster prevention and rescue to improving resilience and ability, provides technical support for the preparation and implementation of emergency system's construction and planning, and enhances the rationality of China's emergency system. (iii) It grasps the laws of emergency evolution and emergency decision-making, puts forward the integration principle and method of emergency decision-making's technical support system, realizes the transition from simple administrative management to data-driven, complex and systematic governance, and provides support for the national emergency platform. (iv) It provides the World Health Organization with a scheme of emergency platform, takes the lead in formulating international standards for emergency capability's assessment, has designed and built public security system and platform for Ecuador, Venezuela and other countries, and thus has enhanced China's voice in international emergency management.

The major research achievements of this major research plan mainly include the following three aspects: (i) Focusing on prevention and emergency preparedness, the theories and methods of risk management, scenario construction, incident evolution modeling and simulation have been formed, and the transition from incident prevention to risk management, and incident prediction to scenario construction has been realized. The "One Case, Three Systems" has been improved, and the emergency preparedness theory and method based on "scenario-task-ability" have been put forward, which provides theoretical and technical support for the construction of national emergency response system. (ii) Around monitoring and early warning, the theories and methods of online hazard monitoring, online perception and data mining based on the Internet of Things have been formed, and the early warning theories and methods based on the evolution law of incidents and the law of human behavior have been put forward. The monitoring and early warning method and system of typical incidents have been constructed by using cloud computing technology to innovate the methods of comprehensive judgment and information release. (iii) Focusing on emergency rescue and handling, grasping the change laws of personnel injury, psychology and behavior, the theory and method of emergency command and coordination, intelligent decision-making and control supported by information have been formed. Besides, the core emergency capabilities of personnel search and rescue, emergency medical treatment, psychological and behavioral intervention, crowd guidance and control, engineering rescue, isolation and epidemic prevention, etc. for typical unconventional emergencies have been improved.

1 Prevention and Emergency Preparedness

1.1 Laws of Incident Evolution

1.1.1 Theory and Method of Unconventional Emergencies' Simulation and Modeling

Huang Quanyi's research group of Tsinghua University put forward the concept of "meta-action" of emergencies, and realized the generalized expression of the mechanism of action of different types of emergencies. The action form of each type of incident is decomposed into basic "meta-actions," including physics, chemistry, biology, information, society, etc. The attribute set and meta-action set of disaster-causer, disaster-bearing body and disaster-incubation body are put forward. Considering the characteristics of mutual transformation among disaster-causer, disaster-bearing body and disaster-incubation body, a chain effect model of emergency incidents (Fig. 3.1) is constructed to describe the relationship among disaster-causer, disaster-bearing body and disaster-incubation body.

Emergencies can easily lead to multiple and multi-level secondary derivative incidents, forming a chain structure relationship, that is, effect of incident chain. The Yuan Hongyong research group of Tsinghua University studied the law of chain effect of emergencies, provided decision-making support for emergency rescue, and effectively reduced losses and impacts. Based on the proposed chain effect model, the characteristics of transition of incidents has been analyzed, the formation mechanism of secondary derivative incidents has been studied, and the disaster chain of "5/12" Wenchuan Earthquake and Great East Japan Earthquake have been analyzed, and the disaster mechanism of earthquake's disaster chain has been obtained (Fig. 3.2).

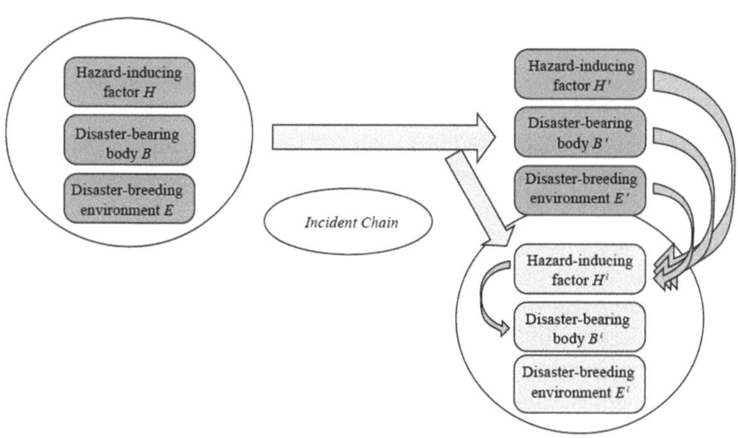

Fig. 3.1 Chain effect model of emergency incidents

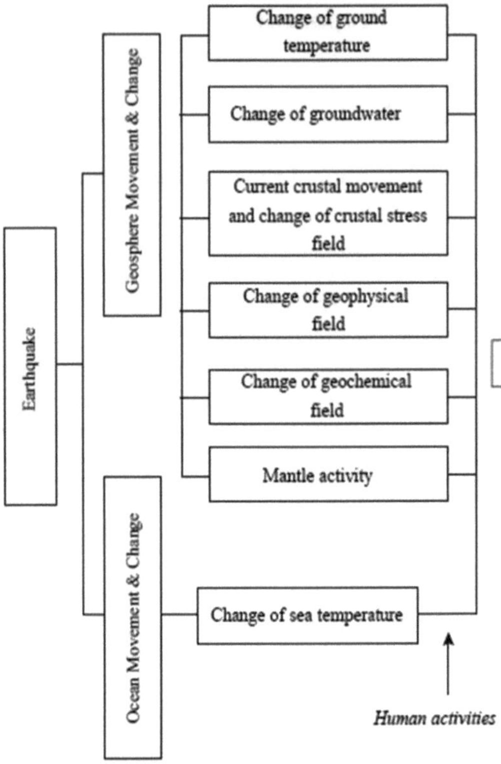

Fig. 3.2 Disaster-formation mechanism of earthquake disaster chain

1.1.2 Theory and Method of Artificial Society and Parallel Computing Experimental Modeling

Based on the mechanism of occurrence, development, transition and evolution of unconventional emergencies and the demand of "scenario-response," Qiu Xiao-gang's research group of National University of Defense Technology systematically proposed artificial society's construction method, computational experiment method and parallel implementation method under the guidance of parallel implementation theory and "triangle model" of public security system, and it has made breakthrough in the construction and operation of large-scale artificial society, multi-paradigm model's integration, real-time situational awareness of incidents, and heterogeneous computing acceleration of large-scale artificial society simulation, and data mining and visualization of artificial society's computing experiment. A dynamic simulation and computing experimental platform for parallel emergency management has been designed. (Fig. 3.3). The platform can assist in the development of basic models needed for parallel emergency management of various emergencies, and

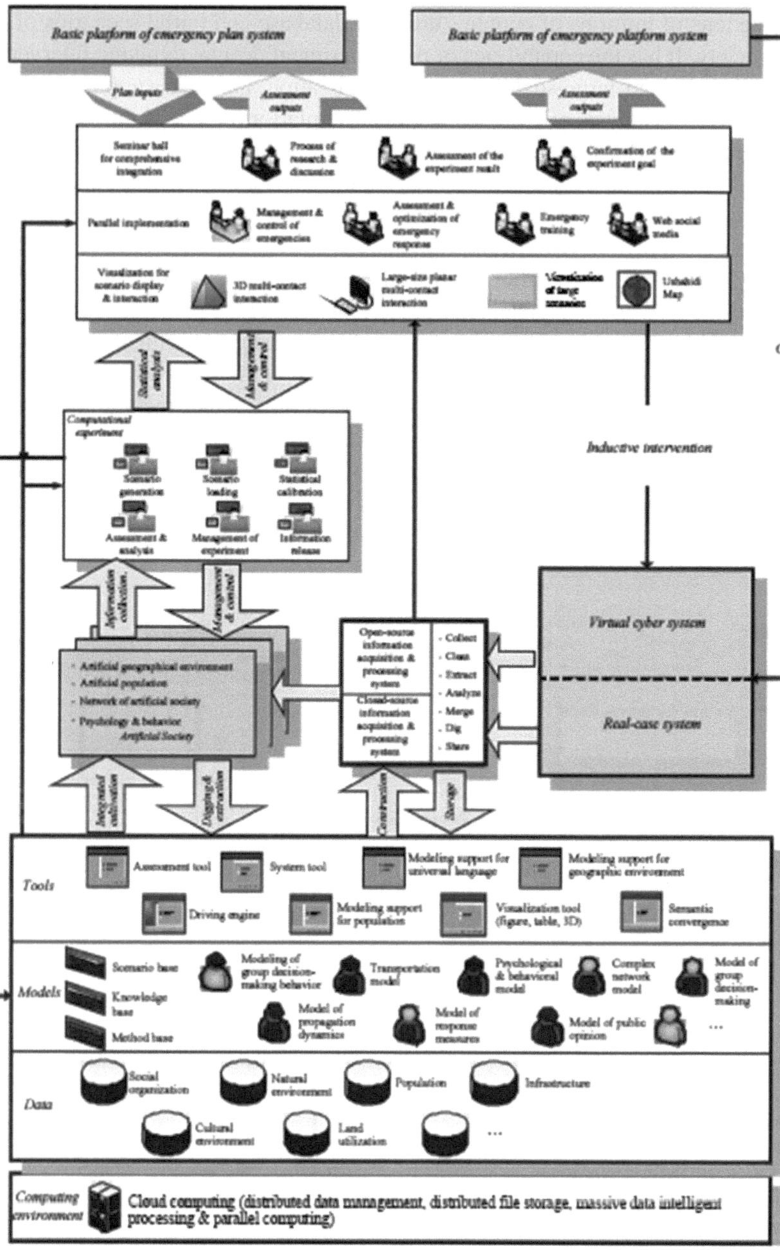

Fig. 3.3 Dynamic simulation and computing experimental platform for parallel emergency management

generate tens of millions of people's artificial database and initial scenarios of artificial society. It has the capabilities of pre-experiment design, dynamic intervention in experiments and large-scale scenario visualization, and also supports real-time acquisition and analysis of emergency data. The platform realizes the separation of model and data, and the separation of model and behavior rules. It is open, extensible and customizable, and can be integrated with other emergency management platforms. The platform's operation process is shown in Fig. 3.4, and the platform software's composition is shown in Fig. 3.5.

Zeng Dajun's research group of Institute of Automation, Chinese Academy of Sciences, focused on the parallel emergency management of public health and social security incidents, and established an artificial computing experimental platform for Beijing, which is composed of agent, environmental entities, social networks, infectious diseases, artificial population database of Beijing, experimental management and control module, display model, etc., and can carry out computing experiments for public opinion and epidemic emergencies of different types, scales and places, and can also support simulation and modeling researches containing multi-agents such as evacuation, psychological warfare and economy.

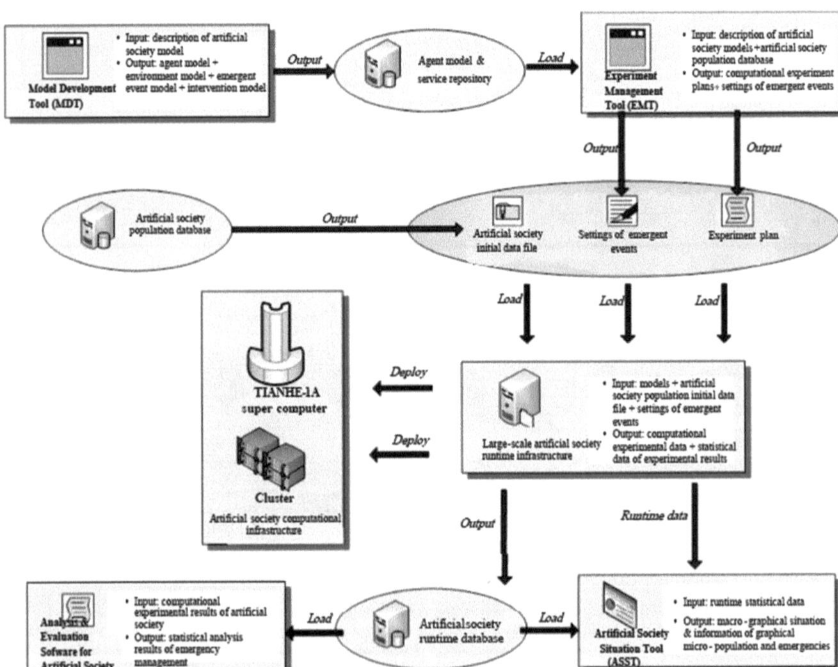

Fig. 3.4 Operating process of the platform

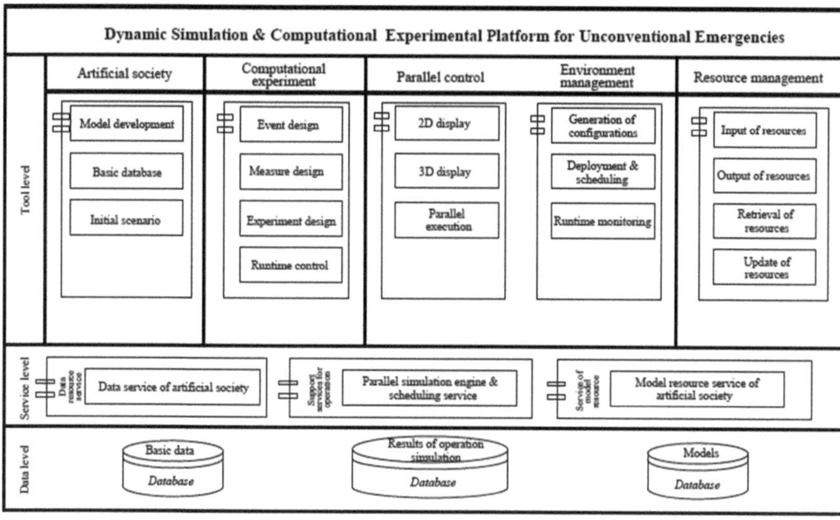

Fig. 3.5 Composition of platform software

1.2 Theory and Method of Risk Management

1.2.1 Theory and Method of Risk Assessment for Unconventional Emergencies

(1) **Comprehensive risk assessment method of urban public security**

Based on the relevant standards of risk assessment at home and abroad, Li Husheng's research group of China Academy of Safety Science and Technology put forward a comprehensive risk assessment method for urban public security: systematically identifying risk sources, scientifically analyzing the possibility and consequences of disaster, comprehensively considering the ability of risk bearing and controlling, evaluating risk levels and choosing risk control strategies. Taking Shenzhen as an example, 60 risk sources were identified in 11 industries, of which high risk sources and extremely high risk sources accounted for 46.67% (Fig. 3.6).

(2) **Analysis method on vulnerability of lifeline system network**

Liu Xiao's research group of Shanghai Jiao Tong University analyzed and identified the vulnerable links, key points, bottlenecks and other vulnerable areas of lifeline system network, and designed the vulnerability analysis model of lifeline system network based on the multi-level network model. Taking Shanghai water supply system as the research object, the research group studies the interdependence between power system and water system, investigates the network structure of water system and its key nodes and paths, considers the risks to emergency response system, establishes the vulnerability assessment model of Shanghai water supply system, designs

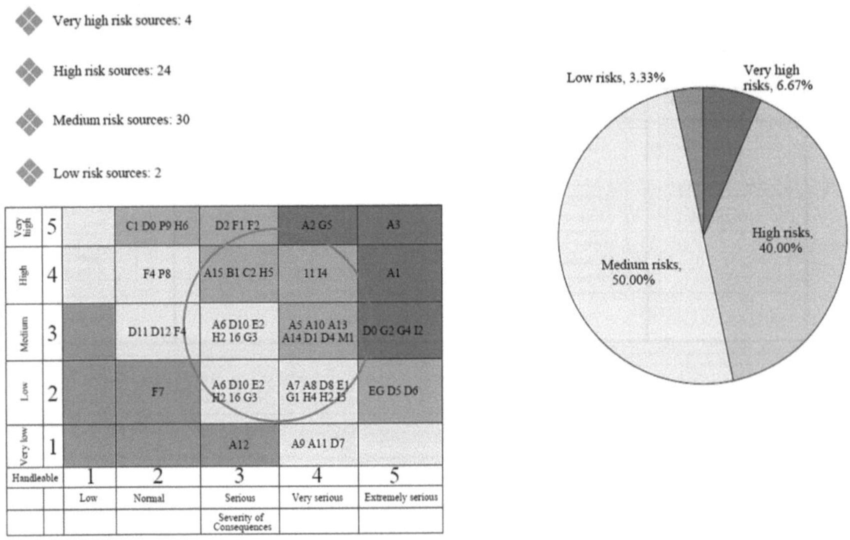

Fig. 3.6 Conclusions of risk assessment

the evolutionary algorithm for solution, and expands the vulnerability research within and among systems.

(3) **Risk- based vulnerability assessment method of urban critical infrastructure**

Li Xiangyang's research group of Harbin Institute of Technology put forward a multi-dimensional evaluation model of urban critical infrastructure's vulnerability, and the model uses state index to calculate the endogenous vulnerability of components, introduces network as a weight and degree of losses as variables to calculate the exogenous vulnerability of components, and considers natural risk factors to calculate the vulnerability of infrastructure network. Taking Shenzhen as an example, the research group analyzes the vulnerability of power grid's components and structures, and puts forward targeted strategies and countermeasures of risk prevention and control, thus improves the scientific nature of urban risk management.

1.2.2 Theory and Method of Constructing Unconventional Emergency Scenarios

In order to better cope with unconventional incident scenarios, an effective method is to set unconventional incident scenarios in advance according to the historical cases and future development trends of related incidents in relevant countries and regions, and prepare corresponding emergency plans, so as to make emergency preparations in advance and improve comprehensive emergency response capability, thus improves response speed and handling effect. Li Husheng's research group of China

Table 3.1 Application demonstration of scenario construction for serious emergencies

Beijing emergency scenarios' construction	Construction of emergency scenarios in petrochemical industry
Destructive earthquake	Leakage of hazardous substances from major hazard sources in chemical industry parks
Extraordinary flood disaster	Blowout accident in offshore oil mining and development
Large-scale blizzard disaster	Leakage and explosion of oil and gas long-distance pipeline
Serious water pollution incident	Fire and explosion in large storage tank area
Large-scale power outage	Blowout and diffusion of toxic gas
Gas stop incident in multi-gas station	Toxic gas leakage incident in petroleum refining and chemical plant
Large-scale network information security incidents	
Incidents of major infectious diseases of unknown origin	
Large-scale mass incidents	
Aviation air crash	

Academy of Safety Science and Technology put forward the theory and method of scenario construction for serious emergencies, including scenario screening, evolution process' scenario construction, consequence evaluation and simulation, analysis of response action, scenario description and presentation, scenario application and update, scenario-based analysis of emergency resources and insufficiency of capacity, setting and planning of objectives of capacity construction, evaluation and improvement of contingency plans, emergency training and drill planning. Taking Beijing and petrochemical industry as examples, typical unconventional emergencies are screened, incident scenarios are constructed (Table 3.1), the contingency plan system is improved, and the insufficiency of emergency resources and capabilities is analyzed, which lays a scientific foundation for emergency preparedness and injects scientific connotation into weaving an all-round public security network based on bottom-line thinking. Relevant achievements have been affirmed by relevant departments of the country and relevant regions, and have been popularized and applied in the field of emergency management.

1.2.3 Communication Obstacles of Crisis Risk and Its Strategies of Solving

(1) **Influence of emotion on risk decision-making**

Uncertainty can easily lead to immediate negative emotional reactions such as anxiety and depression, which are common psychological reactions in disasters. Disaster scenarios can cause acute situational emotional reactions, such as fear, sadness, depression, disappointment, and even reactive consciousness' disorder and emotional disorder, which can seriously lead to behavioral disorder, loss of interest in life and self-care ability. Acute situational and emotional reaction may continue after the emergency, that is, involuntarily fall into the situational reaction at that time. And it includes daydreaming, nightmares and so on. Zhang Kan's research group of Institute of Psychology, Chinese Academy of Sciences, applied international standard emotion pictures to conduct emotion stimulation experiment, and obtained voltage amplitude of different emotions. The results show that emotions (whether positive or negative) will strengthen people's early perception (detection) of uncertain problems; Negative emotions will enhance people's perception of risks; Emotion (whether positive or negative) will reduce people's sense of decision-making's comfort (that is, increase the difficulty of decision-making).

(2) **Word-of-mouth dissemination mechanism based on multi-agent**

Mastering the herd effect is the key to controlling the group behavior (that is, controlling the group's crisis spreading or spreading behavior), and word-of-mouth dissemination plays a core driving role in the formation of herd effect. Information is the core element that determines people's attitude, decision-making and behavior. After an emergency, the environment is full of all kinds of true or false information, which may eventually evolve into a huge word-of-mouth dissemination network interwoven with word of mouth, short messages and online media (including E-mail, forums, Weibo, blogs and other channels). Considering the influencing factors of word-of-mouth dissemination (Fig. 3.7), the research group of Ma Qingguo from Zhejiang University constructed a multi-agent-based online word-of-mouth dissemination network's simulation model, and used NetLogo software to simulate it.

The simulation results of word-of-mouth communication reveal the following conclusions. (i) Word-of-mouth information dissemination is divided into two stages: initial dissemination and re-dissemination. The initial dissemination is completed

Fig. 3.7 The influencing factors of the word-of-mouth dissemination

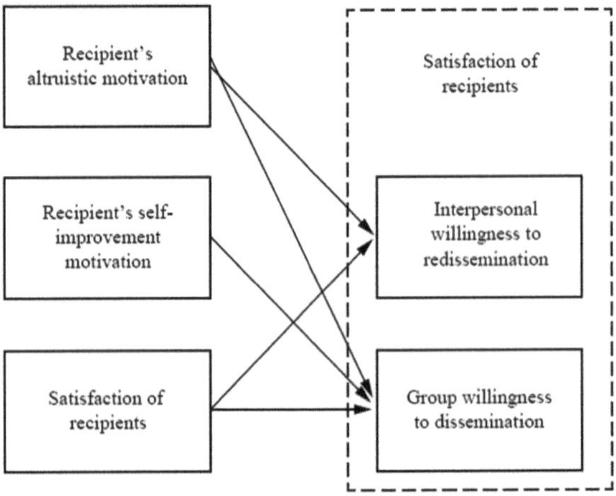

by the manufacturer of word-of-mouth information, while the re-dissemination is completed by the receiver of word-of-mouth information. Individuals with high probability of group dissemination are the first key nodes in online word-of-mouth dissemination network, while individuals with more dissemination times are the second key nodes in online word-of-mouth dissemination network. Therefore, in order to make online positive word-of-mouth dissemination faster and wider, if there are more critical individuals that spread the information, the effect will be better. As for online negative word-of-mouth dissemination, its effect is to the contrary. (ii) Numerous nonlinear effects among individuals are the foundation of the formation of word-of-mouth dissemination network. The slight change of attribute values of a few subjects in the network can cause a long-term and huge chain reaction of the system, and make the whole network show great differences.

(3) **Combination of intervention mechanism and management countermeasures**

Zhang Kan's research group of Institute of Psychology, Chinese Academy of Sciences, started with the extraction and analysis of characteristics and elements of unconventional emergencies and the simulation of group behavior, applied data mining and group behavior simulation respectively, comprehensively analyzed the relationship between characteristics and elements of unconventional emergencies, dynamically tracked the evolution of incidents, and put forward targeted management countermeasures.

The extraction and analysis of characteristics and elements took the severe snow disaster in South China in 2008 and the Wenchuan earthquake on May 12 as examples, collected data and classified them by using text mining technology, and analyzed the correlation among group response and incident state and government response. The results show that there is a strong correlation among group reaction and incident state and government response, and the government can guide group reaction properly through regulation and control and effectively respond to emergencies, thus reducing the damage and loss caused by events. The characteristics and elements and intervention measures are shown in Table 3.2.

The simulation of group behavior takes group behavior as the research object, takes the large-scale Spring Festival travel rush under snow disaster as an example, builds the society model, and uses NetLogo software for simulation research. According to the changing process of group behavior with time, combined with the simulation results, it probes into the movement law of group behavior after emergency, and puts forward the corresponding management countermeasures.

Table 3.2 The characteristics and elements and intervention measures of unconventional emergencies

Unconventional emergencies	Characteristics and elements	Intervention measures
Public health incidents	Highly contagious	Decisive isolation
	Excessive perception of time danger	Release incident information
	Overreaction	Using a few pioneers to guide most people back to normal (herd effect)
Earthquake	The facilities for survival were completely destroyed	Transfer and resettlement of victims
	Traffic paralysis	Emergency repair (team, equipment)
Flood	Being besieged	Transferring, evacuating and resettling the victims
	Overreaction	Using a few pioneers to guide most people back to normal (herd effect)
Snow disaster	Traffic paralysis	Emergency repair (team, equipment)
	Overreaction	Using a few pioneers to guide most people back to normal (herd effect)

1.3 Theory and Method of Emergency Preparedness

1.3.1 Theory and Method of Emergency Preparedness and Planning for Unconventional Emergencies

(1) **System and structure of emergency preparedness**

Li Husheng's research group of China Academy of Safety Science and Technology analyzed the relationship among various functional elements in the emergency preparedness system, and constructed the emergency preparedness system's structure from the aspects of emergency preparedness objects, involved application fields, subjects of responsibility, evaluation and improvement, construction of emergency capability units, integration and configuration, application and verification of emergency capability, and social, political, economic, legal and cultural foundations (Fig. 3.8).

The objects of emergency preparedness include not only conventional natural disasters, accidents, public health incidents and social security incidents, but also unconventional emergencies with low probability but serious consequences. The mission fields involved in emergency preparedness include emergency prevention, monitoring and early warning, emergency response, recovery and reconstruction, as well as disaster reduction and preparation, which are two kinds of actions to

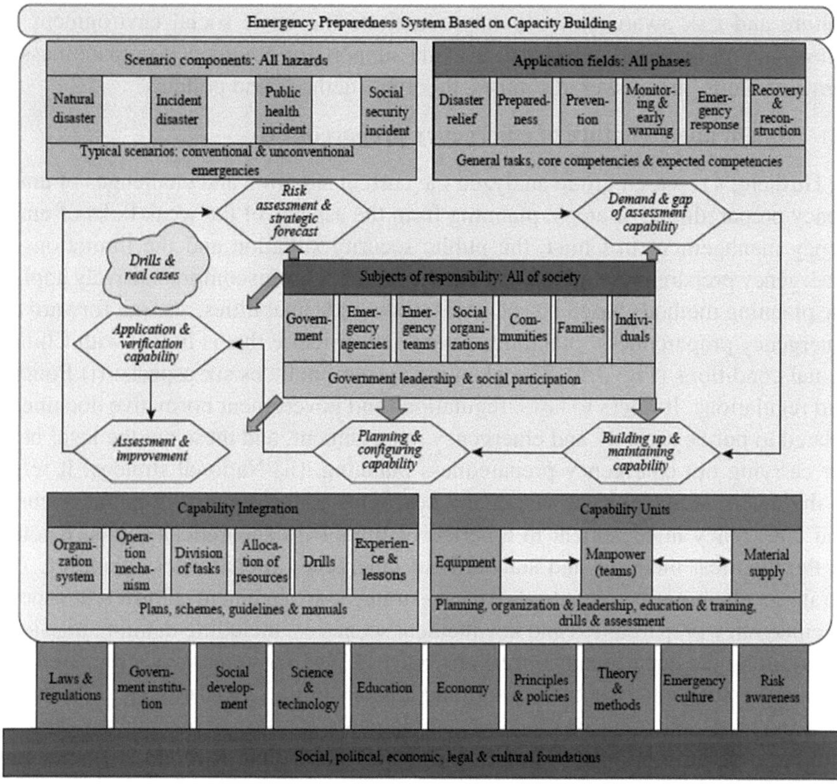

Fig. 3.8 Structure of emergency preparedness system

reduce the loss of events. The subjects of responsibility in emergency preparedness include government, emergency agencies, emergency teams, social organizations, communities, families and individuals. The construction of emergency handling capability's units include basic capability and material elements such as equipment, manpower (team) and materials, and various emergency capability units are formed by combining with non-material elements such as planning, organization and leadership, education and training, and drill evaluation. The integration and configuration of emergency response capability is to integrate the units of emergency response capability with specific organizational system, operational mechanism, division of responsibilities, resource allocation, actual combat drills, experience and lessons, etc., and form systematic arrangements for dealing with certain emergencies. These arrangements include various plans, schemes, guides, manuals, etc. The application and verification of emergency response capability is to test the emergency response capability through scenario simulation and emergency response. Social, political, economic, legal and cultural foundations, as well as laws and regulations, government system, social development level, scientific and technological level, education level, economic strength, principles and policies, theoretical methods, emergency culture and risk awareness, not only constitute the basic social environment for emergency preparedness, but also provide support for emergency preparedness in terms of funds, manpower, resources, theories, methods and policies.

(2) **System and structure of emergency preparedness**

Li Husheng's research group analyzed the current situation and challenges of emergency preparedness' strategic planning from the aspects of the weak links of emergency management in China, the public security situation and the limitations of emergency preparedness' planning, and the research group comprehensively applied the planning methods based on scenarios, tasks and capabilities, and put forward the emergency preparedness' planning system and structure that is in line with China's actual conditions (Fig. 3.9). The planning system includes six aspects: (i) Policies and regulations. It refers to laws, regulations and government normative documents related to public security and emergency management, and these are the legal basis for carrying out emergency preparedness planning. (ii) National strategy. It refers to the macro strategy, basic vision and key fields of the country in public security and emergency management in a period of time. (iii) Theoretical norms. It refers to the theories, methods and standards of emergency preparedness planning. (iv) Strategic planning. It refers to the macro-strategic arrangement of different aspects of emergency preparedness and key incident scenarios, including defining missions, determining the departments of responsibility in Fig. 3.9's emergency preparedness planning system and structure, describing responsibilities, determining key capabilities and goals, etc. (v) Construction and planning. It refers to planning of capacity building and development planning. (vi) Emergency plan. It refers to operational plan and scheme.

(3) **Framework of dealing with general tasks**

The Emergency Response Law of the People's Republic of China divides emergency response activities into emergency prevention and preparedness, monitoring and early warning, emergency response and rescue, and post-incident recovery and reconstruction. Foreign emergency management theory and practice divide emergency management into prevention, preparation, response, recovery and disaster reduction stages. Li Husheng's research group comprehensively analyzed the advantages and disadvantages of classification methods at home and abroad, and divided the mission areas of emergency management into prevention, preparation, disaster reduction, monitoring and early warning, emergency response, recovery and reconstruction, etc., covering all aspects of emergency management activities. The general task framework for emergency response is obtained by applying the analysis method of task fields (Fig. 3.10).

(4) **Structural logic and cultivation strategy of emergency culture**

Han Chuanfeng's research group of Tongji University discriminates the cultivation elements of China's existing emergency culture from three dimensions: depth, width and evolution, as shown in Fig. 3.11.

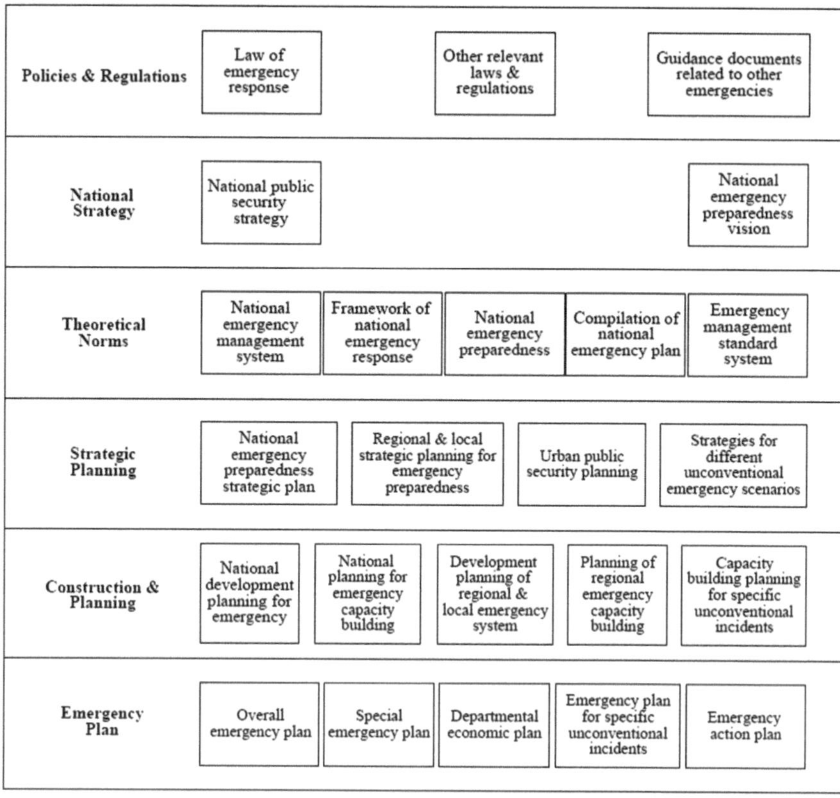

Fig. 3.9 Structure of planning system for emergency preparedness

(i) The depth of emergency culture is manifested as the hierarchical structure of emergency culture, including object, specification and potential basic assumption. The object refers to the objects and behaviors of the emergency culture and their mutual connection, which reflects the deeper content of the emergency culture. The specification is not only a fixed form suitable for material culture, but also the main mechanism and carrier for shaping spiritual culture. The potential basic assumption is the nervous system of the whole emergency culture, and the spiritual pillar and soul of the emergency culture.

(ii) The width of the emergency culture, namely the factors that national and regional emergency management involves, and it involves individuals, organizations and society (group). Individual factor mainly focuses on personal emergency awareness and emergency behavior; Organizational factor involves the main responsibility body of the emergency management, that is, the national management system, relationship among organizations and emergency culture level of social organizations that participate in the emergency management in all stages; Social factor focuses on the overall aspect of society, including the

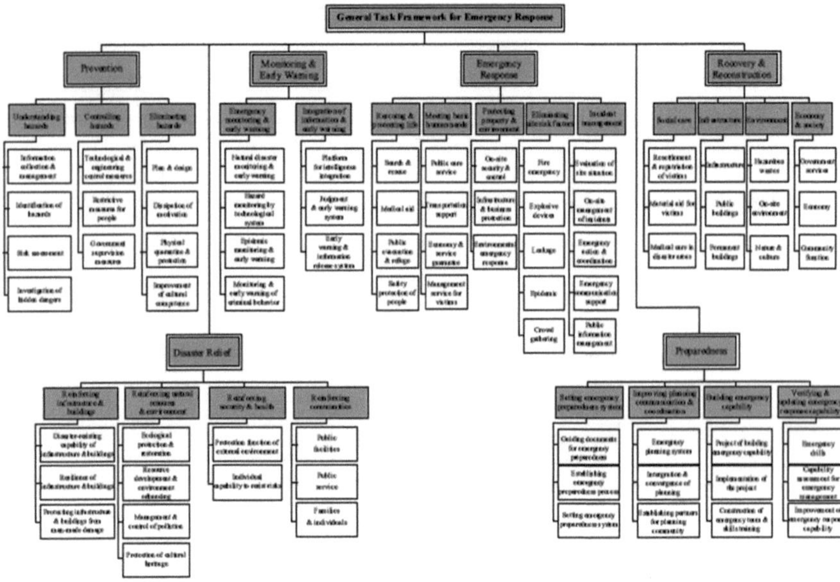

Fig. 3.10 General task framework for emergency response

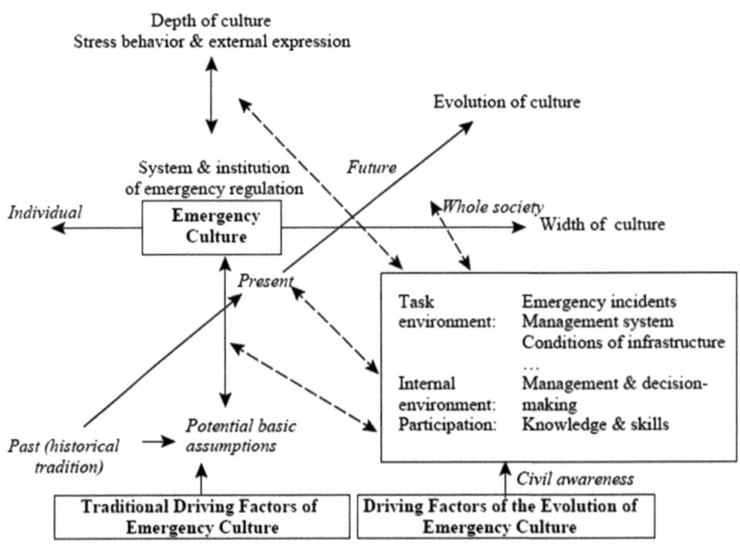

Fig. 3.11 Dimensions and elements of cultivating emergency culture

natural field, the economic field, the political system, and the impact of the cultural system on emergency culture.

(iii) The evolution of emergency culture is formed and developed in the process of the subjects' solving the problems of emergencies' external adaptation and internal integration, and it consists of different processes and states.

Han Chuanfeng's research group analyzed the cultivation mechanism of emergency culture according to the traditional driving factors of emergency culture called "potential basic hypothesis" and the evolution driving factors of emergency culture. The efficient and scientific cultivation mechanism of emergency culture system should include capacity improvement mechanism, goals' achievement mechanism, system integration mechanism and value introduction mechanism (Fig. 3.12). These mechanisms cover the functions of the four cultivation elements of emergency culture: "potential basic hypothesis," "task environment," "internal environment" and "participation degree." They are mutually infiltrated, supplemented and promoted, and it is difficult to study separately. From the aspects of laying the material foundation of emergency, optimizing the emergency system and shaping scientific emergency values, the research puts forward the cultivation strategy of emergency culture, so as to maintain the potential model of value, maintain the basic model of social common values, and keep it institutionalized in the system, and finally achieve the goal of emergency management and promote the harmonious development of society.

The above research fruits have comprehensively formed the theoretical framework of emergency preparedness and planning, provided direct technical support for the

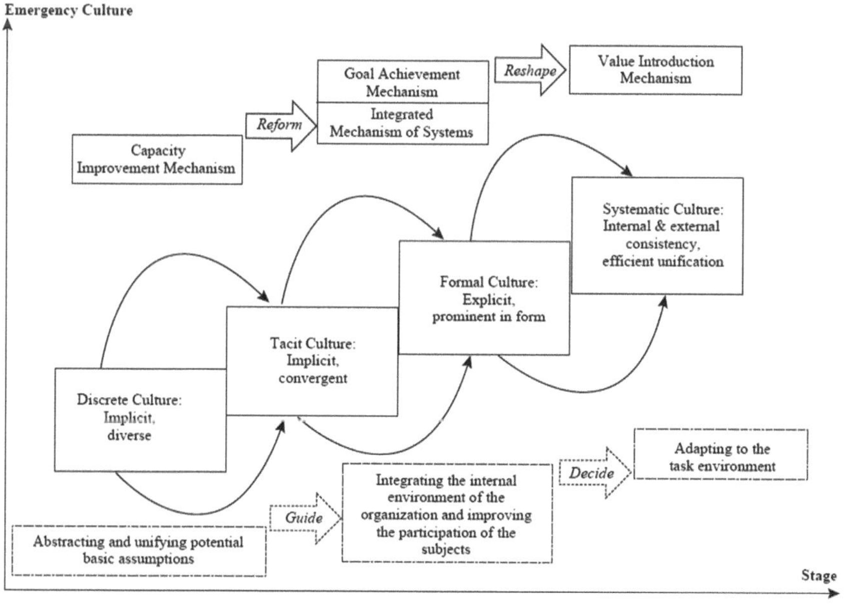

Fig. 3.12 Cultivation mechanism of emergency culture

preparation of the national emergency response system of "11th Five-Year Plan," "12th Five-Year Plan" and "13th Five-Year Plan," and improved the scientificity of China's emergency response system's planning and construction.

1.3.2 Theory and Method of Emergency Plan's Optimization for Unconventional Emergencies

(1) **The emergency plan system's structure and classification**

Li Husheng's research group of China Academy of Safety Production put forward the overall structure of emergency plan system which is suitable for China's national conditions from the aspects of structural prototype, management method, compilation guide, important infrastructure protection plan and work continuity. Through taking the national overall plan and action plan, local governments' comprehensive emergency action plan, enterprises', institutions' and communities' action plans as objects, considering the differences in the influence scope, severity and complexity of emergencies, respectively analyzing the functional characteristics of emergency plans at different levels, and proceeding according to the uncertainty of emergencies and the certainty of emergency responsibilities, the idea of constructing a robust emergency plan system is put forward. The emergency plan system's structure and classification are shown in Fig. 3.13.

(2) **Management model of plan based on life cycle**

Rong Lili's research group of Dalian University of Technology believes that the life cycle of the plan system depends on its entropy increase and entropy decrease. In the research, entropy is used to express the mess degree of the plan system, and the change trend of the plan life cycle's curve is calculated by entropy value. Based on the plan's life cycle curve, the plan's life cycle curve based on information entropy can be established (Fig. 3.14). Because there are differences in the life cycle of different plans (Fig. 3.15), it is necessary to identify the key driving factors, including time and non-time driving factors, to realize the continuous improvement of emergency plans according to requirements of different plans. After that, it is necessary to establish a multi-variable plan system's evolution model, and calculate the planned and the actual average revision time of plans (Fig. 3.16). The results show that the effectiveness of the plan system decreases with the increase of the probability of promulgating plans, and the number of laws and regulations that are removed when plans are revised increases first and then decreases.

(3) **Revision strategy based on internal influencing factors of plans' effectiveness**

Rong Lili's research group identified the internal influencing factors of the effectiveness of the plan from the aspects of laws, regulations, exercise effect, etc., defined the life cycle of all kinds of plans and the dynamic management process of plans, collected and analyzed the case data of revision of plans and exercise cycle, and put

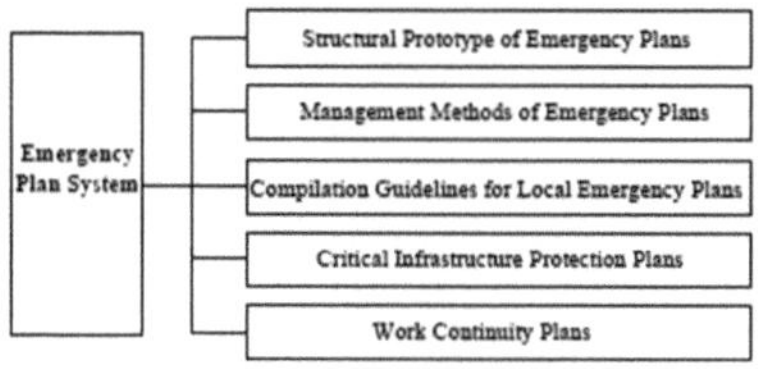

System	Characteristics				
	Targets	Application units	Basic tasks	Elastic demands	Refinement
National overall (strategic) plan	Catastrophe & other crises	The State Council and its constituent departments	Coordinating and unifying major support actions	Limited flexibility not concerned with specific activities	High uncertainty and only functioning as principle requirements
Integrated action plans of local governments (provinces and cities)	Disasters or emergencies with large impact	Local governments at or above the county level and their subordinate departments and units	Coordinating and unifying command and organizing on-site emergency response	Certain flexibility and specific arrangements for evacuation, protection and other functions	Uncertainty and clear behavior route
Stress behavior plan of villages, towns, communities, businesses and institutions	Common accidents or incidents	Grass-roots organizations or businesses and institutions	Organizing on-site rescue activities as the "first response"	Much flexible in dealing with random situations	Clear task and meticulous arrangement
Emergency plan of government and its departments	Overall emergency plan	Stipulating basic principles, organizational system and implementation mechanism of emergency response, the overall arrangement of first-level emergency support, and clarifies the responsibilities and tasks of all parties concerned			
	Special and departmental emergency plans	State level	Provincial level	Municipal & county level	Township level
		Response principles, command mechanism, early warning grading and information reporting requirements, reflecting policy and guidance	Organization and command mechanism, information reporting requirements, graded response, material support and mobilization	Command mechanism, risk assessment, monitoring and early warning, information report, disposal measures and material mobilization procedures	Distribution of early warning information, early disposal, self-help and mutual aid, information collection and report, temporary resettlement of personnel, etc.
	Protection plan for critical infrastructure and lifeline projects	Confirming risk information and preventive measures, forecasting and early warning, information report, emergency disposal, and emergency recovery			
	Guarantee plan for major activities	Focusing on clarifying activities' safety risk information and relevant preventive measures, forecasting and early warning, information report emergency disposal, and organization of evacuation and route			
	Joint emergency plan	Focusing on clarifying emergency linkage mechanisms such as information notification, connection of disposal measures, and sharing of emergency resources between neighboring local people's governments and their departments			
Emergency plans of units & grass-roots organizations	Focusing on clarifying the main bodies of emergency response, risk information monitoring, information report, early warning, emergency response, organization of evacuation and route, and emergency resources that can be called or mobilized for assistance, reflecting the characteristics of self-help and mutual aid, information report and early disposal				

Fig. 3.13 Emergency plan system's structure and classification

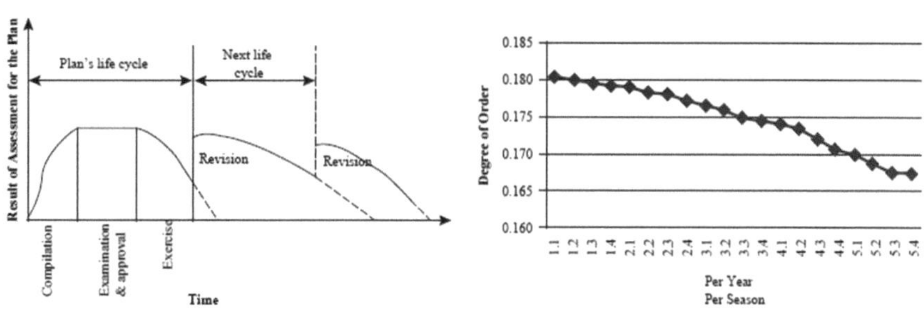

(a) Life cycle curve

(b) Life cycle curve based on information entropy

Fig. 3.14 Plan's life cycle curve and the plan's life cycle curve based on entropy

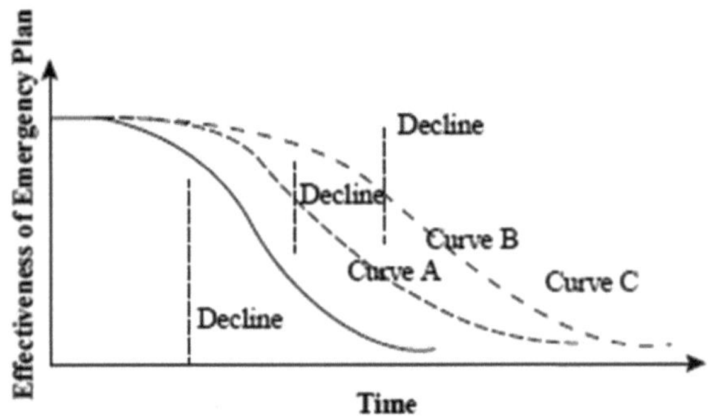

Fig. 3.15 The difference of plan's life cycle curve

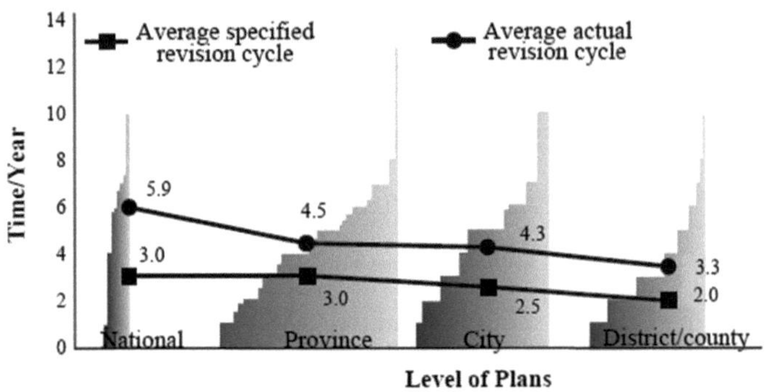

Fig. 3.16 The planned and the actual average revision time of plans

forward the strategy of plan's revision. The revision strategies consider the changes of laws and regulations, and the effect of drills and all internal factors are shown in Figs. 3.17, 3.18 and 3.19 respectively. The results show that the relevant plans must be revised in time after the laws and regulations referenced in the preparation of the plan eventually change; After the exercise effect reaches a certain threshold, the plan needs to be revised; To determine the revision cycle of a single emergency plan, it is necessary to comprehensively consider all internal factors affecting the effectiveness of the plan.

(4) **Digital model of emergency plan**

Rong Lili's research group paid attention to the emergency decision-making and response needs of emergencies, and comprehensively constructed the hierarchical network model of plans (Fig. 3.20), considering the main body layer, theme layer, task layer and resource layer. At present, the release of plans in natural language is characterized by discreteness and centralization, which will easily cause difficulties in finding the information of the plan. Therefore, a prototype system of plan management based on hierarchical network model is constructed. The system has

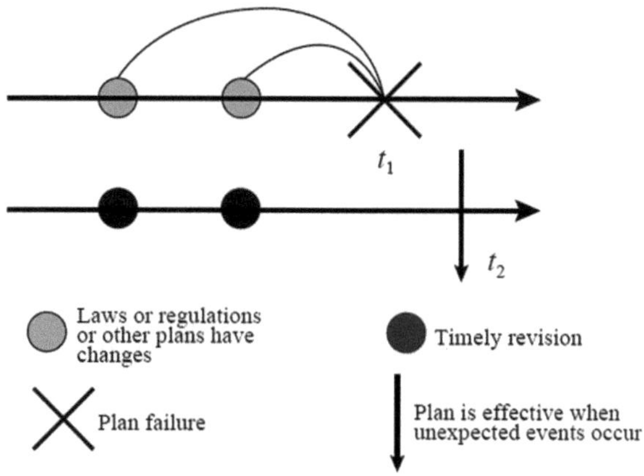

Fig. 3.17 Revision strategy considering changes in laws and regulations

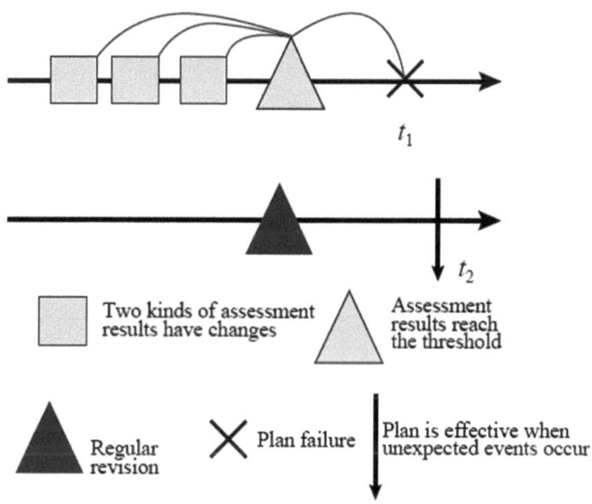

Fig. 3.18 Revision strategy considering exercise effect

the following functions: realizing the digital management and query in the fields of plans, laws and regulations, disaster cases and deduction models; according to the four major emergencies, realizing the deduction of emergency response plan; according to the model stored in the system, the development and evolution of disasters can be deduced; Establishing plan system of effectiveness evaluation and providing revision suggestions. The system promotes the investigation and evaluation of the national emergency's plan system, puts forward opinions and suggestions on further improving China's emergency's plan system, and provides theoretical basis and technical support for the General Office of the State Council to compile and issue the Management Measures of Emergency Plans.

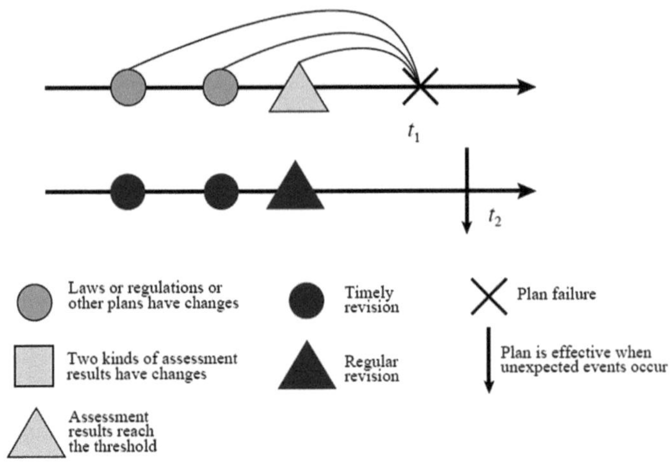

Fig. 3.19 Revision strategy considering all internal factors

Fig. 3.20 Hierarchical network model of plan

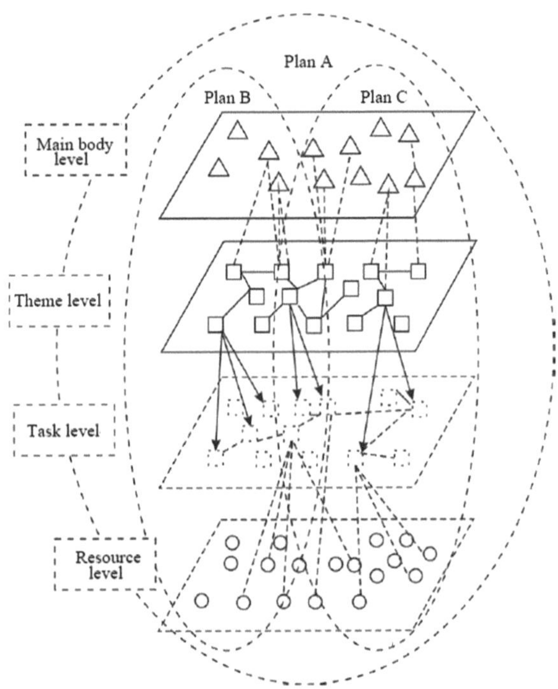

1.3.3 Top-Level Design Theory of Emergency Management System with Chinese Characteristics

(1) **Experiment and computing method of China's emergency management system**

Han Chuanfeng's research group of Tongji University clarified the hierarchy of emergency management organization's system, and grasped the laws of flow of information, resources, funds and other elements among different levels. In terms of the coordination level, based on NK model, the model with great adaptability and of emergency management and organization can be constructed, and the adaptability characteristics of organizational system with centralized, decentralized and hierarchical decision-making modes in different situations, the influence mechanism of organizational culture, decision-makers' ability, information flow and other factors were studied by the group, so as to optimize organization model and organization relationship's coordination. As for the security layer, the scenarios' characteristics of urban waterlogging were analyzed, and the emergency resources' allocation model was constructed. The location and resource allocation of rescue center were optimized by considering the probability of waterlogging's occurrence and the uncertainty of rescue needs, as well as logistics cost and relevant price. The coordination of emergency organizations can be transformed into the functional coordination corresponding to emergency organizations, the emergency functions of different organizations can be abstracted, the stochastic programming model can be established, and the capital investment should be coordinated, so as to improve the functional output of organizations in response to natural disaster. The ultimate goal of organization functions' coordination is to optimize the system. As for the handling level, Markov decision-making model was constructed by the group. Facing the uncertain situation characteristics of emergency management, a dynamic decision-making model based on Markov decision-making model was proposed to support emergency management and decision-making. With comprehensive integration of emergency management's model and method, development of dynamic decision-making support's prototype system, clear logic structure and operation process of prototype system, optimization of system and strategy selection, we can achieve optimal resource allocation and support for managers' decision-making. The research framework of China's emergency management system's calculation experiment is shown in Fig. 3.21.

(2) **Reconstruction of emergency management system in China**

Based on the prerequisite of the emergency management system, Han Chuanfeng's research group systematically analyzed the system's constructive role in the legal system, mechanism and plan, explored and determined the due status of the legal system and the catalytic function of the mechanism and plan, and clarified the connotation, function and interaction relationship of "One Case, Three Systems." This paper analyzes the dynamic structure characteristics of emergency management system in America and China, and puts forward the dynamic structure analysis

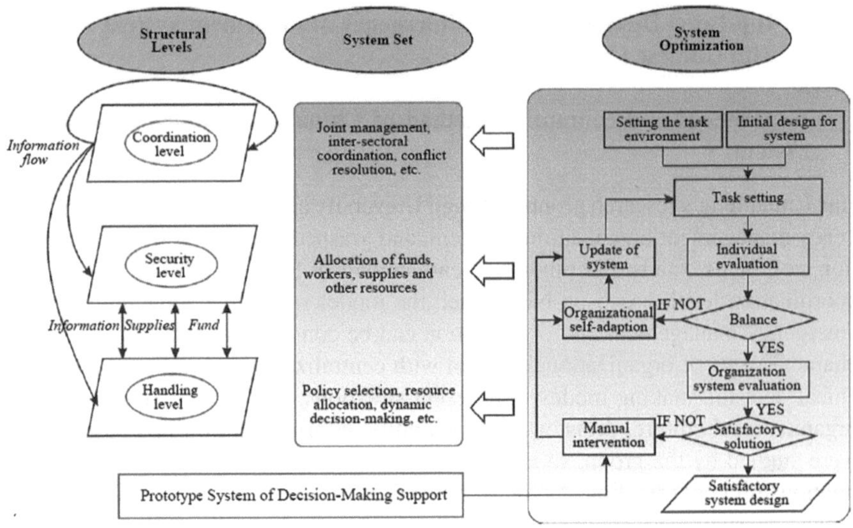

Fig. 3.21 Research framework of computing and experiment of emergency management system in China

framework of emergency management system, which provides a new perspective for the research of emergency management system.

"One Case, Three Systems," namely, emergency system, emergency mechanism, emergency legal system and emergency plan (Fig. 3.22), is an institutional system established under certain rules to clarify the main bodies of action and their role orientations, behavior mode and behavior boundary. In essence, "One Case, Three Systems" is an abstraction of system activities, which concentrates on the core contents of emergency management system, and clarifies the main bodies, norms, methods and procedures of emergency management, which are related and complementary to each other and jointly constitute an organic whole with specific functions. (i) Emergency system. It is the basis of system operation, emphasizes the relationship among different main bodies, takes power as the core and organizational structure as the main content, and clarifies the organizational form and functional division of emergency main bodies. It is equivalent to the hardware in man–machine system and plays a prerequisite role. (ii) Emergency mechanism. It mainly refers to the standardization and institutionalization of management methods and measures of emergency legal system and emergency system. It emphasizes the process and efficiency of main bodies' interaction, takes operation as the core and disposal methods as the main content, and solves the problems of organization, coordination, information communication, resource scheduling, etc., and it is equivalent to the software in man–machine system and plays a supporting role. (iii) Emergency legal system. It mainly refers to the standardization of the operation of emergency management's organization system, which emphasizes the constraints and boundaries of main bodies'

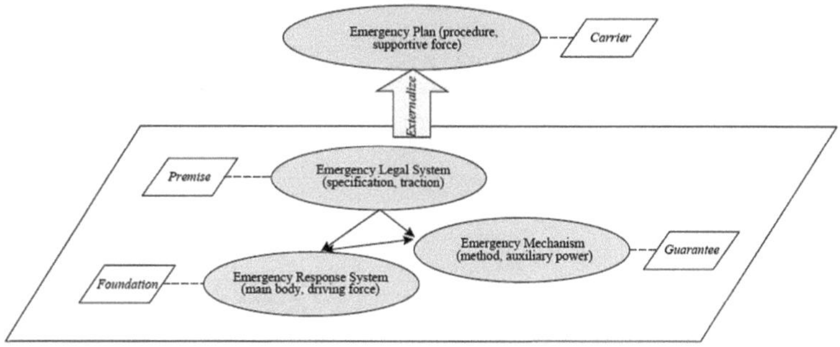

Fig. 3.22 The internal logic of "One Case, Three Systems"

behavior in emergency management, stipulates the setting and functions of institutions, clarifies command and coordination mechanisms such as command transmission and information communication, regulates the behavior of individuals and organizations, regulates the operation of the system, and plays a guiding role. (iv) Emergency plan. It is the externalization of the functional relationship among system, mechanism and legal system, which constitutes the carrier of system's operation, emphasizes the micro-level operation and execution of the main bodies, refines the emergency management process according to the requirements of legal system and the reality of system and mechanism, and promotes the orderly and efficient emergency management. The emergency system is the foundation and the driving force; The emergency mechanism is a guarantee and an auxiliary driving force; Emergency legal system is the premise and plays a role as traction; Emergency plan is the carrier and the supporting force. "One Case, Three Systems" is the internal driving force for the operation and evolution of emergency management's organization system.

The existing emergency management system has exposed a series of structural defects, such as dislocation of emergency handling's subjects, poor relationship and poor mechanism, and these must be improved and upgraded by top-down design. It is found that the reconstruction of emergency management system should pay attention to the following issues: (i) the concept issue, that is, the positioning and interaction among government, market and society; (ii) institutional issue, namely the horizontal and vertical relations in the emergency management's organization system; (iii) mechanism issue, that is, paying attention to the operational efficiency of the emergency management system; (iv) tool issue, it is necessary to make clear the choice of policy tools for emergency management. Therefore, it is necessary to combine the new trends of theoretical presuppositions' renewal, change of governing way, social structure's change and social risks' aggravation faced by China's emergency management, and it is necessary to adhere to the concept of pluralistic governance, take institutional innovation as the core, and put forward the model principles, functions and reconstruction strategies of China's emergency management system from the aspects of reshaping emergency management's organizational culture, changing emergency management's organizational structure, improving emergency management's organizational mechanism, focusing on emergency management's personnel

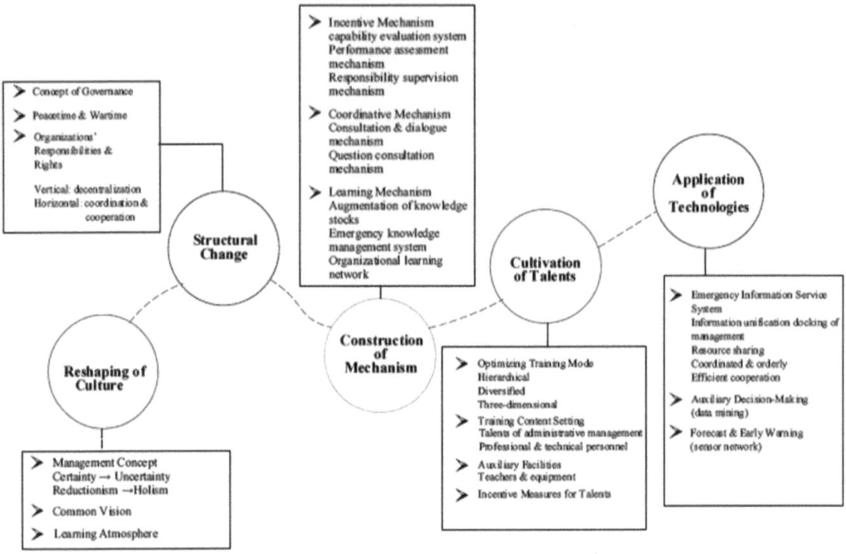

Fig. 3.23 Reconstruction strategy of China's emergency management system

training, and strengthening the application of modern information technology, etc. The reconstruction strategy of China's emergency management system is shown in Fig. 3.23.

1.3.4 Theory of Emergency Organization Design for Unconventional Emergencies

(1) Dynamic reconstruction mechanism of emergency management's organization structure

Unconventional emergencies make the internal and external environment of emergency organizations constantly change, so it is urgent to grasp the law of organizational restructuring in changeable scenarios. Based on fractal theory, Zhao Qiuhong's research group of Beihang University studied the dynamic reconstruction mechanism of "scenario-response" organization of emergency management. From the aspects of emergency management's system, mechanism, legal system and emergency plan, this research group analyzed the present situation of China's emergency management system, and put forward a framework for emergency management system based on collaborative thinking, which has improved the existing emergency management's organization structure in China to highlight the professionalism of decision makers at all levels and the synergy of multiple departments. The three-level emergency organization structure is shown in Fig. 3.24.

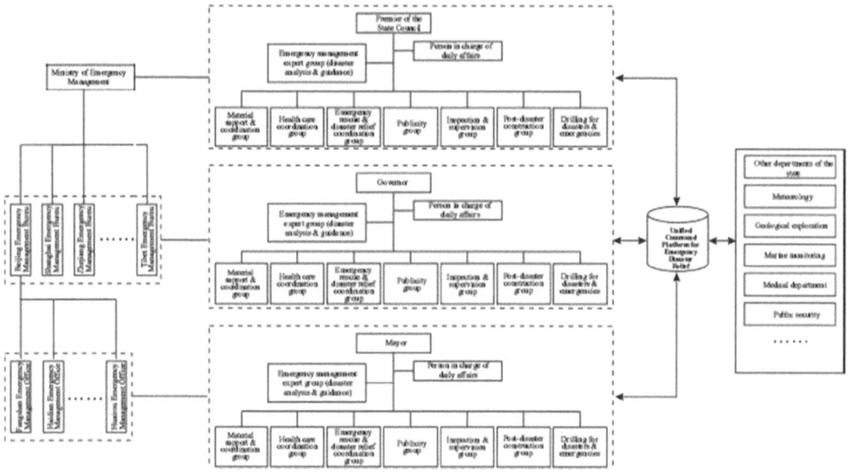

Fig. 3.24 Three-level emergency organizations' structure

From the perspective of complex system, the organizational system of unconventional emergencies is studied by using the fractal system methods of relationship-driven and project-oriented. Under the framework of collaborative emergency organization, taking the relationship between supply and demand and "entrustment-execution" among emergency units as the main line, the fractal element of emergency organization is generated, the system of fractal emergency organization is constructed, and the dynamic reconstruction mechanism of emergency organization in different scenarios is put forward. Taking "8/12" Tianjin major production security accident as an example, this research group studied the reconstruction process of fractal emergency organization from three dimensions: emergency task, emergency handling capability and emergency command, and the group verified the efficiency and flexibility of emergency organization's reconstruction under fractal cooperation.

(2) **Ability and performance evaluation of emergency organization**

Zhao Qiuhong's research group summarized the theoretical research and practical experience related to emergency response's capability evaluation, analyzed the factors affecting capability of emergency response in different stages of unconventional emergency management, constructed the evaluation index system in different stages of unconventional emergency management according to the principles and steps of construction evaluation index system, and put forward evaluation methods. The ability and of emergency organization and relevant performance evaluation include the completeness evaluation of emergency plan, the order evaluation of emergency organization, the ability evaluation of emergency monitoring and early warning, the ability evaluation of emergency response, the performance evaluation of emergency organization during disaster, and the perfection evaluation of post-disaster emergency management system.

(i) Completeness evaluation of emergency plan. The advantages and disadvantages of the emergency plan are related to the final result of the whole emergency management. The existing emergency organization analyzes and evaluates the unconventional emergencies' support ability of the emergency plan from four dimensions: material support, personnel support, infrastructure support, policy and non-governmental organization support, and constructs an evaluation index system for the completeness of the emergency plan.
(ii) Evaluation of emergency organization order. Using structure entropy method, the order degree of emergency organization in flood control in Beijing and Shenzhen was evaluated. The results show that the timeliness and quality of information circulation are very important in the organization structure of flood control. When the emergency response system is put forward, there are too many levels of information transmission, and the management span between levels is too large, which easily leads to problems such as delay and distortion of information transmission. In addition, aiming at the problems of material supply and management in China's rescue, new material resources' distribution organizations have been put forward, and the order degrees of the old and new organizations have been compared by using structure entropy method.
(iii) Evaluation of emergency monitoring and early warning ability. Taking unconventional emergencies and their secondary and derivative incidents as monitoring objects, the evaluation index system of emergency monitoring and early warning ability of unconventional emergencies is constructed from four dimensions: monitoring ability, identification ability, loss assessment ability and early warning ability.
(iv) Evaluation of emergency response capability. Emergency response capability refers to the ability of decision makers to make response plans and issue instructions in emergency rescue in time by using various information provided by relevant departments. The evaluation index system of emergency response ability of unconventional emergencies is constructed from four dimensions: disaster assessment ability, command and coordination ability, emergency rescue ability and information transmission ability.
(v) Performance evaluation of emergency organization in disaster. Using the method of establishing indicators of Balanced Score Card for reference, this research group constructed the performance evaluation index system of emergency organization in disaster from four dimensions: citizen (rescue ability), input control, internal operation and relevant improvement. Based on the influencing factors of performance evaluation of emergency organization during disasters and the feedback relationship among them, taking the "7/21" torrential rain incident in Beijing as an example, the system's dynamics simulation was carried out. The results show that improving the level of organizational division and coordination has the greatest impact on comprehensive performance, followed by the adequacy of rescue workers, the sufficiency degree of resources and the consultation with experts; Under the disaster environment, considering the influence of various factors and constraint of budget, the performance of emergency organizations can be maintained at an appropriate level;

The research is helpful to establish a real-time performance monitoring system and assist managers to monitor the changes of comprehensive performance.

(vi) Evaluation of post-disaster emergency management system's perfection. The success or failure of unconventional emergency management also involves many factors such as disasters' loss assessment ability, post-disaster compensation ability, emergency plan's maintenance ability and post-disaster reconstruction ability. Drawing lessons from the experience and research results of emergency management in developed countries such as the United States and Japan, an evaluation index system for the perfection of post-disaster emergency management system is constructed.

(3) **Structure and operation mechanism of China's emergency response's cooperation network**

Han Chuanfeng's research group of Tongji University collected 4,270 special reports of Wenchuan earthquake, Yushu earthquake, Ya'an earthquake and Dingxi earthquake from Sina.com, www.ChinaNews.com, Sohu.com and ifeng.com, with a total of 420,000 words. Based on the vertical subordinate relationship among governments' emergency organizations at all levels, using the multi-layer network model of exponential random graph, facing the hierarchy and duality of inter-organization relationship, the relationship matrix of governments' emergency organizations in the above four earthquakes was established, and the cooperation network model of governments' emergency organizations was constructed (Fig. 3.25). The structure's logic of cooperation network was analyzed to explore the consistency of cooperation between higher and lower levels of governments' emergency organizations. The results show that the number of national emergency organizations is directly proportional to the number of people affected by disasters (number of victims + number of injured + number of missing persons); National emergency organizations can increase the concentration trend of the whole emergency network. The larger the proportion is, the faster the information will be transmitted in the whole network and the more frequent the information communication between organizations will be; Organizations with resources such as information and materials are more likely to cooperate with each other, and county-level emergency organizations are closely related to each other, but they are on the edge of cooperation network; Among the top ten central organizations, the national emergency organizations account for the largest proportion.

Han Chuanfeng's research group analyzed the influence of affiliation relationship on the multi-layer cooperative network's structure, and studied the overall construction mode of emergency organization cooperation. Using the secondary assignment procedure, the research group analyzed the correlation among decision transmission network, information communication network, resource flow network, and the level, task and nature of the emergency organizations in Wenchuan earthquake. Combined with organizational relationship theory, this research group analyzed the interaction mode of multi-dimensional relationship among emergency organizations, and analyzed the structural logic of multi-dimensional network interaction and

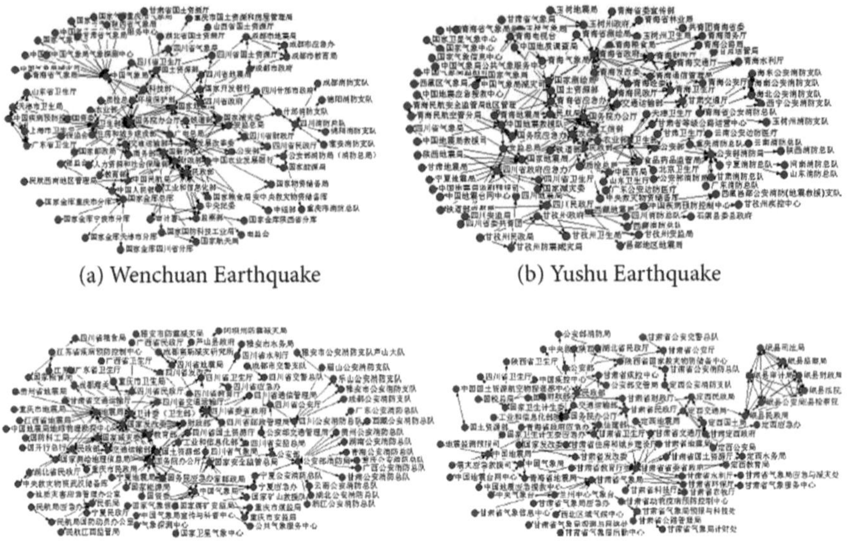

(a) Wenchuan Earthquake (b) Yushu Earthquake

(c) Ya'an Earthquake (d) Dingxi Earthquake

Fig. 3.25 Cooperation network's model of government emergency organization for four earthquakes

the function of emergency organizations' attributes by using multi-network interaction, method and technology. From three aspects (decision transmission, information communication and resource flow), this research group explored the macro-operation mechanism of emergency organizations' cooperation.

Taking Wenchuan earthquake as an example, Han Chuanfeng's research group simulated and analyzed the command's transmission network, information communication network and resource flow network of Wenchuan earthquake's emergency organization. The results show that in the command's transmission network, the centralized power is a prominent characteristic, and the command's transmission is obviously centralized, and there are key delivery organizations; Network closure is not quite obvious, and there is no distributed decision-making; The results verify the homogeneity theory, and the command's transmission among emergency organizations with the same level, task and nature is obvious. In the information communication network, the information's interaction relationship is remarkable; There is obvious concentration, with certain degree of power centralization; High-order network's closure is obvious, the path is shortened, and it is locally gathered. There is no connection among organizations that send information to multiple identical organizations, but there is connection among organizations that receive information from multiple identical organizations. Command's transmission among emergency organizations with the same level and task is obvious, but the nature has no significant effect on information transmission. In the resource flow network, centralization is a prominent characteristic, command's transmission is obviously centralized, and there are key transmission organizations; Network closure is not obvious, and there is no distributed decision-making; Organizational attributes have no significant effect on inter-organizational resource flows.

1.3.5 Legal Guarantee System for Emergency Management of Unconventional Emergencies

(1) **Reconstruction of emergency legal system**

According to Ma Huaide's research group of China University of Political Science and Law, the existing emergency law system with Emergency Response Law of the People's Republic of China as the core is obviously different from the requirements of unconventional emergency response. In order to meet the urgent needs of dealing with unconventional emergencies, it is necessary to reconstruct norms for China's emergency legal system with the main goal of enhancing the adaptability of the law. The specific measures include: designing the organizational system and command organization to ensure the maximum capability of emergency response; making institutional arrangements for ensuring and raising emergency resources; providing enough power for emergency decision-making of unconventional emergencies; standardizing the end of unconventional emergencies and the restoration of normal legal order; clarifying the functional orientation of emergency law in emergency management.

(2) **Perfecting the legal system of risk regulation**

Ma Huaide's research group studied the current situation, problems and perfection of China's legal system of risk regulation based on the law enforcement's inspection of the Standing Committee of the National People's Congress in the field of emergency risk regulation, and held that "Emergency Response Law of the People's Republic of China" regard prevention as the first principle of emergency response, but the effective prevention of emergency needs to rely on the perfection of the risk regulation system under normal conditions, and put forward some suggestions on propelling legalization of risk regulation activities from the aspects of reaffirming the government's security function, strengthening the responsibility of regulatory agencies and improving regulatory measures.

(3) **System design of administrative laws for emergency decision-making**

Ma Huaide's research group believes that in modern society, when the effective response of government departments to the needs of the public becomes the legitimacy basis of their management, the administrative measures formulated by them to prevent extreme incidents sometimes will be irrational. In order to ensure the legitimacy of administrative measures, it is necessary to reform the administrative legal system, including establishing a long-term risk management and evaluation mechanism to reflect, evaluate and filter the public's requirements. And it is necessary to realize the transition of administrative measures through rules. The application period of administrative measures should be stipulated by the Sunset Clause, and it is necessary to ensure justice through compensation mechanism, and at the same time increase the rationality of administrative measures.

1.3.6 Theory and Model of Emergency Resources' Guarantee and Coordination Optimization

(1) **Forecasting method of emergency resource demand based on system dynamics**

Reasonable evaluation and prediction of the types and quantities of rescue materials needed is the premise of efficient distribution and transportation of emergency materials. Emergency material demand for unconventional emergencies is characterized by complex influencing factors, inconsistent dimensions of various factors and lack of historical data. It is necessary to construct emergency material allocation's decision-making scenarios for large-scale emergencies according to disaster attributes, economic and social indicators of the affected areas and already-delivered materials. Wang Xuping's research group of Dalian University of Technology applied three kinds of methods from the perspective of similarity to measure the distribution scenarios of materials in each disaster area, and established a scenario-based assessment method for the demand proportion of emergency materials. The results of analyzed cases show that, compared with the existing methods, the proposed method can not only deal with the mixed scenarios in which accurate values, interval values and fuzzy values exist at the same time, but also reflect the preferences of decision makers.

Facing the problem of earthquake material allocation with dynamic changes of road conditions, based on the prospect theory, considering the rescue center and the disaster places, the risk perception functions of different bodies of responsibility are constructed to measure the risk degree of insufficient and delayed material supply of bodies of responsibility. Besides, a system with dynamics simulation of the whole process of material allocation considering the behavior of bodies of responsibility is established, including road capacity evaluation, material flow, decision-making process, material demand and other modules. Considering the causal relationship of the whole distribution and transportation of relief materials, the research has put forward the prediction method of materials demand of post-earthquake emergency from the perspective of emergency materials' supply-distribution-consumption system, which can comprehensively and systematically simulate the coordination and decision-making effect of earthquake disaster's relief materials, deduce the emergency materials demand in disaster-stricken areas, and display the supply, inventory and transportation of materials in real time. The results of case analysis show that when the decision-making attitudes of the rescue center and the disaster-affected subjects are inconsistent, the decision-making preferences of the disaster-affected subjects have a great influence on the joint decision-making; If the subjects are optimistic, the delivery frequency of rescue center and the material supply frequency of disaster relief areas are higher; if the subject is pessimistic, the single delivery volume and inventory of rescue center are relatively higher.

(2) **Dynamic optimization model and algorithm of emergency resources**

When the rescue resources are restrained, timely, effective and dynamic scheduling of emergency materials is the key to rapid rescue response. Wang Xuping's research group aimed at minimizing the loss of victims and the cost of vehicle dispatching, built a mixed integer programming model, optimized the decision-making processes such as the route selection of vehicles and the allocation of emergency materials under the constraint of transportation capacity, applied a hierarchical solution strategy to narrow the solution space, improved the coding method based on customers, and designed a genetic algorithm and solution model. The results of case analysis show that when the capacity changes in a certain range, the rescue effect is continuously optimized with the increase of the capacity of the material distribution center, the total penalty cost is reduced, and the delivery time is reduced; If too many rescue vehicles are put into use, the rescue time will not be shortened and the operating cost will increase.

(3) **Integrated zoning method for the support of emergency resource**

Under the background of unconventional emergencies, Huang Jun's research group of Chinese Academy of Sciences studied the layout of emergency resources' security zoning and central reserve, and established an integrated regional security system of emergency resources to ensure supply of resources. Firstly, considering the administrative region, geographical location, disaster type, traffic conditions, climate type and other factors, the coordination index of zoning was constructed, and the initial zoning scheme of China's emergency resources' guarantee is obtained by using clustering method. Then, considering the requirements of timeliness, reliability and balance of resource guarantee, the concepts of multi-coverage degree for satisfaction rate and multi-coverage degree of demands were put forward, and considering resource demand, worst scenario and advantage scenario, the minimum cost's layout model based on satisfaction rate of coverage and multi-coverage degree, the maximum coverage's layout model based on multi-level support in demands and the coverage model based on reliability of emergency logistics network were established. According to the calculation results, the layout adjustment scheme for central reserve was given. Finally, taking the overall situation, reliability, timeliness, balance and economy as indicators, the Data Envelopment Analysis Method was used to compare and analyze the layout schemes of the central reserve, and finally it was determined that the central reserve warehouses should be located in Tianjin, Shenyang, Harbin, Hefei, Zhengzhou, Wuhan, Changsha, Nanning, Chengdu, Xi'an, Kunming, Urumqi, Jinan, Hangzhou, Nanchang, Shijiazhuang and Hohhot.

(4) **Design and coordination's optimization model of blood security system**

Ma Zujun's research group of Southwest Jiaotong University analyzed the characteristics of emergency blood support in unconventional emergencies from the aspects of need for blood, blood collection, blood supply, blood's clinical use and blood security. According to the current situation of blood security work at home and abroad and the construction requirements of emergency blood's security system, the framework of emergency blood's security system in China was designed. Based on the emergency command system, the national, provincial and local command systems of emergency blood support were established, and the coordination optimization's model and algorithm of unconventional emergencies' blood support were put forward. Firstly, according to the characteristics of different stages of emergency blood support, the emergency blood demand's forecasting model based on Logistic curve characteristics, the model of emergency blood's information updating based on Grey Envelope-Markov Chain and the model of emergency blood's combination forecasting based on Grey-Artificial Neural Network were proposed to optimize the emergency blood's demand forecasting in unconventional emergencies. Secondly, the location-allocation model of national and strategic blood reserve, two-stage heuristic algorithm based on Tabu Search algorithm, Blood Reserve Bank's rotation and update strategy, emergency blood's dynamic collection model and genetic algorithm are put forward to optimize emergency blood's reserve and collection plan. Finally, the emergency blood banks' choosing-allocation model and Vector Coding Genetic Algorithm, the optimization model for choosing the blood banks and transportation routes and Genetic-Tabu hybrid algorithm for emergency blood distribution, the bi-level programming model and Greedy Heuristic Algorithm of emergency blood's allocation considering blood types' substitution, the optimization model and analytical algorithm for backlog blood's emergency transportation based on blood bank age were put forward to optimize the emergency blood's allocation model.

2 Monitoring and Early Warning

2.1 System and Model of Monitoring and Early Warning Technologies

2.1.1 Internet of Things Technology and Method for Unconventional Emergencies

(1) Internet of Things technology and method for active perception and emergency command

Although the Internet of Things technology has been widely valued and gradually become the research focus, the active perception and emergency command of the Internet of Things system for unconventional emergencies are still in the initial stage, and only focus on the sensor layer, network layer and servers' hardware layer. The research on the processing and intelligent analysis of massive heterogeneous perception data is rare. Thus, it is difficult to meet the requirements of unconventional emergency management. Ding Zhiming's research group of Institute of Software,

Chinese Academy of Sciences, combined the Internet of Things technology and emergency management technology, applied individual perception, group statistical perception, prediction perception through model and other technologies to analyze emergencies and make relevant early warning. Through unified management and status tracking of emergency resources, as well as optimal scheduling of multiple resources and tasks, the coordination of emergency command was realized, and new methods and new ideas were put forward for emergency management of unconventional emergencies. The framework of active perception and emergency command of unconventional emergencies based on Internet of Things technology is shown in Fig. 3.26.

(2) **Mass data storage technology of Internet of things compatible with "key-value" query and logical condition query**

In the fields of sensors' data management, spatio-temporal database and mobile objects' database, some scholars have studied the storage technology of spatio-temporal data collected by a single type of sensors, such as global positioning system, GPS) and radio frequency identification (RFID). However, there is no feasible scheme for unified representation, query and interoperability of massive heterogeneous sensors' streaming spatio-temporal data. Ding Zhiming's research group proposed a method of unified database representation, query and processing for all

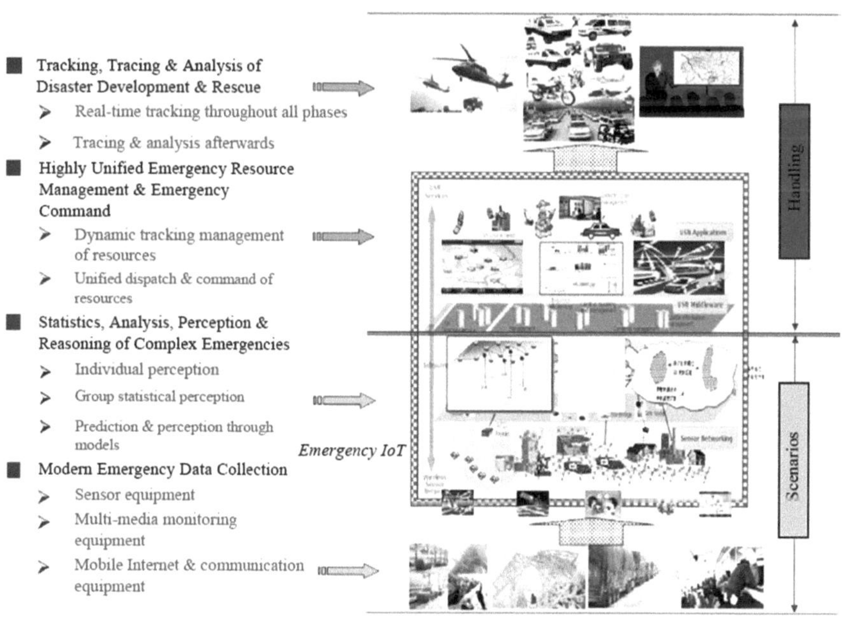

Fig. 3.26 Active perception and emergency command's framework of unconventional emergencies based on Internet of Things technology

kinds of heterogeneous sensors' streaming spatio-temporal data, constructed spatio-temporal data types, query operation and query language, and data stream's spatio-temporal index mechanism, and unified representation, processing and query of all kinds of sensor data sequences.

Existing researches mostly apply the "Top-level Reconstruction" method. Based on the original "Key-Value" data management layer, the research group added one layer to support relational database's query language. Because of the complexity of model's transformation, the "Top-level Reconstruction" method was inefficient. Ding Zhiming's research group proposed a massive data storage mechanism of Internet of Things that is compatible with "Key-Value" query and logical conditional query. By taking the advantages of the underlying model combined with relational database and "Key-Value" database, and by supporting and optimizing the global data distribution, global index and global query mechanism, the research group met the requirements of efficient management query and processing of massive data of Internet of Things in unconventional emergencies. A "Massive Data-Cloud Processing" collaborative storage, query and processing model for massive sensing data of Internet of Things has been proposed, which effectively improves the system's big data storage capability.

(3) **Distributed perception mechanism of emergencies based on massive data flow and incidents' perception rules of Internet of Things**

Emergency Internet of Things contains a large number of sensor nodes and computer nodes. In order to meet the principle of minimum business radius and ensure the efficiency of task processing, incidents' perception rules need to be stored and executed in a distributed way, but it is still paid little attention to in the existing researches. Ding Zhiming's research group put forward a distributed active detection mechanism for unconventional emergencies. A large number of early warning rules were implemented in the sensors' access layer. The reasoning rules of the emergency perception and command layers completed complex data analysis, ensuring the efficiency of incidents' perception; A kernel-level statistical analysis model of Internet of Things' perception data based on database cluster was proposed to support the statistical analysis of Internet of Things; An emergency detection technology based on dynamic perception data of Internet of Things was proposed to realize accurate perception of incidents. A hierarchical model of social sensor network was proposed, which effectively improved the emergency perception performance of the system; The research group also put forward the technology of measuring the influence of emergencies based on massive perception data, which improved the ability of detecting and warning for emergencies; The evolution model of groups' abnormal situation was put forward, and the group behavior's perception was realized. The disaster tracking technology based on the evolution of incidents' situation was proposed, which improved the emergency handling ability of the system. The distributed perception process of emergencies based on massive data streams and incidents' perception rules of Internet of Things is shown in Fig. 3.27.

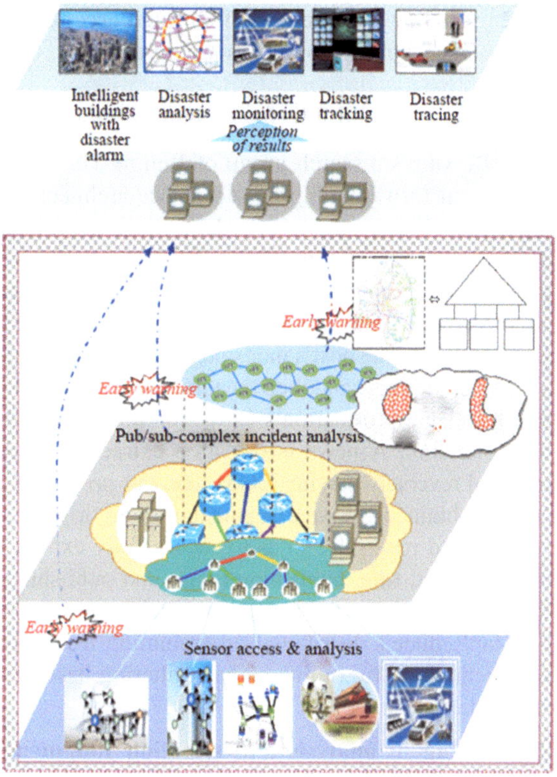

Fig. 3.27 Distributed incidents' perception process based on massive data flow and incidents' perception rules of Internet of Things

(4) **Unified emergency command and dispatching mechanism based on time and space logic**

Ding Zhiming's research group studied and put forward the unified time–space and state management methods for emergency resources with fixed positions such as warehouses and rescue agencies, emergency resources with mobile positions such as rescue vehicles and emergency logistics vehicles, multi-channel positioning methods for emergency resources with mobile positions (such as GPS and RFID), and position tracking methods based on unified time–space trajectory model; The research group also put forward the hierarchical and batch representation model of emergency resource management to realize the effective management of massive emergency resources; An intelligent and continuous dynamic navigation mechanism based on traffic conditions, meteorological conditions, disaster situations and other factors was proposed. A spatio-temporal multi-resource and multi-task matching method based on space cost and time constraint and a heuristic resource scheduling strategy based on multiple metrics were proposed to optimize scheduling of emergency resources. The research group also put forward the dynamic traffic guidance and evacuation technology based on the development trend of incidents, so as to improve the ability of emergency command and dispatch.

2.1.2 Theory and Method of Online Perception and Early Warning of Unconventional Emergencies

(1) Extraction and filtering technology of online crisis intelligence

Fang Binxing's research group of Beijing University of Posts and Telecommunications put forward a high-performance architecture of multi-channel online crisis intelligence's acquisition. In order to meet the requirements in scale and flexibility of intelligence information's acquisition, a distributed directional acquisition architecture of "master-subordination distribution and independent coordination" was adopted. The active collection technology facing Weibo, forums and social networks was proposed. The post's index page (column page) can be located by man–machine assistance strategy, and then the target post's link can be accurately located by accurate webpage information's extraction algorithm, which can completely eliminate the noise link. A news information extraction method based on page segmentation and reversely parsing DOM tree was proposed. The main areas of news pages can be obtained by page segmentation algorithm, and their DOM structures can be analyzed reversely, so as to successfully extract key news information such as news content, title, publishing house and publishing time. An efficient text classification technology for intelligence information's extraction was proposed. Based on MapReduce's distributed programming framework, an efficient text classification technology for online garbage filtering based on vector space model has been realized.

(2) Intelligent analysis and situation judgment technology of multi-source crisis intelligence

Fang Binxing's research group proposed a mining method of atomic event's evolution pattern. Based on spectral clustering method, atomic events (including time, place, people and other features) were extracted from the text corpus. According to the similarities in content and time of events and the co-occurrence frequency of events in multiple documents, the correlation degree of events was judged and the evolution relationship of events was mined. A multi-dimensional vector-based Weibo emotion analysis model was proposed. Emotion words were extracted from the perspective of clinical psychology to form an emotion vectors' multivariate model. Based on online Weibo data stream, emotion vectors were monitored in real time, mainstream emotion patterns were mined, and users' emotion evolution was detected by analyzing temporal characteristics. The research group put forward an emotion evolution's analysis method for security incidents, defined the relative strengths of positive and negative emotions, depicted the emotional evolution state, confirmed that the main factors affecting the emotional evolution are emotional factors, social influence factors and dynamic factors, designed calculation methods, and monitored public emotion fluctuations based on security-related data.

(3) Uncertainty analysis of network events' propagation and evolution process

The complexity and suddenness of network events make it difficult to analyze the uncertainty of the propagation and evolution of network events. Therefore, we should study the characteristics of different types of network events and the uncertainty of their distribution, and model and calculate the trueness of network events. It is necessary to propose a method to measure the uncertainty of network events based on feature distribution. Firstly, the factors affecting the uncertainty of network events should be analyzed from the aspects of events' source website, source webpage and events' attributes. Secondly, it is necessary to use matrix operation, analyze the calculation process, and put forward the simulation algorithm that simulates factors affecting the uncertainty of network events. Finally, the effectiveness of the network events uncertainty's measurement algorithm can be verified.

(4) **Early detection and early warning technology of online abnormal signs**

Fang Binxing's research group put forward the multi-sequence emergencies' detection technology based on Hidden Markov Model, applied the improved Markov transformation model and the model with two observable time sequences to predict the emergencies' outbreak, fitted the outliers by introducing jumping points, found the most possible hidden variable sequence by optimizing methods, and made early warning of the occurrence time of potential emergencies. An early warning technology for abnormal security incidents based on spatio-temporal scanning statistics was proposed. All kinds of emergencies can be counted by spatio-temporal scanning, and relevant events can be monitored in real time. The significant difference between local and global Poisson Distribution was compared, and unknown risk areas and periods with spatio-temporal abnormal aggregation were found to realize early warning of emergencies.

The research results have been widely used in the analysis and early warning of emergencies such as violence and terror, covering more than 5,000 source channels of social networks such as Weibo, blogs and forums. And it fully support online crisis intelligence's collection, intelligent analysis and situation judgment, online and offline interaction mechanism, anomaly detection and early warning, dynamic intelligence navigation and human–machine interaction, design and optimization of online crisis command and control mechanism, and providing online emergency perception, intelligence analysis and early warning, crisis intelligence's navigation and other functions (Fig. 3.28 Structure of Application Demonstration).

2.1.3 Multi-source Data's Mining and Integration Theory of Unconventional Emergencies

Zhang Hui's research group of Tsinghua University proposed the overall framework of emergency data's management system for emergencies (Fig. 3.29), and it dynamically collects, cleans, integrates data in real time. The overall framework of emergency data's management system is divided into three layers: (i) Data source layer. It needs to effectively manage and integrate multi-source data from relevant departments, Internet and on-spot collection. (ii) Data processing layer. It includes three key technologies: data extraction, data integration, and storage for entities and their relationships. (iii) Application layer. It develops and applies entities and

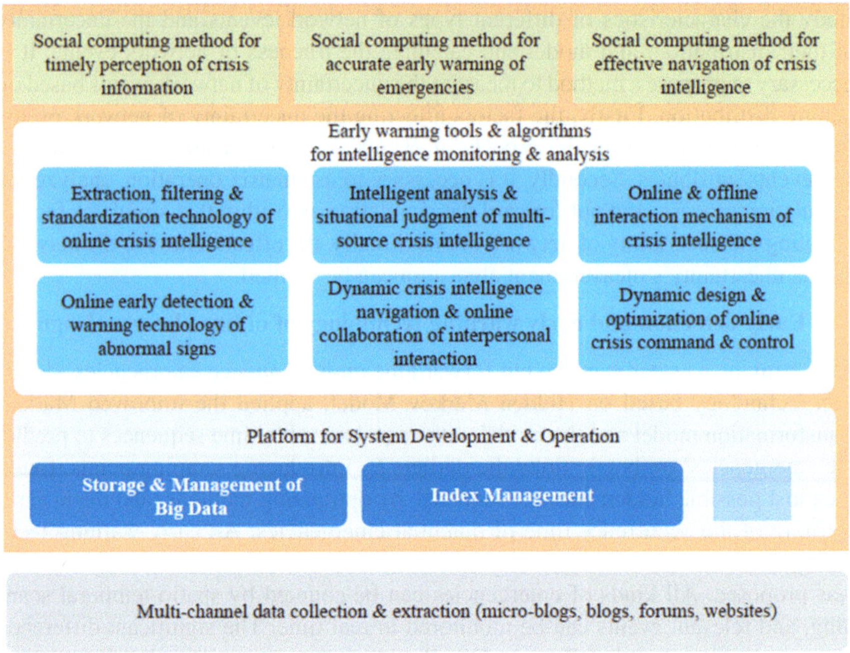

Fig. 3.28 Structure of application demonstration

their associated databases obtained from the data processing layer, so as to meet the demand of decision makers in using Web data.

Sensation and mining technologies of sources of emergency data include data extraction, classification methods and Web information's credibility identification methods, etc. According to the requirements of emergency management scenario data's integration, multi-source data's integration technology was adopted, data sources were analyzed, description files were summarized and established, and domain-oriented Web data's integration technology was applied to integrate semi-structured and unstructured data sources into structured data sets.

(i) Data storage and query technology. Zhang Hui's research group proposed a framework of data storage and management system based on "dual-core" and cloud, developed a prototype system of cloud data management, TaijiDB, proposed a query and scheduling algorithm for backup data to improve query parallelism and ensure load balance among nodes, proposed an efficient and extensible multi-dimensional index architecture, proposed an online aggregation system based on MapReduce in cloud environment, and developed an online aggregation system, COLA.
(ii) Sharing and using standards of emergency data. To solve the problem of data sharing in unconventional emergencies, PAVLL open-source scheme is adopted to realize the construction of distributed virtual network. Based on the service

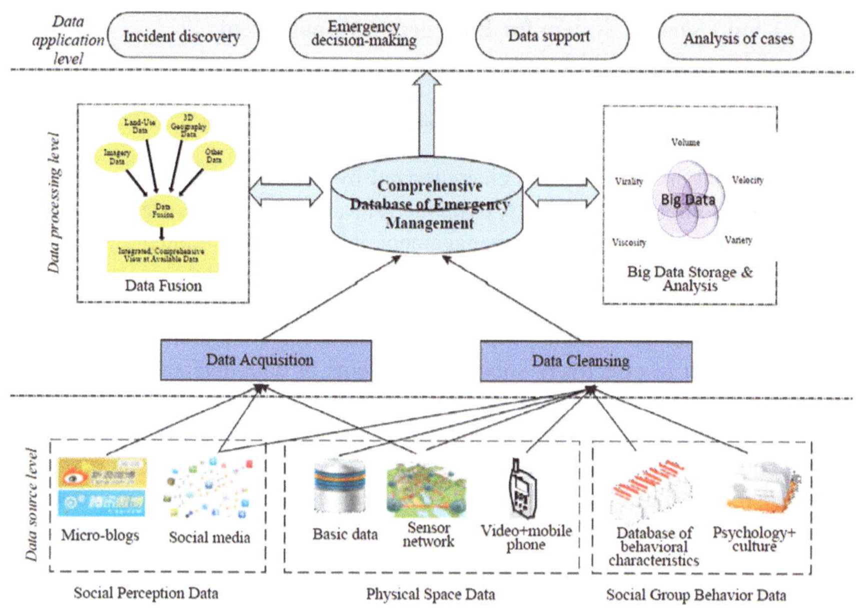

Fig. 3.29 Overall framework of emergency data management system

support of cloud computing technology, an experimental platform for unconventional emergencies' data processing and analysis is initially established and designed, which provides large-scale intelligent computing methods and computing support for relevant subjects. Parallel computing services based on Internet architecture for optimizing models and remote shared and parallel computing resources for some models of the platform are provided.

(iii) Trajectory data's privacy protection technology. To solve the problem of leakage of personal privacy in trajectory data's releasing, the research group proposed trajectory data's privacy protection technology based on different attack models, including trajectory privacy's protection methods that distinguish location sensitivity, trajectory privacy's protection methods based on prefix tree, and attacks through trajectory reconstruction that can infer private locations.

(iv) Mobile Internet query technology. Considering the characteristics of mobile Internet such as openness, sharing, collaboration and innovation, aiming at providing users with efficient and accurate query results and high-quality services, three new keyword query technologies on Internet have been proposed: collaborative "space-tag" keyword query, multi-keyword route query with limited access sequence, and query through mobile Internet users based on trust mechanism.

2.1.4 Cloud Service System for Emergency Management of Unconventional Emergencies

(1) Emergency management's cloud service architecture

Unconventional emergencies are characterized by suddenness timeliness, massive information, complexity, diversity, dynamic relevance and high uncertainty. When we make decisions on emergency management, it is necessary to integrate data from various fields, and consider various information characteristics and complex scientific issues throughout the life cycle. The existing emergency management systems are mostly target-oriented systems, which are aimed at the combination of functions, and it is difficult to effectively solve the problems of data diversity and technical complexity of unconventional emergency management.

Starting from the needs of typical emergency management cases such as economy, finance and foodborne diseases, Li Jianhui's research group of Computer Network Information Center designed a universal and dynamic extensible emergency management cloud service system, formed a complete theory for emergency management cloud service system and methods suitable for China's national conditions, provided extensible service interfaces for different emergency needs, and provided a series of emergency management services based on cloud computing technology for monitoring, early warning, information reporting, emergency handling, aftercare and improvement of emergency plans. The emergency management cloud service architecture (Fig. 3.30) mainly includes three parts: infrastructure, platform and software, and it can quickly allocate resources and support real-time emergency-making and emergency management.

Infrastructure as a Service (IaaS), serves emergency management systems, researchers and decision-makers by virtualizing hardware resources such as processing, storage and network, that is, it can deploy and run any software, including operating systems and applications. Platform as a Service (PaaS) provides an open application programming interface (API) or development platform for researchers in emergency management and emergency management systems, including emergency data, emergency models and visualization services. Software as a Service (SaaS), provides Internet-based emergency software and application services for researchers, decision makers and the public. Users can enjoy the service only through browsers/handheld devices, and it can favor multiple clients to share the same application. It also designs interface-related specifications, which connects the interfaces among different levels of cloud service system, including cloud service resource pool's specifications and protocols in service interface, physical computing resources' application protocols and access interfaces, virtual computing resources' application protocols and access interfaces, storage resources' application protocols and access interfaces, and network resources' application protocols and access interfaces. In this way, it provides standards and specifications for cross-level communication.

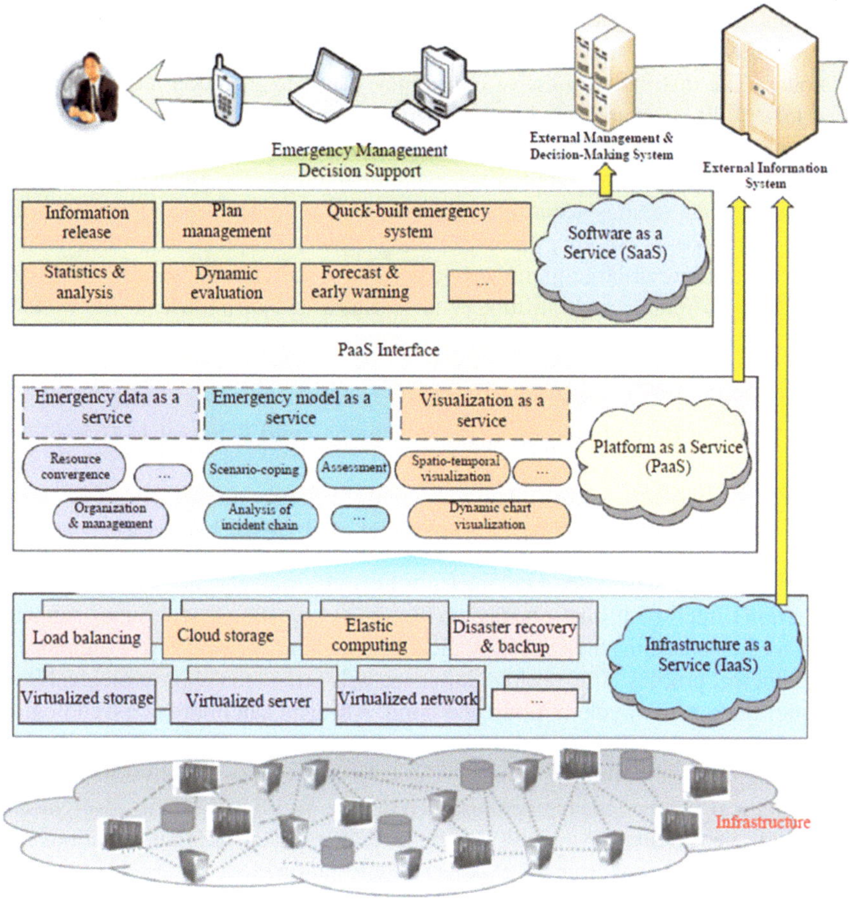

Fig. 3.30 Emergency management cloud service architecture

(2) **Key technologies of emergency management cloud service system**

Li Jianhui's research group proposed a hierarchical and modular design and description language, DVDL (Data Visualization Description Language), which combined with the model-driven theory and End-user Development Theory to realize the platform of data visualization. The platform supports 10 types of data sources, provides 11 kinds of visualization models including maps, and supports a variety of interactive components. Through WYSIWYG (What You See Is What You Get) configuration, visualized results can be generated and released just with one click. This platform can not only provide users with data analysis, display and other services, but also provide support for rapid decision-making of unconventional emergency management.

The unified management and scheduling method of storage and computing resources has invented the core technology of unified management and scheduling of distributed cross-domain heterogeneous storage and computing resources, solved problems such as network security, cross-domain delay, local caching, and realized rapid deployment of various basic supporting applications for tens of thousands of nodes. The distributed massive emergency data's aggregation technology puts forward a multi-agent and collaborative data aggregation mechanism, which realizes automatic detection for targets, updating of research data, time and space partitioning, periodic scanning, intelligent screening, intelligent acquisition of metadata, and caching data entities according to demand. A big data processing framework, NoPar, which does not modify the model to segment data, is proposed, which realizes parallel computing of large-granularity data. The reading performance (800 MB/s) of the 100 TB remote sensing images is 10 times that of traditional file systems (MooseFS, HDFS).

2.1.5 Laws of Individual and Group Psychology and Behavior in Crisis

(1) **Model verification of physical and mental interaction effectiveness—from body to mentality: Self-motivation effect of altruistic behavior**

Xie Xiaofei's research group of Peking University innovatively put forward the "Self-motivation Effect" of altruistic behavior (Fig. 3.31), revealing that altruistic behavior can make helpers adjust their psychological resources and bring positive influence to psychological process. This positive influence will extend to the physiological level and start the positive cycle of body and mind. Altruism can improve psychological and physiological utility, and can offset the reduction of external utility caused by the cost and consumption of helpers. Three basic variables in physiological and psychological perception, temperature perception, feeling of bearing weight and expression sensitivity, have great significance on survival in crisis situations and have been selected to explore the concrete manifestation of altruistic self-motivation effect. Through a series of strictly controlled laboratory experiments and on-spot experiments based on real crisis incidents, and it is confirmed that altruism can effectively improve temperature perception and expression sensitivity, and reduce the feeling of bearing weight. The research reflects the typical results of interaction between physiological system and psychological system, and verifies its internal path from mentality to body and the special adaptability of altruistic behavior in crisis situations.

(2) **Model verification of physical and mental interaction effectiveness—from body to mentality: pain and sense of unfair feeling**

Pain is a basic variable that plays an important role in crisis response. Based on the existing pain experimental paradigm, combined with behavioral and functional magnetic resonance imaging techniques, Xie Xiaofei's research group explored the interaction between individual altruistic behavior, sense of fairness and pain perception. The results show that altruistic behavior can effectively reduce subjective pain, and can predict unfair experience through difference of individual pain sensitivity. Experiments have proved that physiological system can affect mental state, and

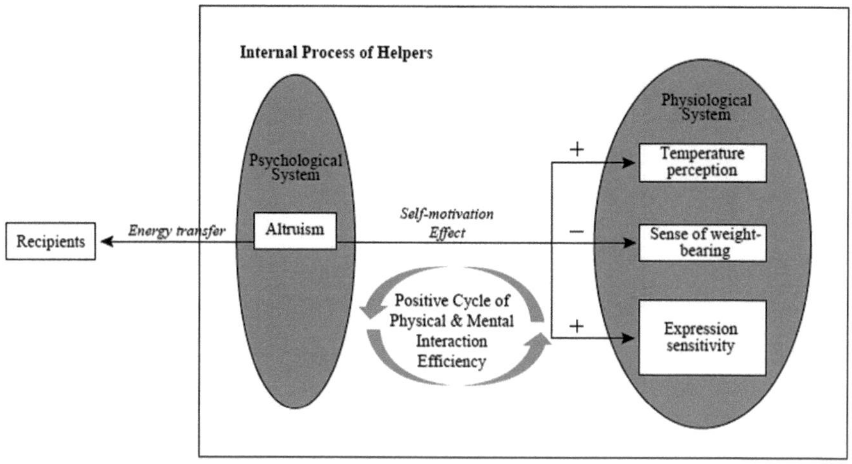

Fig. 3.31 Self-motivation effect of altruistic behavior

positive psychology has a positive feedback effect on physiological system, and has an important effect on relieving pain in daily life and crisis situations. In a crisis situation, physical pain and psychological pain are easy to form a negative cycle. Thus, creating a fair environment is conducive to preventing the negative cycle (the physical-mental interaction efficiency model is shown in Fig. 3.32).

(3) **Special laws of cognition and behavior in crisis situations**

Threat to security, time pressure and high uncertainty are three typical characteristics of crisis, which easily make individuals show special laws of cognition and behavior. Exploring altruism based on crisis scenario is a unique perspective to solve the worldwide difficulty of "human's nature." The special laws of cognition and

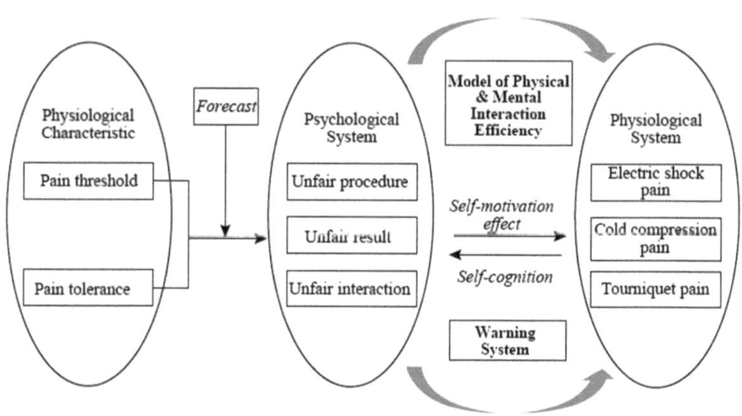

Fig. 3.32 Physical-mental interaction efficiency model

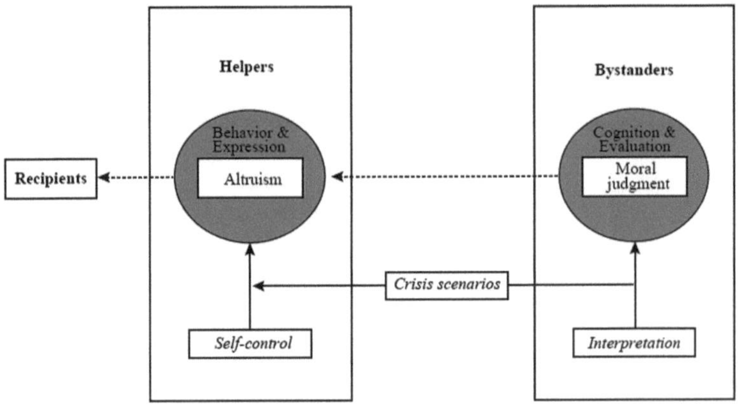

Fig. 3.33 Special laws of cognition and behavior in crisis situations

behavior in crisis situations are shown in Fig. 3.33. Xie Xiaofei's research group found that when the level of individual self-control is low, there will be automatic and instinctive response. If human's nature is good, when the level of self-control is low, human's moral instinct will produce more altruistic behaviors; On the contrary, it will lead to more egoistic behaviors. In addition, people's moral judgments on altruism and egoism are different because of the crisis degree and interpretation level of the situation. When perceiving at a high level of explanation, bystanders' evaluation of negative events is more acute. In the crisis situation, the degree of crisis and the level of interpretation have interactive influence on the moral judgment of bystanders. From the perspective of evolution, altruistic behavior has a positive impact on individual's coping with crisis, which helps to improve the adaptability of altruists and make them survive in natural selection. Thus, it is a virtuous circle.

(4) **Detection indicators of stress and its prediction of brain functions and behaviors**

Zhang Kan's research group of Institute of Psychology, Chinese Academy of Sciences, focused on the characteristics that unexpected events can lead to people's stress state, studied the detection indicators of stress and the influence of stress on behaviors, and took special police, soldiers, post-disaster residents and other key groups in China as objects, and examined the influence of long-term stress caused by sudden acute stress and daily events on people's behaviors, and the research group also studied the measurement and judgment of stress state. Combining with the analysis of neurological, biochemical and behavioral data, the research group found the sensitive indexes of judging stress state represented by cortisol reaction. It was found that under stress state, people's cognitive function, especially executive function, will be reduced, while their emotions are more excited, but it will lead to the reduction of people's emotional control ability and lead to irrational behaviors; Using stress-related physiological and biochemical indicators such as cortisol in the morning, we can predict the individual's brain function and behavior control ability. This is due to the reduction of neuroendocrine resources, which reduces the prefrontal control function and leads to the decline of individual's ability to control their own behaviors. The research results provide an important theoretical basis and practical method for the measurement and intervention of group psychology and behavior when emergencies occur.

(5) **Cognition's processing efficiency of basic cognition function under emotion state in disaster stress**

Based on the theory of post-disaster crisis ossification and cognitive processing efficiency, Xu Yan's research group of Beijing Normal University verified the impact of disasters' negative emotion immersion on the effectiveness and efficiency of non-automated cognitive processing. The blood flow in bilateral prefrontal regions (the target brain region of Stroop Effect) was examined by fNIRS technique, which was used as an index of psychological energy. Compared with the group with neutral emotion, there was no difference in error rate and response time in the group under the situation of disaster with stress emotion, but the target brain regions (channels 7, 17 and 18) was more active (Fig. 3.34). The results show that under the emotion state of disaster stress, individuals have to pay more psychological energy to complete the same tasks. The Near Infrared technology was applied for the first time in this study, which indirectly explained the existence of psychological energy and verified the loss of psychological energy after the disaster. In the state of disaster emotion immersion, individuals can ensure the execution quality of cognitive activities by consuming more psychological energy, and the effectiveness of cognitive processing has no obvious change, but the processing efficiency (cognitive effort) will decrease. It shows that under the interference of negative emotions, processing efficiency will be more sensitive than effectiveness of cognitive processing.

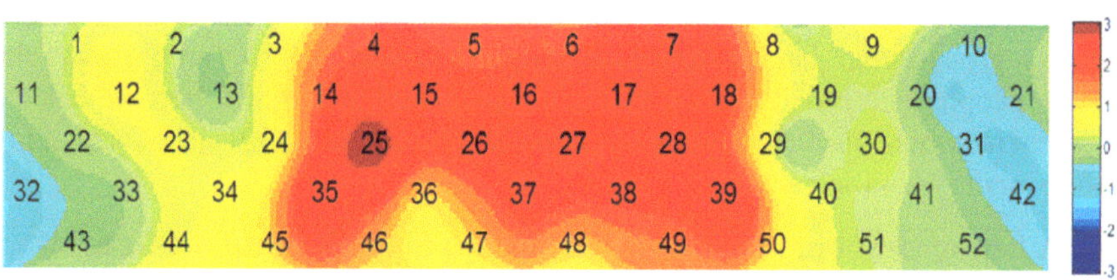

Fig. 3.34 Topographic map of stroop effect difference between group with disaster stress emotion and group with neutral emotion

2.2 Monitoring and Early Warning Methods of Typical Incidents

2.2.1 Monitoring and Early Warning Theory and Method for Major Network Public Opinion Incidents

(1) **Mining key nodes of public opinion's transmission in social network**

Unconventional emergencies can easily lead to rumors, which can be spread through the Internet, mobile phone short messages, and traditional face-to-face and mouth-to-mouth methods. Through the network, rumors spread most rapidly and do the greatest harm. Key node identification is the research focus and hotspot in the field of network science. Key nodes refer to special nodes in the network that can influence the structure and function of the network more obviously such as nodes located in the center of the network, or the most influential transmission sources of public opinion/epidemic. Accurate and efficient finding of crucial network nodes is helpful to better discover opinion leaders, predict the development of the situation, control the spread of rumors, restrain the outbreak of epidemic, and discover important pathogenic genes. Shang Mingsheng's research group of University of Electronic Science and Technology of China proposed a LeaderRank method for mining opinion leaders in social networks. Compared with PageRank algorithm, LeaderRank method has the advantages of high accuracy, good robustness and no parameters.

(2) **The formation mechanism of opinion leaders in information dissemination**

Refuting rumoring is one of the important means to control public opinion, and it is of great significance to dispel rumors by intentionally cultivating opinion leaders. Controlling opinion leaders can control public opinion obviously. In addition, through theoretical analysis of the formation mechanism of this structure, Shang Mingsheng's research group found that the formation of opinion leaders follows an obvious rule, that is, people with broad interest and better judgment ability will become leaders of social networks. And the finding has important reference significance for shaping online media opinion leaders.

(3) **A new model of information dissemination in small-world network**

Compared with disease transmission, information transmission has the following characteristics: Firstly, information transmission has the characteristic of memory, that is, after multiple rounds of transmission, the probability of being "infected" increases; Secondly, it has the characteristic of " being strengthened by the society," that is, a message may be suspected at first, but it may be believed after multiple rounds of transmission; Thirdly, each link is usually used only once. In view of this, Shang Mingsheng's research group put forward an information transmission model: every step of time, every individual is in a state of not knowing information, knowing information, confirming information, and it is an exhausting state. Through the experimental research in regular network, small-world network and random network, it

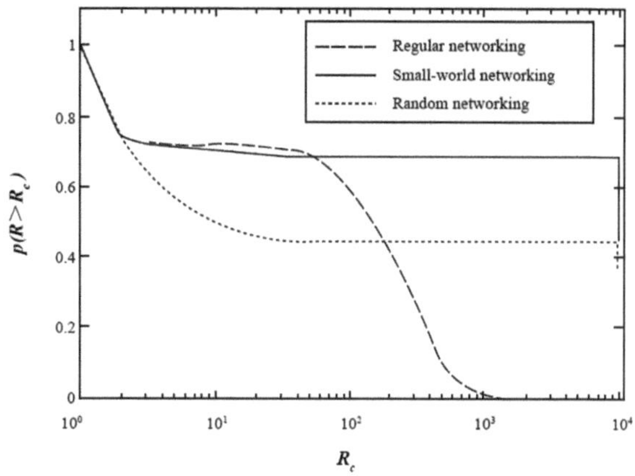

Fig. 3.35 Information's transmission speed in regular network, small world network and random network

is found that when the probability of believing just after receiving the information is small, regular network spreads faster and wider than random network, and small-world network has the best transmission effect. The transmission speed of information in regular network, small-world network and random network is shown in Fig. 3.35.

(4) **Visualization method of public opinion information**

Existing public opinion systems often use text, list, graphics and other forms when presenting results, which cannot directly reflect the relationship between public opinion and region. Shang Mingsheng's research group put forward a visualization method of Internet public opinion map, which can not only reflect the geographical distribution of Internet public opinion and the degree of attention in the distribution area, but also show the possible geographical distribution trends in the future. This method can improve the accuracy and immediacy of public opinion's prediction and early warning. By studying the overall dynamic evolution process of public opinion, it can provide decision-making support for the formulation of control strategies.

2.2.2 Monitoring and Early Warning Theory and Method of Crowded Stampede

(1) **Crowd tracking method based on scenario model of crowd**

Sun Jinhua's research group of University of Science and Technology of China fully analyzed the normal and abnormal behaviors of people, and combined with the motion model in the tracking field, the Social Force Model in crowd simulation, optimization issues and the inherent characteristics of video tracking, proposed a crowd scenario model that can be used for tracking and simulation. The framework of crowd behavior's analysis system based on crowd scenario model is shown in Fig. 3.36.

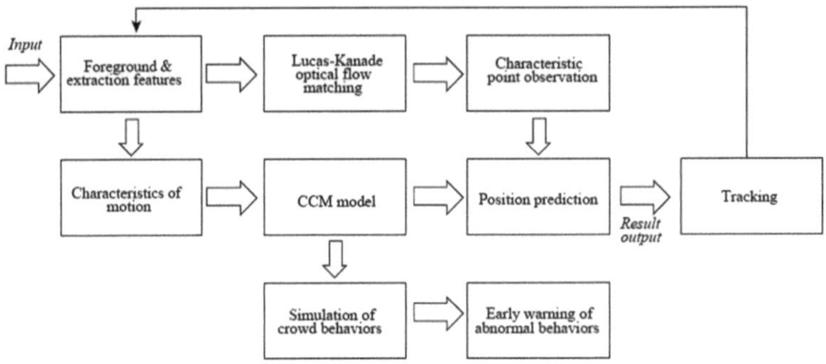

Fig. 3.36 Framework of crowd behavior's analysis system based on crowd scenario model

Combining feature point tracking and crowd simulation model of optical flow method, a crowd tracking method based on crowd scenario model is proposed. According to the idea of Kalman Filtering, the motion features in the video will be captured, the model will be driven to give a reasonable prediction, and then the prediction results can be corrected through observation. Therefore, the interaction between the model and the actual observation can be realized, and different scenarios have been selected for testing. The results show that the crowd tracking method can better cope with the loss and confusion of targets caused by obstructing. The actual scenario and simulation scenario are shown in Fig. 3.37 and 3.38 respectively.

(2) **Detection of abnormal crowd distribution based on fractal dimension**

Natural disasters, traffic accidents and social security incidents will cause abnormal crowd distribution and movement patterns. With the wide application of GPS and the popularity of smart phones, it becomes increasingly easier to obtain people's physical location. Yang Su's research group of Fudan University gridded the city. Facing the non-uniformity of crowd distribution on the grid, a method of anomaly detection of crowd distribution pattern based on fractal dimension was proposed. The large-scale crowd is regarded as a particle swarm, and the distribution of particle swarm is regarded as an image. The feature of particle swarm is extracted by using the idea

Fig. 3.37 Actual scenario

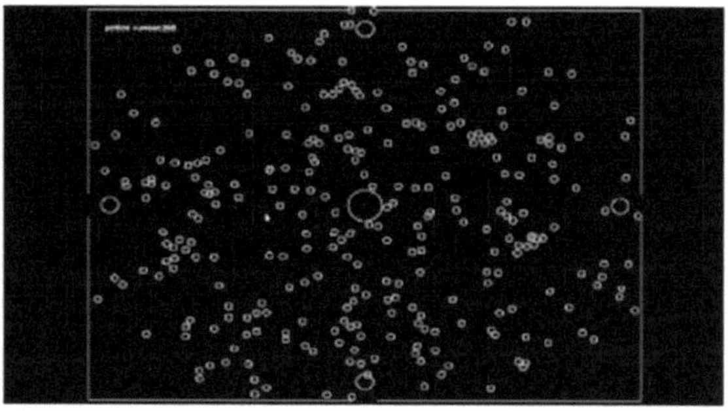

Fig. 3.38 Simulation scenario

of image texture analysis. The spatial distribution of moving points and groups in Levyflight is used to represent the crowd distribution state, and the fractal feature of point-group aggregation degree is used to describe the spatial distribution of points and groups. In order to remove the interference factors from the feature vector and keep the main related information, principal component analysis (PCA) is used for dimension reduction, and an adjacent points' accumulation algorithm is proposed to detect the abnormal distribution of particle swarm in relevant feature space. This method can effectively streamline massive data and realize monitoring and early warning of abnormal distribution and movement of large-scale people flow in cities.

(3) **Traffic abnormal pattern's detection based on hidden markov urban dynamics model**

Yang Su's research group regarded the distribution of urban traffic flow at a certain time as a state, and thought that its state's transition conforms to the probability model in normal state, so the state evolution of urban traffic flow can be dynamically modeled by using hidden Markov model. When abnormal incidents occur, such as when a large number of people gather in a certain area, the evolution of system state will deviate from the original model and can be manifested as a small probability incident of the Hidden Markov model. Using the artificial data test model, the logarithm of probability of abnormal traffic flow at different time intervals can be obtained, as shown in Fig. 3.39. Based on the traffic flow data collected by more than 4,000 traffic coils released by Minnesota Traffic Management Center from January to September, 2010, the abnormal time of traffic flow has been detected, and the time mostly corresponds to extreme weathers such as heavy snow, hurricane, as well as hot summer, festivals, large gatherings and road closures.

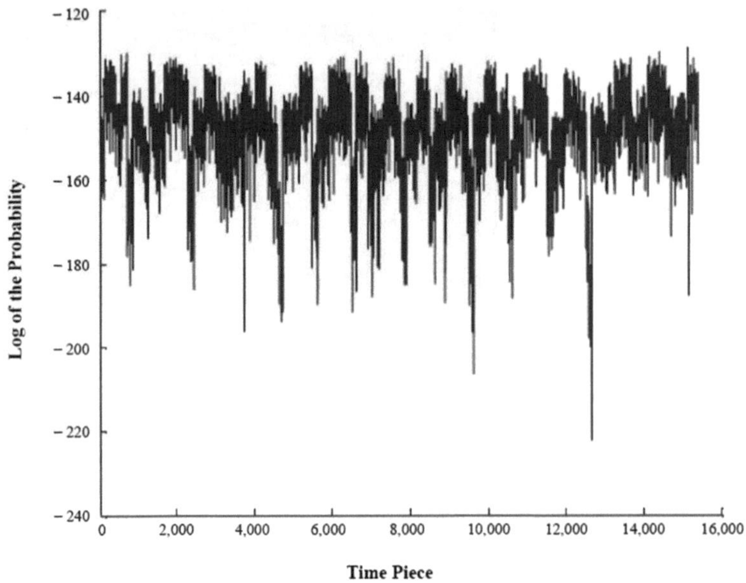

Fig. 3.39 The logarithm of probability of abnormal traffic flow at different time intervals

2.2.3 Monitoring, Early Warning and Handling Strategies of Disastrous Meteorological Incidents

(1) Temporal and spatial distribution laws of typical disastrous meteorological incidents in China

Based on the meteorological data of 756 ground observation stations in China from 1951 to 2010, Huang Quanyi's research group of Tsinghua University studied the temporal and spatial distribution characteristics and laws of persistent extreme precipitation incidents, regional heavy precipitation incidents, heavy rain, heavy snow and strong wind in China. The main conclusions are as follows. (i) Persistent extreme precipitation incidents in China mainly occurred in the central and eastern regions (26–34 N) and South China (south of 26°N); There are few persistent extreme precipitation incidents in North China (north of 34°N). (ii) Since 1961, the frequency of regional heavy precipitation incidents has showed a weak increasing trend, and the regional heavy precipitation incidents were more frequent from late 1980s to 1990s. (iii) The areas with daily precipitation greater than 25mm are mainly in the southeast of China, and the precipitation days decrease from southeast to northwest. (iv) Heavy snow is mainly distributed in the north of 25°N in China, while heavy snow is scattered in the west and regular in the east. (v) Xinjiang, central and eastern Inner Mongolia, eastern Heilongjiang, Qinghai-Tibet Plateau, Hexi Corridor, southeast coast of China and other places are the main areas affected by strong winds. (vi) Persistent low-temperature rain and snow mostly occur in the south of China, especially in the southwest, east and south of China; Persistent freezing rain mainly occurs in the south of the middle reaches of the Yangtze River; The duration of persistent low-temperature rain and snow and persistent freezing rain incidents in the northern Yunnan-Guizhou Plateau is relatively long; Precipitation is the limiting condition for the occurrence of persistent low-temperature rain and snow and persistent freezing rain in the north of China, and extreme low-temperature is the main factor for the occurrence of persistent low-temperature rain and snow and persistent freezing rain in the south of China.

(2) **The refined risk assessment model for meteorological disaster ontology**

Previous studies were mostly based on natural attributes or economic dimension, and used enumeration method to classify disaster-bearing bodies. This kind of method accords with habitual thinking and contributes to the statistics of post-disaster losses, but it is not conducive to the unified management of all kinds of disaster-bearing bodies in disaster prevention and mitigation, nor can it characterize the essential reasons for the damage of disaster-bearing bodies caused by many factors in disastrous meteorological incidents. Huang Quanyi's research group classified disaster-bearing bodies according to the action forms affected by meteorological disasters, and put forward a conceptual model of meteorological disasters ontology based on meta-action. Using the symbols of ontology primitives such as concept, individual and attribute (relationship), the research group formally described the constituent elements of meteorological disasters and the semantic relations among different concepts such as attributes, methods, method types, parameters and data types of each constituent element. The ontology rules of meteorological disaster ontology's conceptual model were realized by SWRL rules. The research group also studied the technical realization of constructing reasoning rules for meteorological disasters' risk assessment based on SWRL. In order to facilitate the compilation of risk reasoning rules and serve the specific reasoning process, a set of reasoning rules for the risk assessment of urban rainstorm waterlogging disaster have been constructed, which can provide support for the refined assessment of waterlogging risk.

(3) **Technologies for meteorological disasters' deduction and pre-evaluation of plans**

Huang Quanyi's research group divided scenario elements into four categories: disaster-causing, disaster-bearing, background and resources, and thought that the occurrence of unexpected incidents was a process of logical combination of the four types of scenario elements. Meanwhile, the connection of multiple scenarios and the combination of mutual relations will form a scenario chain. The evolution of complex scenarios such as series connection, one-to-many scenario, many-to-one scenario, parallel connection scenario and cyclic scenario was constructed. The research group studied scenario matching and chain reasoning such as mutually exclusive relation, inclusion relation, inherited relation and parallel relation. Based on the structured emergency scenarios, the changing rules of structure and parameters between upper and lower scenarios during scenario conversion, meteorological disasters' scenario construction and scenario-driven deduction, the meteorological disasters' scenario expression model based on objects was established. Besides,

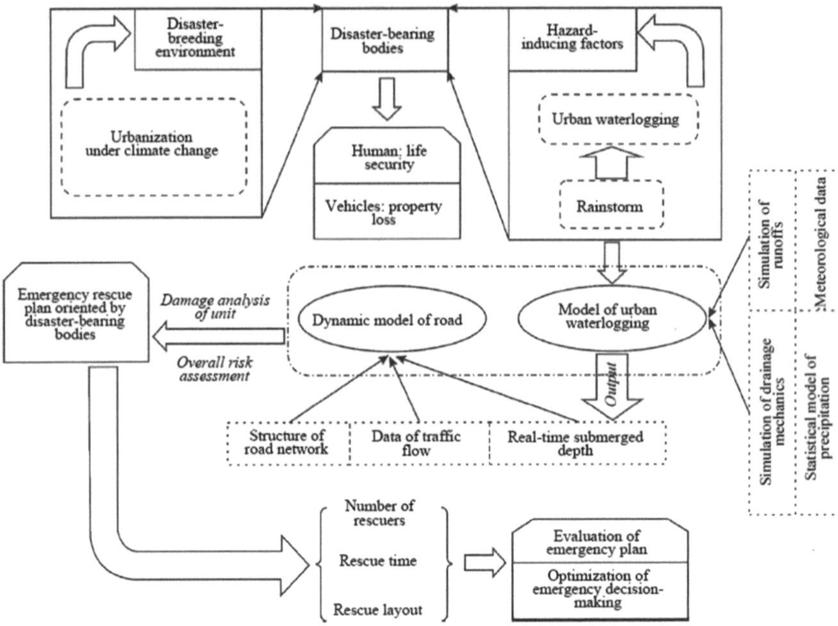

Fig. 3.40 Framework of meteorological disasters' scenario deduction and pre-evaluation of plans

the disasters' scenarios deduction technology based on multi-agent and Bayesian Network's uncertain reasoning was realized, the linear programming method and evaluation method for emergency plan based on analysis of multi-dimensional preference were established, and the dynamic implementation effect of plans was evaluated. The framework of meteorological disasters' scenario deduction and pre-evaluation of plans is shown in Fig. 3.40. The research results have been well received by experts from the World Meteorological Organization, and supported the construction of emergencies' early warning system and early warning's release system in Beijing, Tianjin and Guangxi.

2.2.4 Monitoring, Early Warning and Risk Prevention and Control Model for Major Hazard Sources of Hazardous Chemicals

(1) **System dynamics model of hazardous chemicals incidents**

According to the method of system dynamics, the research group of China Academy of Safety Science and Technology regards the evolution process of hazardous chemicals incidents as the system "flow," and takes the incidents' spread range and direct influencing factors as the system boundary. For example, in the leakage of hazardous chemicals, the cumulative amount formed after leakage of hazardous substances can be set as the Level Variable, and the rate of leakage, diffusion, dilution or elimination can be set as flow rate variable. On the premise of defining level variables and flow rate variables, several level variable systems and flow rate systems can be set up in a system of major incident. There are also needs to make clear the accumulation and distribution characteristics of hazardous substances, and the auxiliary variables, supplementary variables and constants related to incident's handling decision and their mutual relations, and there is also need to establish flow rate equation and auxiliary variable equation. The Level Variable $\mathrm{LEV}(t) = \mathrm{LEV}(t - \Delta t) + \Delta t \times \mathrm{RAT}(t - \Delta t)$. $\mathrm{RAT}(t - \Delta t)$ is the flow rate variable. The flow diagram of hazardous chemicals incidents' evolution is shown in Fig. 3.41.

In Fig. 3.41, C_n ($n = 1, 2, 3, 4, 5$) indicates closing, and R_1, R_2, R_3, R_4 respectively indicate taking mitigation, preparation, response and recovery measures. Major incidents generally start from faults or mistakes, and after improper handling, they will become incidents, and after further development, they produce harmful consequences and cause sequent incidents. Under the influence of internal and external factors, incidents may further intensify and even evolve into crises. In several major development stages, the incident's development can be interrupted by control to restore the normal state, but it mainly depends on the changes of Level Variables and flow rate variables. Besides, it also includes the influence of external information such as media and public pressure on feedback and control.

(2) **Risk assessment model and acceptability standards for major hazard sources of hazardous chemicals**

The research group of China Academy of Safety Science and Technology has studied the theory, method and model of risk assessment for major hazard sources of hazardous chemicals, which provides a scientific basis for China to formulate risk assessment methods and risks' acceptability standards of hazardous chemicals. The risk calculation process is shown in Fig. 3.42.

(i) Individual risk and its acceptability criteria. Individual risk refers to the probability of death of personnel in a specific place caused by hazardous chemicals incidents in a unit time. For any hazard source in the area, the individual risk

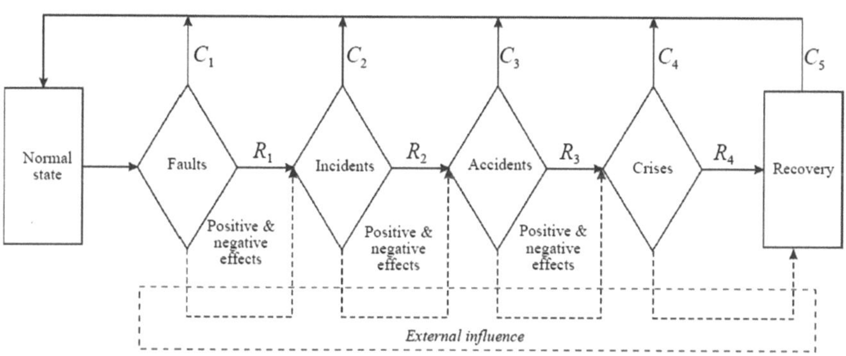

Fig. 3.41 The flow diagram of hazardous chemicals incidents' evolution

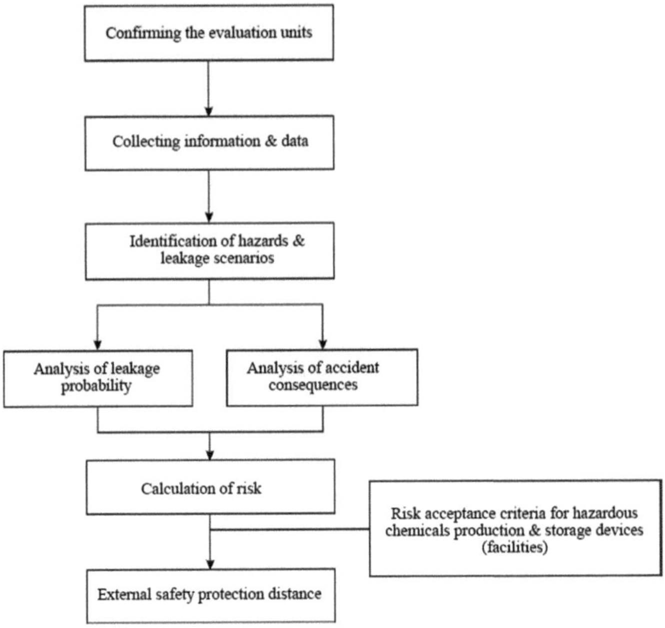

Fig. 3.42 Risk calculation process of hazardous chemicals

caused to the geographical coordinates in the area after the incident can be calculated by the formula $R(x, y) = \sum_s f_s v_s(x, y)$. $R(x, y)$ *indicates individual risk and its unit is* $(\text{man} \cdot \text{year})^{-1}$; f_s is the probability of occurrence of the sth incident; $v_s(x, y)$ is the consequence of the sth incident causing death at the position (x, y). The acceptability standards for individual risks from major hazard sources of hazardous chemicals are shown in Table 3.3.

(ii) Social risk's calculation and its acceptability standards. Social risk refers to the cumulative frequency F with more than N deaths caused by specific hazardous chemicals incidents, that is, the number of deaths per unit time, and the calculation formula is

Table 3.3 The acceptability standards for individual risks from major hazard sources of hazardous chemicals

Protection targets	Acceptable standard of individual risk (probability value)	
	Newly built devices (per year) ≤	Current devices (per year) ≤
First-class protection target	3×10^{-7}	3×10^{-6}
Second-class protection target	3×10^{-6}	1×10^{-5}
Third-class protection target	1×10^{-5}	3×10^{-5}

$$R_s = \sum_{i=1}^{N} P_i \times P_w \times P_D(\geq N)$$

R_s indicates the social risk of the risk source and its unit is (man·year)$^{-1}$; N refers to the number of deaths; P_i refers to the probability of the incident's consequence i; P_w refers to the occurrence frequency of atmospheric stability w; P_D ($\geq N$) indicates the frequency of occurrence of wind direction D which causes N deaths or more.

Based on the principle of as low as reasonably practicable (ALARP), the social risk's acceptability standard is designed, and the social risk map is divided into three areas by two risk dividing lines, namely unacceptable area, area to minimize risk and acceptable area. If the social risk curve falls in the unacceptable area, safety improvement measures should be taken immediately; If the social risk curve falls in the acceptable area, only the existing safety measures need to be maintained; If the social risk curve falls in the area to minimize risk, the social risk is actually at an acceptable level, but it is still necessary to take safety improvement measures if the economic cost is reasonable and feasible. In addition, in order to avoid huge incidents or disasters, the abscissa upper limit of socially acceptable risk benchmark is 1,000 people, that is, incidents exceeding this upper limit are unacceptable regardless of the probability.

(3) **Monitoring and early warning of major incidents of dangerous chemicals**

Based on historical case analysis and risk assessment, there are key time and space nodes in the occurrence, development and evolution of major incidents of hazardous chemicals, namely opportunity time window (Fig. 3.43). According to the research group of China Academy of Safety Science and Technology, there are two typical time windows in the evolution process of major incidents. In the first time window, before the incident, there are signs and destructive energy is accumulating. If correct intervention actions are taken, hidden dangers may be controlled. Otherwise, it may lead to accidents. Therefore, relevant parameters of typical accidents should be monitored, risks should be assessed quickly, and early warning measures should be taken to avoid accidents. The second time window is in the process of incident handling. If correct response measures are taken, the incident's scale can be controlled. In this time window, information on incident's evolution and response should be collected, early warning information should be released in time, and personnels should be evacuated, so as to avoid developing into a major incident.

Based on the evolution law of major incidents and the theory and method of risk assessment, the research group of China Academy of Safety Science and Technology put forward the method, model and system structure for monitoring and early warning of major hazard sources from hazardous chemicals, and developed a monitoring and early warning system taking the toxic gas leakage incident as an application case. The functional modules of the system are shown in Fig. 3.44.

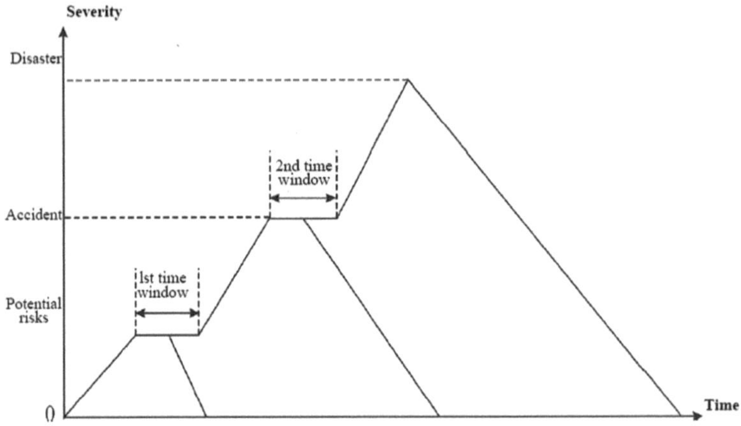

Fig. 3.43 Evolution process and time window of major incidents of hazardous chemicals

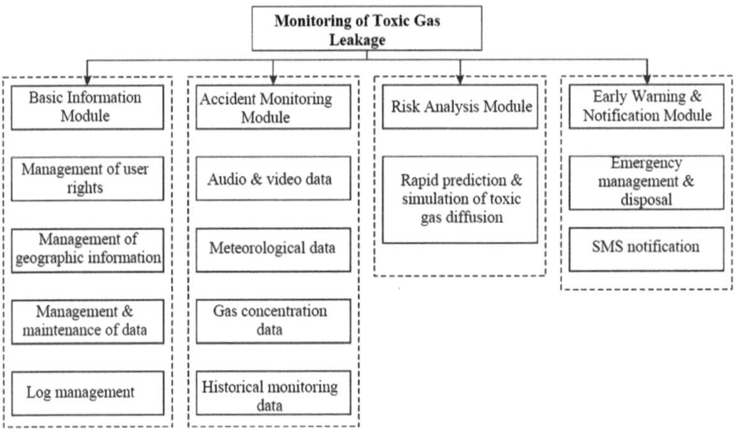

Fig. 3.44 Function module of monitoring and early warning system for major hazard sources

(4) **Construction of incident scenarios for major hazard sources of typical hazardous chemicals**

Using the theory and method of scenario construction, the research group of China Academy of Safety Science and Technology constructed scenarios of typical hazardous chemical incidents, and considered the evolution process and emergency response for the incidents in detail, and obtained the scenarios' summary, consequences, response tasks and handling measures.

Table 3.4 Blowout incident scenario in sulphur-bearing gas field

Items	Contents
Scene	Sulphur gas field well site
Casualties	100 people were killed and 700 were injured
Evacuated population	10,000 people were told to evacuate and avoid risks
Economic loss	50 million yuan
Possibility of multiple incidents at the same time	Relatively low
Recovery period	One month

(i) Blowout incident scenario of sulphur-bearing gas field (Table 3.4). Inaccurate construction design, poor mud performance, improper operation technology, serious downhole leakage and other factors cause blowout which is out of control, and hydrogen sulfide spreads rapidly after leaking from wellhead, which seriously threatens the life of personnels at downwind side. This kind of accident has a large influence range, which is easy to cause poisoning casualties within several kilometers around the well pad.

(ii) Incident scenarios of leakage and explosion in storage areas of hazardous chemicals (Table 3.5). A total of 36 tons of liquid chlorine were stored in the liquid chlorine's storage tank area of a chemical enterprise. The staff's mishandling led to the explosion, which caused the leakage of liquid chlorine and released a large amount of chlorine gas. The gas diffused with the wind. The explosion damaged the security system of liquid chlorine's production line in the chemical plant. In the subsequent accident handling process, many sequent explosions occurred again, and all dangerous places of the incident had to be blown up to release all liquid chlorine.

The research results related to scenario construction have been applied by the National Emergency Rescue Command Center of Production Safety and petroleum and petrochemical enterprises to guide emergency preparedness activities such as incidents' monitoring and early warning, emergency plan's optimization, emergency training and drills, and emergency capability's assessment, and achieved good effect.

3 Emergency Rescue and Handling

3.1 Law of Psychological Behavior

(1) **Psychological effect of social justice based on social investigation**

Tong Hengqing's research group of Wuhan University of Technology studied the relationship between social group incidents and justice, including the relationship

Table 3.5 Scenes of leakage and explosion incidents in hazardous chemicals storage area

Items	Contents
Scene	Hazardous chemicals storage area
Casualties	500 people died, 3,000 people were poisoned to varying degrees, and 30,000 people were treated in outpatient clinics
Damage to infrastructure	The metal near the direct explosion area is completely damaged and the metal in the heavily exposed area is corroded
Evacuated/relocated population	Evacuating 150,000 people to safe areas and evacuating 200,000 people out of affected areas
Pollution situation	The environmental pollution area is mainly near the explosion point
Economic loss	150 million yuan
Possibility of multiple incidents at the same time	Possible
Recovery period	Several months

among social justice, social class and authority legitimacy (Fig. 3.45), and discussed the mitigation mechanism of unfair events. On the mechanism of social justice's influence on authority legitimacy, based on the Theory of Interpretation Level, the research group proposed that social stratum can regulate the interaction between justice of distribution and procedural justice on the perception of authority legitimacy. The results show that for the high class, only in the case of procedural justice can distribution justice significantly affect the level of authority legitimacy; For the lower class, whether the procedure is fair or not, distribution justice significantly affects the perception of authority legitimacy. In terms of alleviating the negative effects of injustice, relevant research analyzes the adjustment mechanism of the effect of justice from the perspective of "environment," and discusses the influence of easy extraction on procedural justice under uncertain conditions and the influence of processing system on unfair response. The results show that when people are in information uncertainty and self-uncertainty at the same time, if they are led to recall more procedural injustice, people will experience a stronger sense of procedural justice; If we guide them to use rational processing system to process unfair information, people will restrain their strong negative reaction to injustice.

(2) **People's cognitive changes caused by deprivation of the sense of control**

Richard Nisbett, an academician of the National Academy of Sciences, United States, put forward a far-reaching theory of cultural psychology: there are differences in thinking modes between oriental and occidental people, that is, East Asians tend to use holistic thinking, while occidental people tend to adopt analytical thinking. Zhou Xinyue's research group of Sun Yat-sen University: By systematically examining the change rule of cognitive model after losing control, it is found that the differences of

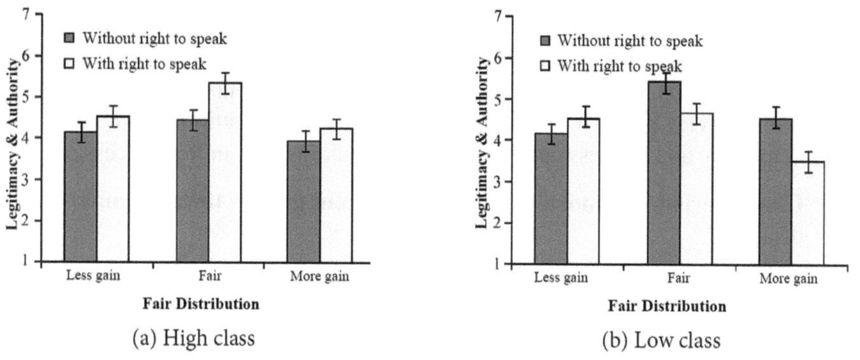

Fig. 3.45 Relationship among social justice, social class and authority legitimacy

holistic and analytical cognitive processing between oriental and occidental people are not absolutely stable, but will be affected by the sense of control. When the sense of control is deprived for a short time, both East and West tend to adopt analytical thinking mode; When the sense of control is deprived for a long time, both oriental and occidental people tend to adopt a holistic way of thinking. Both kinds of changes will lead to the disappearance of the thinking differences between oriental and occidental people. When the subjects are induced to adopt analytical thinking, the sense of control will also increase. This discovery explains the differences in cognitive styles between oriental and occidental people. This study expands Nisbett's theory, and shows that when people face emergencies, they can cope with them by changing their thinking tendency, so as to make up for the negative psychological impact brought by the sense of losing control.

(3) **Nostalgia and psychological mechanism of fair disposal of economic resources**

Zhou Xinyue's research group believes that nostalgia has a significant impact on physiological comfort. Nostalgia can make people feel more warmth from both physiological and psychological aspects. Its practical function lies in that through training individuals or groups' nostalgia ability, it can relieve the uncomfortable feeling of bad weather, help them persist in looking for food and shelter for a longer time, and enhance people's adaptability in unconventional emergencies.

Unconventional emergencies often require the support and distribution of economic resources. As the second stimulus of conditional response, will economic resources cause people's selfish and greedy behavior, or fair and reciprocal behavior? Zhou Xinyue's research group understands the influence of economic resources on people's attitudes and behaviors by manipulating the "cleanliness" of money. The experimental results show that "clean" money and "dirty" money can regulate the relationship between economic resources and behaviors, that is, "dirty" money triggers greed and selfish trading motives, while "clean" money triggers fairness and reciprocity motives which is more consistent with social principles. "Clean" money can not only activate the trading motivation, but also activate the self-concept maintenance mechanism, which requires individuals to treat others fairly and mutually. After unconventional emergencies, the fairness degree of economic resources' distribution will have a positive or negative impact on people's attitudes and behaviors, and reciprocity and fairness can play a positive guiding role in this process.

(4) **Characteristics of moral emotion of different groups toward "man-caused disaster"**

Xu Yan's research group of Beijing Normal University made an emotional analysis of 94,562 related posts on Weibo within 40 days after the "July 23" Ningbo-Wenzhou railway traffic accident, and discussed the moral and emotional characteristics of netizens to the "man-caused disaster," and got the following findings. (i) Internet users had anger, contempt, disgust, sympathy and love for the train accidents. (ii) According to the theory of moral basis, three incidents corresponded to the moral basis respectively: "Little Yiyi" event violated the moral bases of "injury/care" and induced sympathy and love of netizens; The improper remarks of officials violated the moral bases of "authority/subversion" and induced the contempt of netizens; "Italian passengers got high compensation" violated the moral bases of "fairness/deception" and induced netizens' anger. (iii) Men had higher intensity and willingness of expressing emotions such as anger, disgust and contempt, while women were more likely to express love and sympathy. (iv) In virtual network, authenticated users expressed emotions more actively than anonymous users, and anonymous users expressed more anger and negative emotions than authenticated users. (v) Among real-name users, the group users with VIP authentication were more prominent than individual users with VIP authentication, and group users with VIP authentication tended to express positive emotions such as sympathy and love. The corresponding relationship between emotional inflection point and social derivative events is shown in Fig. 3.46.

Focusing on the derivative events accompanied by "man-caused disasters," the network supervision department should take corresponding control and guidance measures according to different moral bases, and lead the focus of public opinion to the injured through mass media to reduce the discussion of the injured; Based on the group differences in moral emotions, the network supervision department should implement different management and control efforts for different groups to improve the supervision efficiency; Social media should develop more different levels of identity authentication, so that most anonymous users can get different degrees of authentication, and in this way, it will help make them take more civilized and rational behaviors in network.

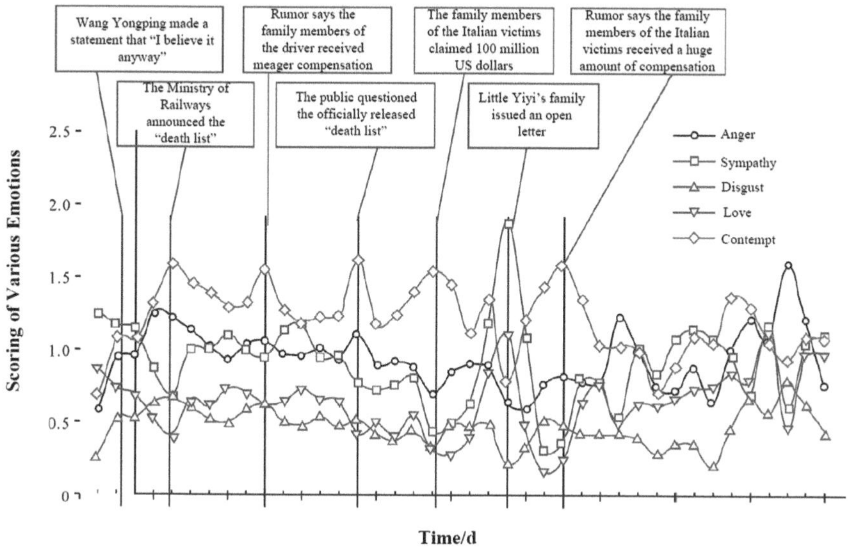

Fig. 3.46 Corresponding relationship between emotional inflection point and social derivative incidents

3.2 Theory and Method of Emergency Command Decision-Making

3.2.1 Theory and Method of Multi-agent and Multi-link Collaborative Emergency Decision-Making

(1) **Emergency decision-making organization's modeling based on multi-agent system**

At present, there are some problems in emergency decision-making's organization structure, such as the lack of emergency decision-making organization's standards. Besides, the optimization of intelligent design of organizational structure has not been done, and there are more emphases on static capacity's building rather than dynamic capacity's building. Therefore, based on the typical unconventional emergency cases in China, Wang Hongwei's Research Group of Huazhong University of Science and Technology analyzed and summarized the characteristics of China's emergency decision-making's organization structure, including complexity, temporality, hierarchy, authority, openness and power change. According to the organization theory, using a systematic perspective, the research group put forward the organizational framework of emergency decision-making that is suitable for China's national conditions (Fig. 3.47).

According to the emergency decision-making's organization structure, Wang Hongwei's research group established an emergency decision-making's organization model based on multi-agent system (MAS) to simulate the interactive process

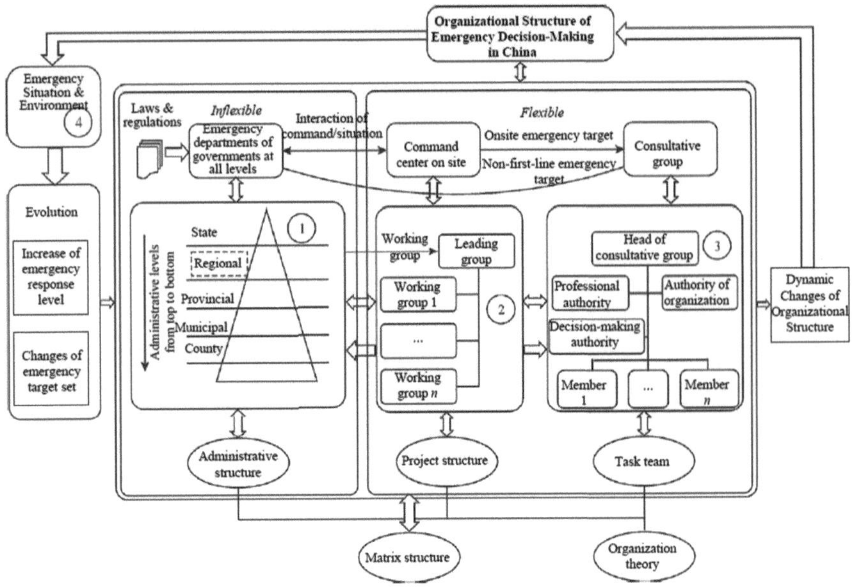

Fig. 3.47 Organizational framework of emergency decision-making suitable for china's national conditions

of emergency decision-making. The research group formally described emergency related laws and regulations, and established emergency decision-making's organization model of governments at all levels. Besides, the group used tuple symbolization to describe the ability, responsibility and rights of emergency decision-making's organization unit, defined social norms, described organization rules and consultation system, expanded AUML (Agent-oriented Unified Modeling Language), and built model for organization structure and interaction mode by using sequence diagrams and class diagram; Based on Protégé ontology modeling tool, the traditional description of emergency decision-making's rules was transformed into ontology knowledge. Considering that departments of emergency decision-making need to have thinking state to make reasoning and decisions, Tropos method was selected and multi-BDIAgent emergency decision-making's organization was realized based on JACK platform (Fig. 3.48).

(2) **Multi-department distributed collaborative planning based on hierarchical task network**

The emergency response process involves multiple departments, so it is difficult to share information completely. Each participating department needs to complete its own task planning based on the local emergency situation, and also needs to realize the coordination and operation of all task planning within the emergency organization through reasonable coordination, and it is essentially a distributed collaborative task planning. Facing emergency tasks, Wang Hongwei's research group first designed

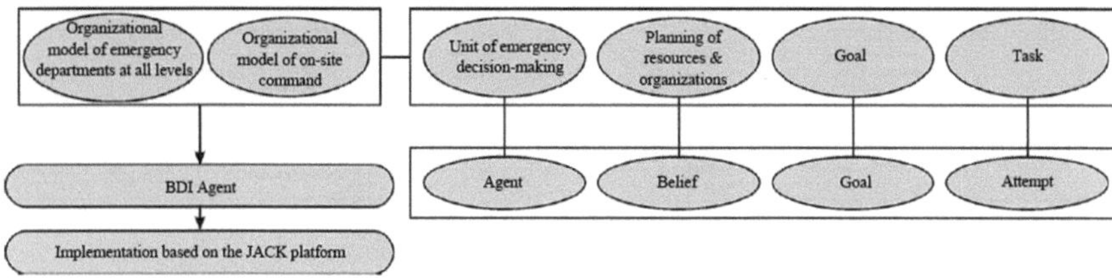

Fig. 3.48 Implementation of Multi-BDI Agent emergency decision-making organization based on JACK platform

a framework of collaborative task planning to ensure the collaboration in the planning process. Secondly, a coordination mechanism was designed to solve the dependency relationship, including conflict relationship, enabling relationship and incentive relationship, which was suitable for the hierarchical task network, HTN) planning process. As a participating department in emergency response, a single HTN planner needs to be able to identify the dependencies of corresponding planning actions, and change the existing planning schemes according to the requirements of cooperation mechanism when necessary. Finally, in the emergency decision-making and planning, in terms of the emergency resources, taking the reusable resources, consumable resources and bandwidth resources as objects, the research group designed coordination method of emergency resources.

The collaborative task's planning framework for HTN planning process designed by Wang Hongwei's research group (Fig. 3.49) is suitable for the coordination between any two agents, and can also be extended to the coordination of Multi-Agent. The main idea of the framework is to embed the task view's sharing and task coordination's cooperation process which are separated from each other in the traditional framework into the task planning process, so that the planner can coordinate atomic actions in time. And there is no task scheduling in the new framework. The node return and re-planning function of HTN planner should be used to integrate the process of task adjustment and modification into the planning process.

(3) Multi-person and multi-criteria emergency group decision-making model based on distance

Emergency decision-making for unconventional emergencies has become an important issue that the government and organizations are increasingly concerned about. It has the characteristics of unified command, multi-department coordination, great pressure on decision-makers, and the need for multidisciplinary knowledge. Based on the characteristics of multiple decision-makers and multiple influencing factors in unconventional emergencies, Yu Lean's research group of Institute of Mathematics and Systems Science, Chinese Academy of Sciences constructed a multi-person and multi-criteria model of emergency group decision-making based on distance, and solved the emergency decision-making problem of unconventional emergencies. Using this model, the decision-making problem of Canadian fire rescue is studied, and satisfying result of decision-making is achieved.

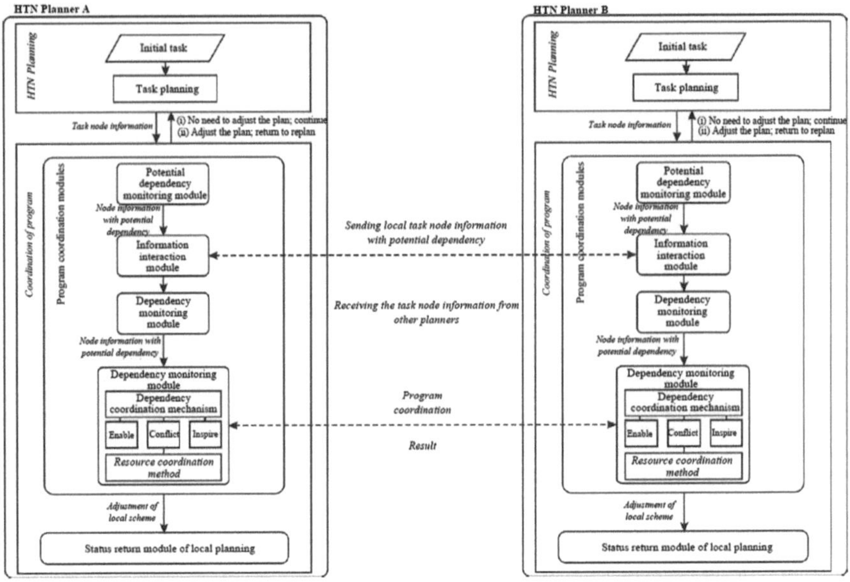

Fig. 3.49 Framework of collaborative task planning for HTN planning process

(4) Multi-party consultation's simulation method and technology

China's emergency management implements the principle of territorial management. When the incident involves multiple provinces and cities, the on-site emergency command shall be jointly organized by the responsible departments of multiple provinces and cities, and cross-regional and inter-departmental coordination shall be implemented. In addition, on-site disaster information, such as longitude and latitude of the incident's location, disaster type, casualty statistics, accident spread range and resource demand, must be sent back to the relevant departments in time. With a powerful database and professional analysis system, after comprehensive analysis, the disaster prediction and early warning results, emergency resources' allocation and scheduling information should be fed back to the front-line command departments in time to assist the emergency response on site. In the process of emergency handling of major emergencies, it is necessary to carry out information interaction and collaborative handling among multiple front-line command centers and rear auxiliary departments, collect and obtain a large amount of emergency information from disaster spot, and distribute the information accurately and timely. It has become the key factor to carrying out scientific and rapid emergency response actions.

Yuan Hongyong's research group of Tsinghua University put forward the application mode and architecture of multi-party collaborative consultation for emergencies based on "Emergency Map." The system adopts B/S and C/S mixed modes, and it is composed of GIS server, collaborative consultation server software and collaborative consultation client software. The initiators are the management departments

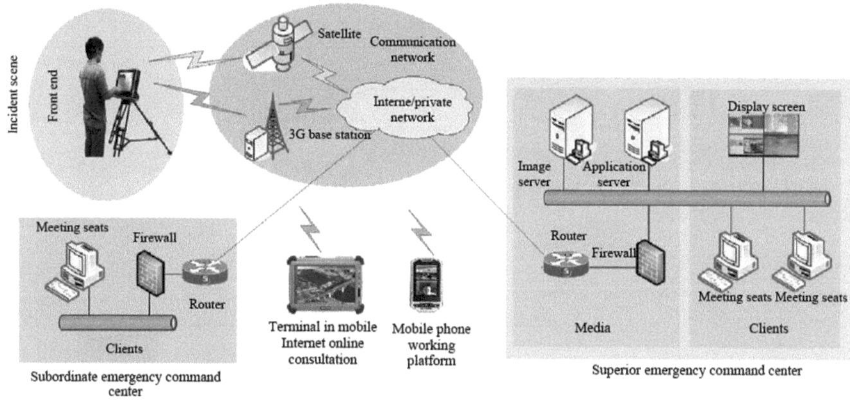

Fig. 3.50 System structure of multi-party collaborative consultation

or responsible departments for emergency response, and the participants are cooperative departments or units. The structure of multi-party collaborative consultation system is shown in Fig. 3.50. Through distributed deployment, the system uses the collaborative consultation client software deployed in the on-site mobile emergency platform, accesses the private network via the Internet, and interacts with all levels of emergency platforms running on the private network, so as to quickly distribute the disaster information of the accident site to the decision-makers of all levels of emergency platforms. And at the same time, it transmits the auxiliary information for decision-making by multi-departments to the on-spot emergency handling departments in time, so as to make emergency consultation and decision such as multi-party collaborative command and coordinated deployment on the same map.

The "emergency map" receives data provided by the initiator of the collaborative consultation, and the participants draw the map based on the map provided by the initiator, and discuss the response measures for disasters through the interaction of text, map, plotting, voice and video. The research introduces the concept of WebServices into the functional design and construction with a sharing mode, establishes geographic information services according to industry standards and relevant interfaces under the computer network environment, and provides users with callable geographic information components. Through computer network technology, we can freely access various geographic information and services distributed in different places, such as maps, images, data set services, geospatial analysis and report. It has the characteristics of standardized system, universal interface, hierarchical service, stability, robustness and transparent service. The geographic information service framework of "Emergency Map" is shown in Fig. 3.51.

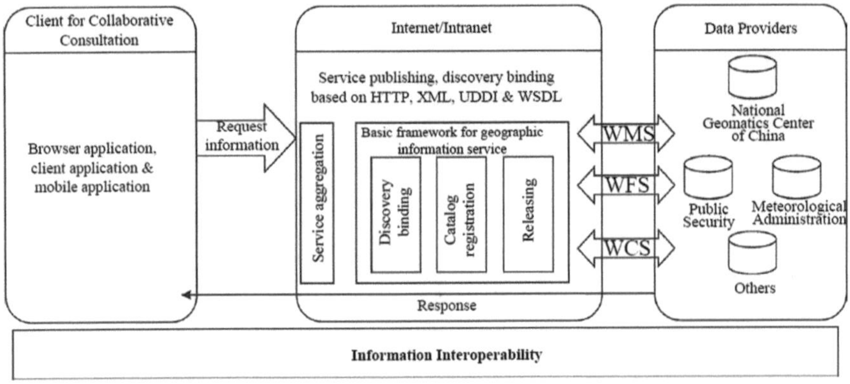

Fig. 3.51 Geographic information service framework of "Emergency Map"

3.2.2 Theory and Method of Dynamic Emergency Decision-Making for Unconventional Emergencies

(1) Structured expression model of emergency cases' content

A complete case needs to describe the space–time coordinates containing all the information of the emergency. In addition to the basic information such as the cause, course and result of the incident, the information should also include emergency response information such as emergency tasks at different development stages of the incident, the executing agencies and personnel of the tasks, and the effects of handling measures. Yu Lian's research group of National School of Administration regarded the environment where the disaster-bearing body is located, emergencies and emergency management as the three record dimensions of the case. Environmental dimension is the environmental background of the occurrence, development and response of incidents, including static and dynamic environmental elements. Static environment should be described at the whole incident level and dynamic environment should be described at the sub-incident level. Incident dimension describes the cause, process and result of unexpected incidents in the order of time, space or incident chain; Management dimension describes the main bodies, actions taken and resources used in emergency management, including the main bodies involved in task management. It is found that although the amount of information and information types of attributes in emergency cases are different, the features can be mainly classified into four categories, namely, digital eigenvalue, symbolic eigenvalue, fuzzy concept eigenvalue and fuzzy number or fuzzy interval attribute value. On this basis, Yu Lian's research group developed a management system for emergency cases (Fig. 3.52). Through on-the-spot investigation and in-depth interview, based on the cases' expression paradigm, more than 40 typical emergency cases were developed and applied to the practical teaching for building up relevant personnels' ability of emergency management in National School of Administration.

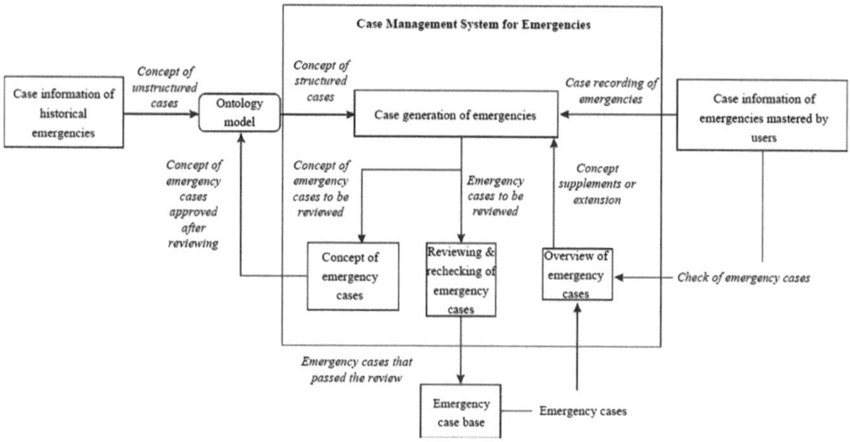

Fig. 3.52 Management system of emergency cases

(2) Decision-making method under risk of emergency response considering decision-makers' behavior

The evolution of emergencies is highly uncertain and risky. Fan Zhiping's research group of Northeastern University considered the behaviors of decision makers, such as reference dependence, loss avoidance and attention to incidents with small probability but high hazard, and introduced prospect theory into emergency response's decision analysis, and put forward a decision-making method under risk of emergency response considering the behavior of decision makers. Firstly, the emergency scenario tree was constructed, and the index value of scenario response's results was calculated according to the value function of the prospect theory. Then, considering the correlation characteristics among indicators such as casualties and property losses in emergency decision-making, Choquet integral was used to integrate the index values of the results, so as to determine the comprehensive value of emergency response's results. Finally, according to the weight function of prospect theory, the weight of each possible response result was calculated, and the prospect value of emergency response's cost was confirmed by integrating the comprehensive value and weight of possible results of response. Compared with the existing emergency decision-making methods, the method proposed by Fan Zhiping's research group considered the behavior characteristics of decision-makers in emergency response and decision-making, and it can make the analysis results more in line with the subjective perception of decision-makers, and it is easy for decision-makers to understand and adopt.

3.2.3 Emergency Knowledge Management Model and Decision-Making's Deduction Method

Wang Yanzhang's research group of Dalian University of Technology, based on knowledge element, integrated systems and other discipline models, discussed the comprehensive simulation and analysis method of complex system driven by knowledge-guided data and model, and put forward the management and decision-making's deduction method based on knowledge, and the method laid a theoretical foundation for emergency response. Its main function is to construct the six-space system of comprehensive knowledge based on duality principle and the formal representation method of scenario model based on knowledge element.

(1) **Six-space system of comprehensive knowledge based on duality principle**

First of all, using the duality principle of systematics, the unconventional emergency related system can be divided into the original image of objective things and the system image of human's cognitive space. The objective things are real and things in cognitive space are virtual. Then, it can be subdivided into the duality of objective things and metadata, the duality of knowledge and knowledge, the duality of formal model and instantiation model, and the duality of metadata and instance data. Human's cognition of objective things can be divided into six categories, or "six spaces," namely knowledge element space, metadata space, data space, entity model space, formal model space and generalized operator space. The duality relationship among information, knowledge and mathematical models and the six-space system are shown in Fig. 3.53.

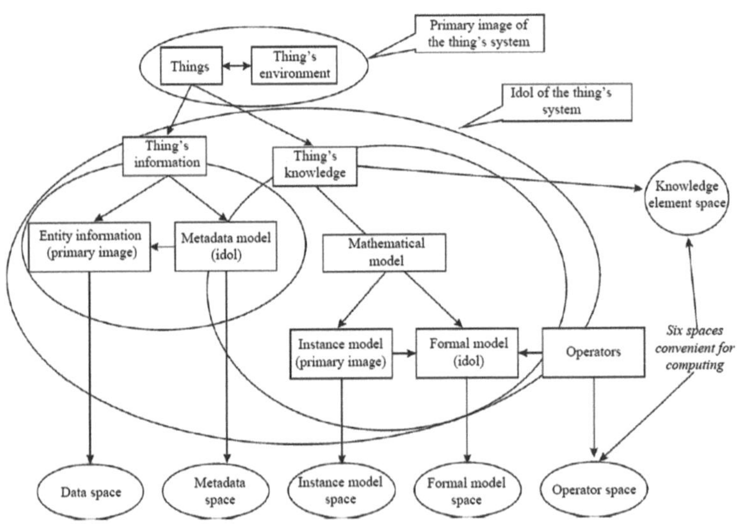

Fig. 3.53 Duality of information, knowledge and mathematical models and six-space system

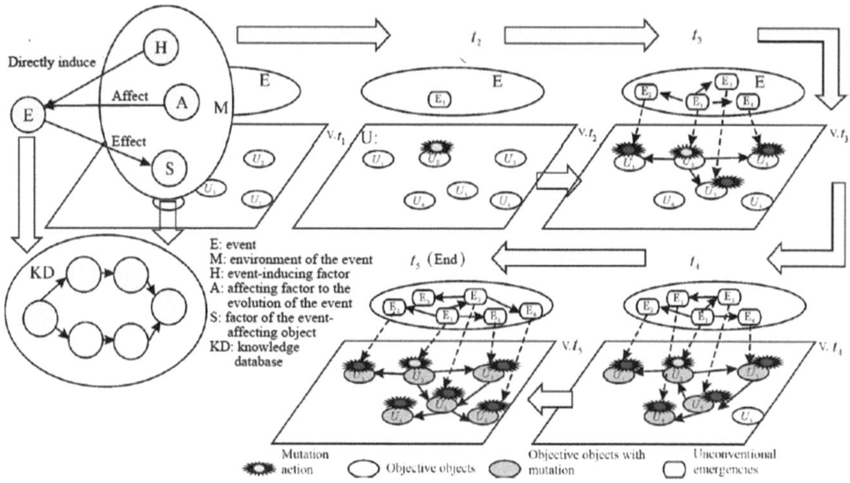

Fig. 3.54 Scenario and conceptual model based on knowledge element

(2) **Formal representation method of scenario model based on knowledge element**

Facing the complexity of unconventional emergencies, Wang Yanzhang's research group put forward a conceptual model of scenarios based on knowledge elements (Fig. 3.54), which combed the elements, granularity, hierarchy, constraints among elements and complex relationships among scenarios. At time t_1, all objective objects were in a normal state; At time t_2, the attribute of the objective object U_2 changed suddenly. At time t_3, the change influenced U_1, U_3 and U_5; At time t_4, U_3 acted on U_5 and U_5 acted on U_4. At t_5, U_5 acted on U_6, and the emergency was about to end. This scenario model can be used to simulate the whole process of incidents from primary incidents to secondary derivative incidents until the end. In the simulation process, based on the knowledge element structure, the data of objective objects (specific emergency scenarios) can be obtained after the knowledge element structure is materialized, and it is conducive to reducing the high uncertainty of scenarios and making effective response plans.

3.2.4 Integration Principle and Method of Emergency Decision-Making Technical Support System

(1) **Integration method of data and model**

By studying the evolution process of unconventional emergencies, Xu Wei's research group of Renmin University of China constructed the ontology model of emergency management, refined the minimum information set needed to support emergency management, and realized the integration and sharing of spatio-temporal data. Based on ontology modeling method and OWL language as modeling standard, an emergency ontology construction method oriented to event-task-model-data is proposed (Fig. 3.55). (i) Given an emergency, through the plan engine, it is necessary to get possible emergency tasks. (ii) For specific emergency task, the corresponding model and data are obtained by inference engine, and the relationship structure between the model and data is automatically given. (iii) According to the above topological structure, it is necessary to find available models and data, replace unavailable resources, and train the involved models through data. (iv) It is necessary to output the results, and give specific emergency solutions through the interpretation engine.

(2) **Technical scheme of integrated platform**

Zhang Hui's research group of Tsinghua University put forward the structure of a comprehensive emergency integration platform for emergency decision-making. Focusing on network integration, computing integration and application systems'

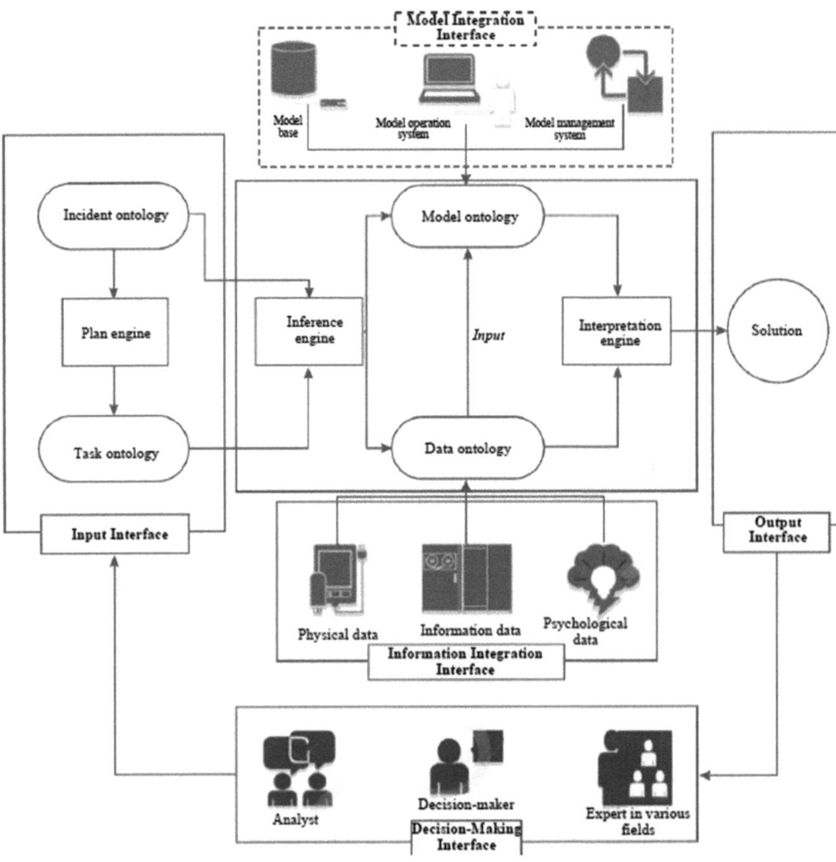

Fig. 3.55 Emergency ontology construction method for unconventional emergencies

integration, an open online sharing integration platform is built by using the self-organization mode of information. In this way, it realizes the collection and sharing of interdisciplinary, cross-regional and cross-fields emergency systems' information and data, and provides data support for the whole emergency service. The research group also applied cloud computing, cloud storage and other technologies to build a integration platform of computing's sharing to meet the interdisciplinary and cross-regional emergency requirements of handling and analysis and provide smarter and more efficient simulation and analysis services; Through cloud application technology, it provides an integrated platform for cross-regional and cross-platform application systems. The overall framework of the integrated platform is shown in Fig. 3.56.

In order to achieve the integration goal, it is implemented in stages and steps according to the requirements. (i) Building a basic platform for emergency management. Under the support of cloud computing, it is necessary to provide operation support and required computing and storage environment. (ii) Realizing the percep-

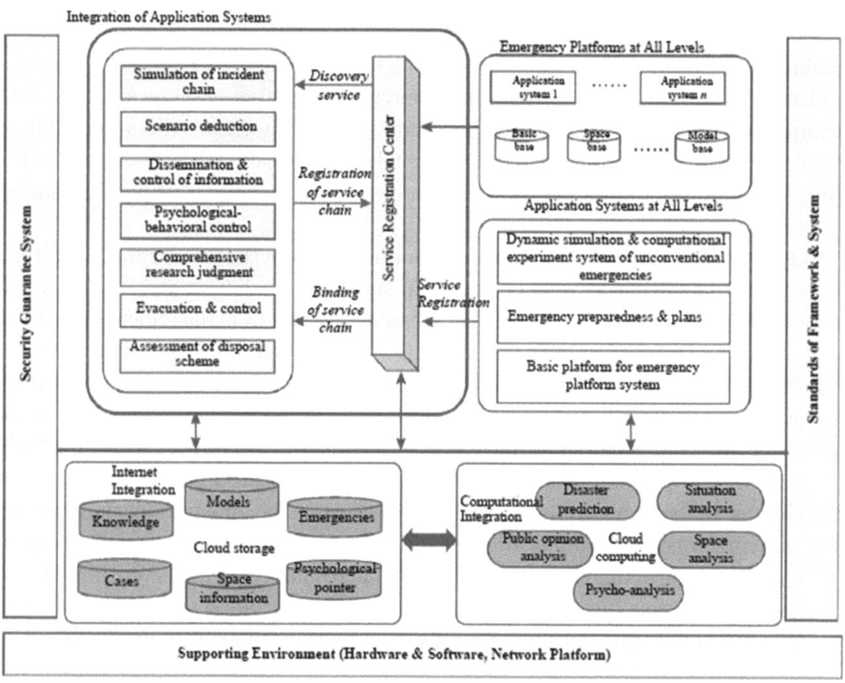

Fig. 3.56 Overall framework of integrated emergency platform

tion and access of emergency resources, models and tools. It is necessary to support the access of various software and hardware resources and the access of service requests, and realize the dynamic collection, storage and preprocessing of models and tools. (iii) Realizing the encapsulation and integration of service. For various data, models and tools, it's necessary to realize service's virtualization, and realize service's integrated operation in various professional fields and capabilities. It is necessary to provide access to various services by using virtualization and other methods to realize the comprehensive management of service's publishing, organization and aggregation, management and scheduling, and service process's supervision. (iv) Application of the model. The platform can realize personalized customization of data, resources, tools, services, etc., and realize real-time response to emergencies.

(3) **Integrated sublimation of Web platform**

Zhang Hui's research group has developed an integrated sublimation of Web platform (Fig. 3.57). The platform includes project space, data center, resource center, topic center, models and tools. (i) Project space. It includes project overview, project progress' introduction, etc., and it supports doc, xls, pdf and other document formats, as well as various pictures, slides, videos, etc. (ii) Data center. It mainly includes establishing data interface to realize seamless connection of data. (iii) Resource center. It mainly includes storage of resources, computing resources, documents, pictures, videos, etc. (iv) Special topic center. It mainly shows the application of comprehensive integration and in-depth analysis of cases. (v) Models and tools. This part mainly includes psychological measurement, analysis of public opinion, evacuation, dispatching and planning.

Based on cloud data management, computing and analysis, the integrated platform can provide data of basic aspects, research, statistics, prediction and it can provide other services such as information management, document management and search for authorized systems (projects) according to the agreed data exchange protocol. Pictures, animations and videos are obtained and integrated by the core algorithm of the model and they will be shown to relevant users in the form of web pages. At the same time, the system can show the research progress of each project according to the "time axis," and provide channels for online communication and the transmission and sharing of project results through the platform. The integration platform can start network transmission and file transmission in other formats according to the content of the control field; It can provide simulated data for the decision optimization system. It can receive the decision of the decision optimization system, deploy it on the deduction platform, and show it on the client synchronously.

Fig. 3.57 Integrated sublimated web platform interface

3.3 *Handling Strategy of Typical Emergencies*

3.3.1 Emergency Rescue and Handling Strategy of Catastrophic Earthquake Disaster

(1) **Evolution of life rescue scenarios in extraordinary earthquake based on case deduction**

Li Shiming's research group of University of Electronic Science and Technology of China took Dongfang Turbine Co. Ltd (referred to as Dongqi Plant) as the evolution case of earthquake's rescue scenario, and divided the rescue scenario into four sections: instinctive escape, local self-help, enterprise self-help and social rescue [Fig. 3.58(a)]. It is not only the logical process and main element of the earthquake's rescue process of Dongqi Plant, but also a typical section and key event of the rescue scenario, with pertinence and typicality. The basic state of each profile of Dongqi Plant's post-earthquake rescue is shown in Fig. 3.58(b).

The main elements of earthquake's rescue scenarios can be divided into four levels: individual, group, organization and society (Fig. 3.59 indicates the disaster-bearing and disaster-resisting bodies' states). As living beings, people take on two

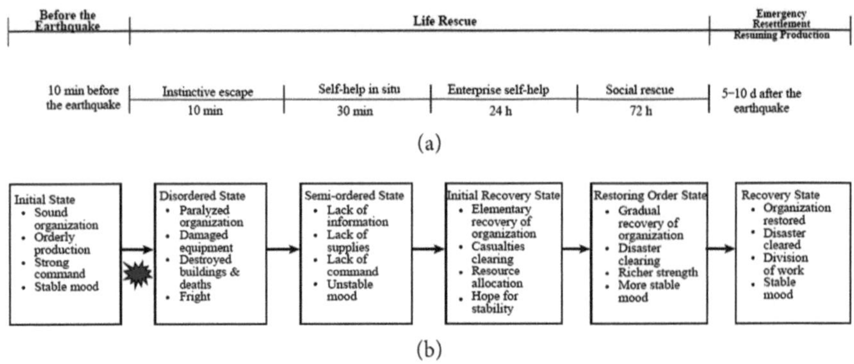

Fig. 3.58 The basic states of each profile of Dongqi Plant's post-earthquake rescue

different roles: disaster-bearing bodies and disaster-resisting bodies. As disaster-bearing bodies, people are generally in a state of panic and bewilderment, while as disaster-resisting bodies, they are mainly responsible for rescue. From individuals, groups, organizations to society, the typical disaster-bearing bodies respectively have the characteristics of escape, hope for rescue, loss of help and disorder, while the typical disaster-resisting bodies have the characteristics of life-saving, self-help, rescue and order.

In the scene's section plane of life rescue, instinctive escape, local self-rescue, enterprise self-rescue and social rescue have obvious correspondence with individual level, group (grass-roots) level, organization (enterprise) level and social level of disaster-bearing bodies. Each level may become a disaster-bearing body, and when certain conditions are met, it may be transformed into a disaster-resisting bodies. Hierarchical transition from disaster-bearing body to disaster-resisting bodies is shown in Fig. 3.60. There are three meanings in the transformation from disaster-bearing body to disaster-resisting bodies: (i) As far as each scenario's section plane is concerned, there is a logical sequence among the four section planes: individual instinct escape, local self-rescue at the grass-roots level, internal self-rescue within

Fig. 3.59 Four levels of disaster-bearing bodies and the states of disaster-resisting bodies

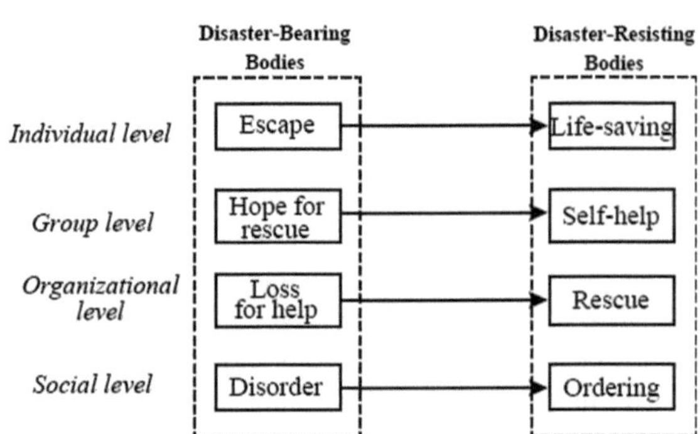

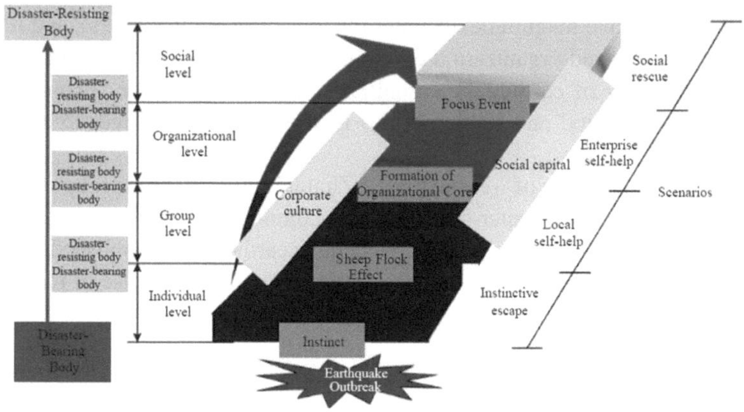

Fig. 3.60 Hierarchical transition from disaster-bearing body to disaster-resisting bodies

the enterprise and external social rescue. The next level's situation is the basis for carrying out the next-step rescue, and life rescue is promoted orderly step by step. (ii) In life rescue, there may be four section planes, namely, individual instinct escape, local self-rescue at the grass-roots level, internal self-rescue at the enterprise and external social rescue, that is, the four sections overlap and cross. (iii) Realizing the transformation from disaster-bearing body to disaster-resisting bodies is the "nuclear fusion" of unconventional emergency's scenario-response. Therefore, researching, creating and cultivating the conditions for the transformation from disaster-bearing body to disaster-resisting bodies has become an important method for emergency rescue.

(2) **The "two-stage" law of the growth of wounded persons by earthquake and the "three-stage" characteristics of the use of rescue forces**

Zhang Lulu's research group from the Second Military Medical University of the Chinese People's Liberation Army focused on the empirical research on earthquake emergency's medical rescue. Focusing on the wounded (the demander) and the rescue forces (the supplier), the research group successively spent eight years in investigating and collecting data from the earthquake-hit areas in Wenchuan, Yushu, Lushan and Rudian, and summarized and proposed the basic "two-stage" law and "three-stage" law for emergency medical rescue under earthquake situation.

Based on the main line of "the sick and wounded," it is found that within 2 weeks after Wenchuan earthquake, there is a rapid growth period of casualties, and after 2 weeks, there is a "stable period" of casualties (Fig. 3.61). Compared with Wenchuan earthquake, the inflection point of cumulative percentage of death and cumulative percentage of wounded in Yushu earthquake is about 7 days and 10 days earlier respectively, as shown in Figs. 3.62 and 3.63. There is an obvious inflection point in the growth curve of the number of earthquake casualties. There is a rapid growth period before the inflection point and a stable period after the inflection point. The occurrence time of inflection point is related to the earthquake magnitude (the larger the earthquake magnitude, the later the inflection point appears), and it is also related to the time and organization efficiency of rescue. Scientifically grasping the time sequence law of earthquake casualties' growth (that is, the "two-stage" law of rapid growth period and stable period) is very important to reduce earthquake casualties.

Based on the main line of "rescue force," Zhang Lulu's research group made an in-depth investigation of 8 disaster-stricken areas, command organizations from 5 levels and 6 types of security organizations in Wenchuan earthquake, established a database of rescue force in Wenchuan earthquake, and proposed that the use of rescue force

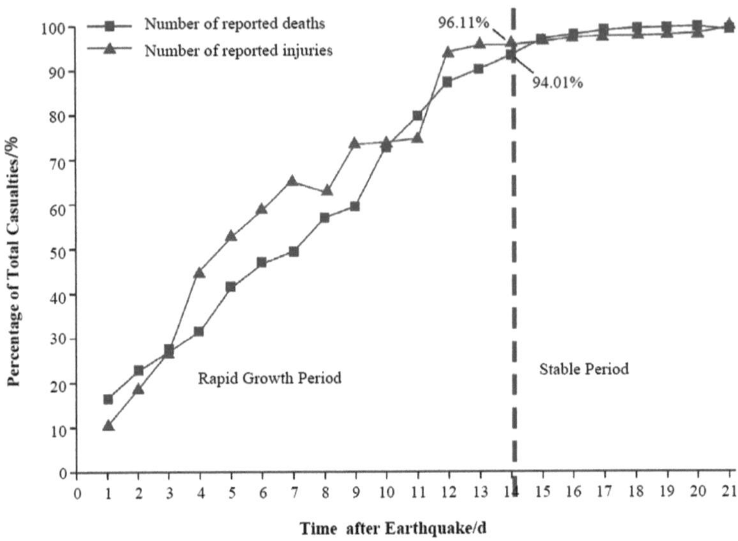

Fig. 3.61 Growth trend of casualties in Wenchuan earthquake

Fig. 3.62 Comparison of cumulative percentage of deaths

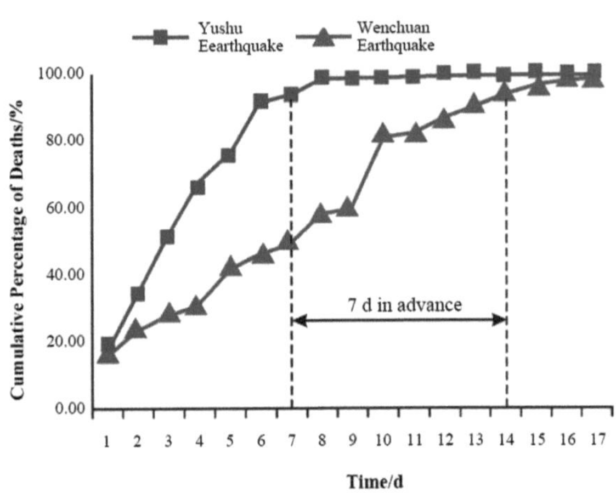

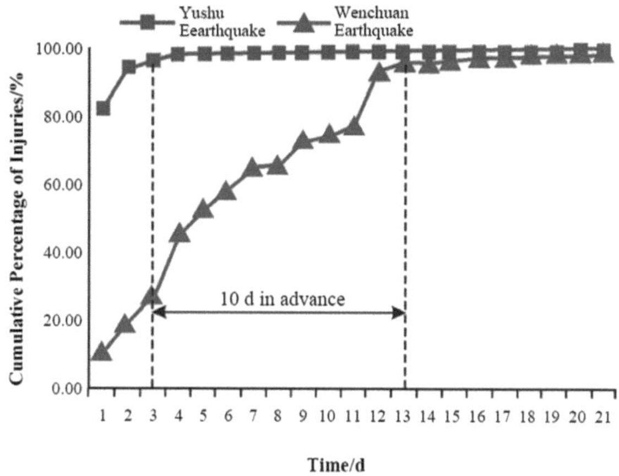

Fig. 3.63 Comparison of cumulative percentage of casualties

has the characteristics of "three-stage," namely, emergency stage, effective handling stage and maintenance section. In the emergency stage, a large number of casualties occurred. The command of medical service was mainly a kind of strategic command, and the medical service forces were quickly selected according to functional modules. The medical service support was mainly medical security, on-site first aid and protecting disaster victims. In the effective handling stage, casualties continued to rise, and the medical service command was dominated by the strategy of detachments. Medical service support turned to medical rescue, health guarantee and epidemic prevention and psychological assistance, in which medical rescue turned to emergency treatment and early treatment (treatment method of controlling damage), and the focus of support changed from treating victims to supporting troops' rescuing work. In the maintenance stage, casualties reached saturation, the command system of the disaster area was formed, and medical rescue turned into specialized treatment. Yushu earthquake's rescue force is also characterized by "three-stage." Compared with Wenchuan earthquake, the duration of each stage of Yushu earthquake is shortened. More than half of the rescue forces were deployed within 72 h after the earthquake, and the military shelter hospital played a great role in treating the wounded and sick after the earthquake. Figure 3.64 shows the relationship between the occurrence of earthquake victims and the use of medical rescue forces in Yushu.

3.3.2 Emergency Rescue and Handling Strategies for Unconventional and Sudden Water Disasters

(1) **Risk identification and scenario construction of unconventional and sudden water disasters driven by big data**

Wang Huimin's research group of Hohai University focused on the problem of depicting the potential trend law and risk control of flood and drought disasters and secondary derivative disasters under the strong coupling of nature and society.

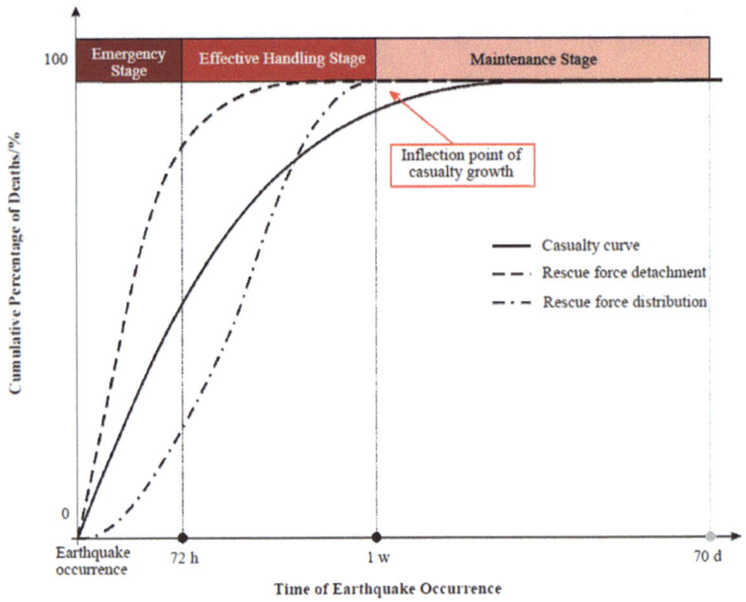

Fig. 3.64 Relationship between the emergence of sick and wounded and the use of medical rescue force in Yushu earthquake relief

Through statistical analysis of hydrometeorological big data, it revealed the evolution law of hydrometeorology and its correlation with hydrometeorological disasters; By interpreting and analyzing multi-source satellites' remote sensing data, a new method of monitoring flood and drought disasters was established. Based on the knowledge Meta-theory, the research group analyzed the relationship among emergency incidents, disaster-bearing bodies, environmental units and emergency activities during the occurrence and development of unconventional sudden water disasters, put forward the knowledge super-network's modeling method in disaster factors, state evolution and transmission, and established the association structure of scenario-disaster factors; The similarity calculation method between nodes and edges of knowledge Meta supernet, the disaster-causing factors based on conditional information entropy, and the time–space distribution's identification method were proposed. The research group also established a multi-level scenario expression model (Fig. 3.65), and realized the formal description of the complex process of incubation and occurrence of disaster.

(2) **Emergency cooperative management method of unconventional sudden water disasters based on scenario response model**

Climate change, water disasters, complex and uncertain water disasters' environment and other objective conditions increase the difficulty of emergency management of unconventional sudden water disasters. At present, there are some problems in the emergency handling of water disasters, such as low efficiency, mutual prevarication

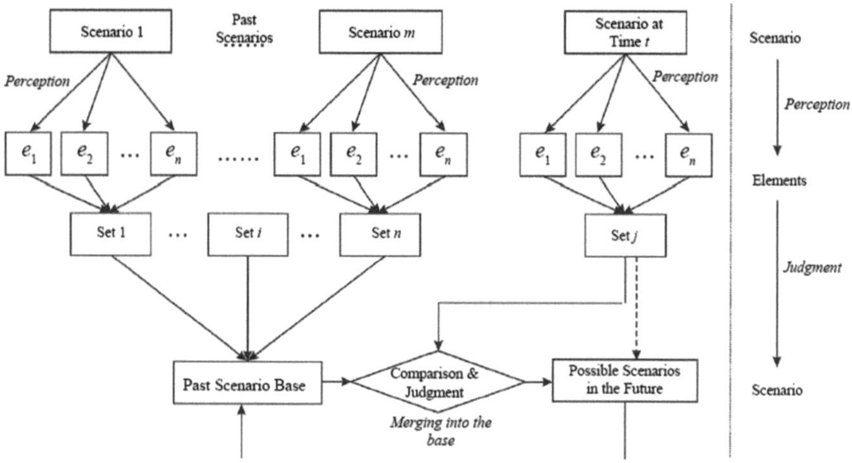

Fig. 3.65 Multi-level scenario expression model

and lack of cooperation. Wang Huimin's research group used the complex system, a science method, and was guided by the concept of harmonious development between man and water to analyze the role, function, adaptability and heterogeneity of the water disaster emergency management's main bodies, and described the behavior law and system's evolution equilibrium of the water disaster emergency cooperation's main bodies. The management system of emergency cooperation of flood and drought disasters based on multi-agent cooperation was constructed respectively, and the emergency cooperative management's process and its efficiency were analyzed. Based on the framework of emergency cooperative management system, the research group put forward a new idea of water disaster's emergency management based on scenario-response and multi-agent cooperation, and the research group constructed systematic models of flood and drought disasters respectively, including macro and micro models of flood disaster's emergency management of cooperation, cooperative storage model and allocation model of water resources for emergency drought disaster.

Flood disasters are frequent in Huaihe River Basin, which is characterized by cross-border and complexity. The research group of Wang Huimin selected Huaihe River Basin as an example to study the behavior pattern of multi-agent cooperation. The emergency cooperation system for sudden floods in Huaihe River Basin is mainly composed of the National Flood Control and Drought Relief Headquarters (hereinafter referred to as the National Defense General) and the Flood Control and Drought Relief Office of Huaihe River Water Resources Committee (hereinafter referred to as Huai Committee). The National Defense General and the Huai Committee play the role of strong reciprocity in the whole process of emergency decision-making, and they formulate the overall emergency plan, supervise the behavior of subjects of emergency handling, and promote the cooperation of subjects with a lower level (Henan Province, Anhui Province, Jiangsu Province) by means of punishment and incentive mechanism. Henan Province, Anhui Province and Jiangsu Province, under the unified leadership of superior strong mutualists, adapt to each other and evolve to achieve Pareto optimality. The results show that when the supervision cost of Huai Committee is too high and the punishment for local government's passive implementation of emergency policy is too light, after the evolution of the game, Huai Committee gradually tends not to supervise, while local government chooses passive implementation; When Huai Committee's punishment for local government's passive implementation of policies increases to more than the cost of supervision, it can propel its enthusiasm for supervision over local governments. However, due to the high cost from local governments' active implementation and the lower invisible income, through long-term repeated games, the limited rational local government finally tends to continue passive implementation.

(3) **Collaborative Decision-making's Simulation System for Comprehensive Flood Control in Three Gorges Area**

Wang Hongwei's research group of Huazhong University of Science and Technology has developed a collaborative decision-making's simulation system for comprehensive flood control in the Three Gorges region and it aims to simulate the collaborative planning process of multi-departments working together to formulate response plans. According to the emergency situation, emergency decision-making entities use coordinated reasoning and task planning methods to dynamically generate response plans through multi-department information's interaction. Through scheduling and simulation control, comprehensive simulation of implementation process of response plans, emergency situation's change (environmental change) process and inter-departmental information interaction is realized, and a general framework for emergencies' collaborative decision-making simulation based on distributed interactive simulation technology is formed. Thus, comprehensively simulating multi-departmental collaborative response process, decision-making mechanism and operation process in flood control in Three Gorges area. Wang Hongwei's research group simulated three typical decision-making processes: discussion and decision-making for the flood in the Three Gorges area, handling by engineering rescue and people's relocation and resettlement. The discussion and decision-making process mainly simulates: when the flood occurs, Office of Flood Control and Drought Relief Headquarters of China, Yangtze River, Hubei and Three Gorges Company determine the flood's dispatching target through negotiation (Fig. 3.66). The handling by engineering rescue is mainly manifested in setting up the on-site headquarters quickly after the emergency occurs in the local area, and formulating the action plan to deal with the emergency by using the HTN planning method under the condition of limited resources. In the process of task's implementation, the emergency changes dynamically, and unexpected incidents will happen during the task's implementation, which will require re-planning (Fig. 3.67). When the water level reaches a certain height and flood diversion and storage areas need to be put into use, the relocation and resettlement of its personnel and property requires the cooperation of departments of civil affairs, transportation and public security. Under such situation, the HTN

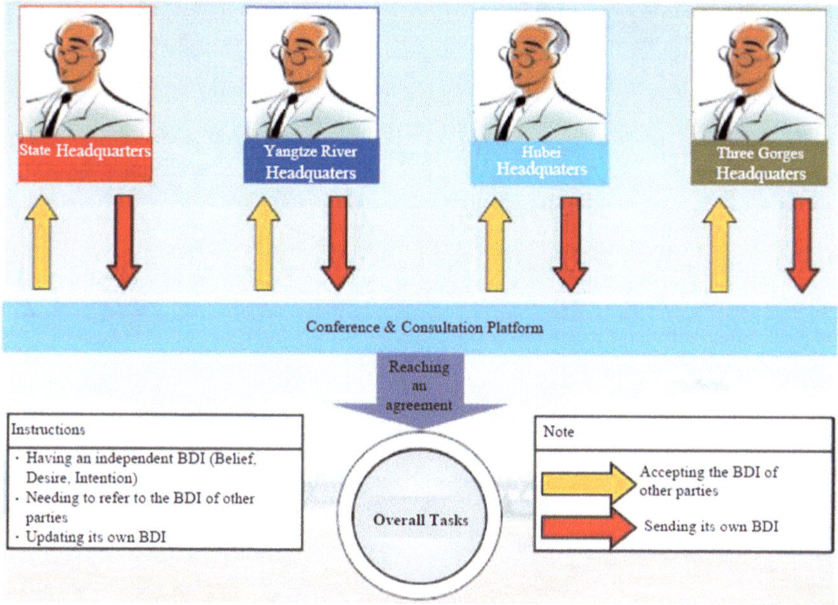

Fig. 3.66 Consultation and decision-making process under flood

collaborative task planning method should be applied to formulate action plans for each department to eliminate conflicts among plans (Fig. 3.68).

3.3.3 Evolution Mechanism and Handling Strategies of Major Infectious Diseases

(1) Study on dynamic characteristics of infectious disease's transmission and contact network

At the beginning of September, 2009, Langfang broke out the largest H1N1 incident in Chinese mainland. The first case appeared on August 27th, and the epidemic ended on September 17th. There were 586 influenza-like cases, of which 226 were confirmed by laboratory tests. Based on the survey data of epidemiological contact history of infectious diseases, Zeng Dajun's research group of Institute of Automation, Chinese Academy of Sciences reconstructed the communication network of infectious diseases among students in this epidemic, analyzed the communication and contact relationship among students by using complex network method, and used Agent modeling technology to build a model of artificial society, and dynamically simulated the dynamic process of the spread and diffusion of infectious diseases among students. The experimental simulation's results show that the simulated cases' growth process is basically consistent with the actual evolution process, and the

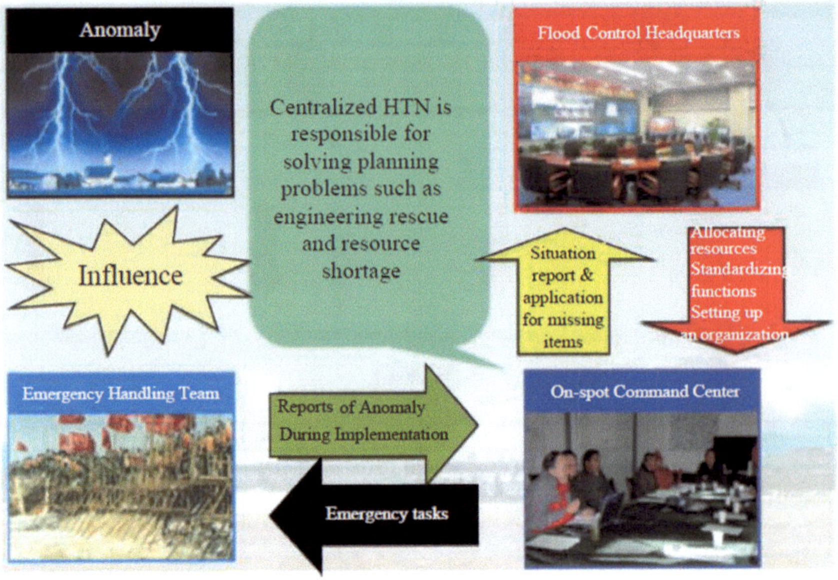

Fig. 3.67 Engineering rescue and disposal

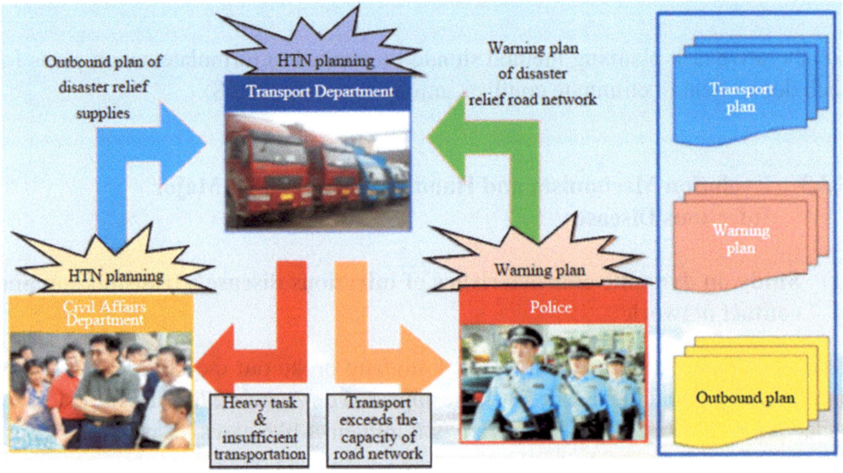

Fig. 3.68 Transfer and resettlement of flood diversion and storage areas

distribution characteristics of infectious cases in the students are consistent with the actual observation results, indicating that Agent modeling can basically describe the internal development mechanism of infectious disease's transmission and contact among the students. Using this artificial society, we can scientifically and quantitatively evaluate the effectiveness of prevention and control measures in infectious diseases. Zeng Dajun's research team also quantitatively evaluated three prevention and control strategies' influence on the evolution of H1N1 epidemic when H1N1 broke out in Langfang (isolated dormitory building; cutting off the communications among dormitories and prohibiting mutual visits; isolation measures for dormitory with H1N1 cases). Before September 3rd, the simulated results were very similar to the real outbreak situation. After September 3rd, the decline trend of the number of simulated cases was slower than the real situation. This is because the model only pays attention to the prevention and control measures taken by schools, but on the personal level, students have strengthened their awareness of prevention (such as wearing masks, reducing going to crowded places, washing hands frequently, keeping indoor ventilation, etc.), and these individual prevention measures have greatly reduced the spread rate of H1N1 among students. Nevertheless, the simulation results also show that the prevention and control measures of cutting off social communications play an important role in controlling the spread of H1N1.

(2) **Quantitative risk assessment method for susceptible people**

Hospitals, engine rooms and other confined spaces are high-risk areas for the spread of infectious diseases. The spread risk is the focus of the research. Weng Wenguo's research group of Tsinghua University has established a quantitative risk assessment method for susceptible people (Fig. 3.69) to study the spread and spread risks of infectious diseases in confined and crowded spaces. This method is mainly divided into three parts: description of infectious sources, analysis of air flow distribution and assessment of infectious risks. Firstly, the source of infection is described quantitatively, the particle size distribution's characteristics of exhaled droplets are analyzed, and the physical model of exhaled infectious substances of patients with disease source is established; Secondly, according to the established disease source model, considering the ventilation conditions and thermal boundary conditions in the space, the diffusion and spread process of infectious substances of respiratory infectious diseases in the space is quantitatively simulated, and the spatial and temporal distribution of infectious substances in the indoor space is calculated as the input condition of the risk assessment process; Finally, according to the pathological characteristics of the disease, it is necessary to determine the relevant parameters of the risk assessment process, assess the infection risk of the susceptible people, obtain the distribution of infection risk, and confirm the correlation between the amount of inhaled infectious substances, residence time and infection possibility.

(3) **Infectious diseases' transmission model based on complex network**

Weng Wenguo's research group of Tsinghua University has established an infectious diseases' transmission model based on complex network (Fig. 3.70), and studied the transmission mechanism of infectious diseases under the situation of large-scale crowd flow. (i) Population structure model: According to the existing administrative divisions in China, the population of 31 provinces, autonomous regions and municipalities directly under the Central Government is divided into three levels of sub-groups with hierarchical structure, and sub-groups at the same level can be connected with each other through actual or virtual transportation networks. (ii)

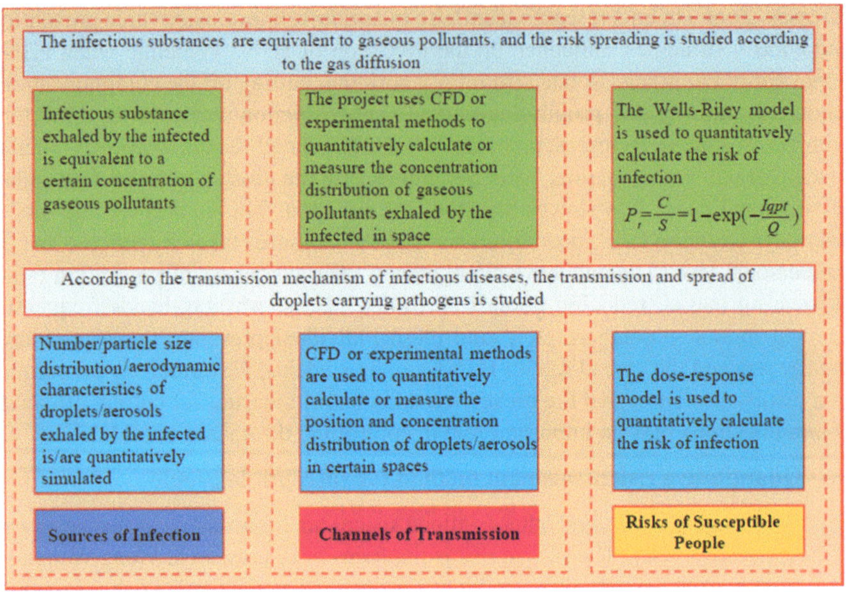

Fig. 3.69 Quantitative risk assessment method for susceptible population

Personnels' travel model: The daily travel process of personnels is simulated through the transportation network, so that people from different sub-groups also have contact opportunities. (iii) Model of stochastic SEIR and partial part of the transmission: It describes the transmission process of infectious diseases in the lowest sub-groups (also called community). Through the coupling of this partial part of the transmission process and the travel process of people, infectious diseases may spread to the whole population (taking SARS as an example, the transmission process and containment strategy are simulated, and the effect is good).

(4) **Online public opinion during the epidemic and online-offline action laws**

According to Zeng Dajun's research group of Institute of Automation, Chinese Academy of Sciences, there is a significant positive spatial correlation between the number of posts posted by netizens participating in the topic's discussion of influenza A (H1N1) in China in 2009 and the incidence of influenza A (H1N1). Taking 31 provinces, autonomous regions and municipalities directly under the Central Government as spatial scale units, the spatial pattern of the number of posts and the number of patients is close, and the correlation coefficient is 0.848 ($p < 0.01$). Taking 18 districts and counties in Beijing as spatial scale units, the spatial pattern of the number of posts and the number of patients is close, and the correlation coefficient is 0.901 (*p*

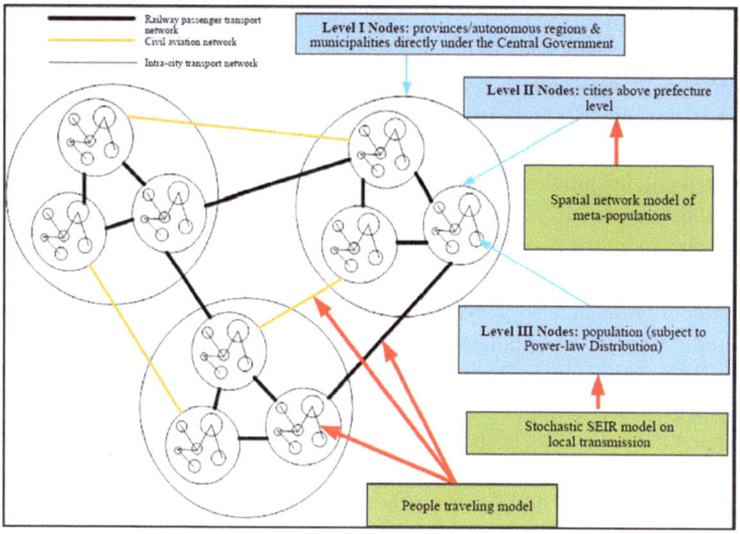

Fig. 3.70 Infectious diseases' transmission model based on complex network

< 0.01). Therefore, it is a reliable method to predict the risk of infectious diseases' spread in time and space by using open source information on the Internet.

(5) **Real-time monitoring and early warning platform for major epidemics on the Internet**

Zeng Dajun's research group has developed a technical prototype system for crisis information's monitoring, collaboration and emergency response. It is called Internet Real-time Monitoring and Early Warning Platform for Major Epidemics, and it integrates traditional closed-source data, network open-source data (such as news, forums, blogs, Weibo, etc.), mobile data and active call emergency data to realize online perception, online call for help, proactive early warning, group emergency response and parallel management and control of crisis incidents. By collecting, processing and analyzing open-source information on the Internet, we can provide crisis information navigation services for the public and relevant departments, such as forecasting, early warning and auxiliary analysis of crisis information. Through real-time linkage of maps and information, the platform monitors the time, place, development, influence scope and public reaction of crisis information.

The Internet epidemic real-time monitoring and early warning platform includes eight functional modules: home page, network epidemic, epidemic hot spot, time and space analysis, hot spot tracking, network report, epidemic topic and data retrieval. (i) Home page. It is a comprehensive display of all sub-modules, providing users with macro-global information, including geographical distribution of online epidemic monitoring, statistics of monitored data, latest epidemic monitoring results and official announcements, epidemic topics and cloud system of public health information. (ii) Network epidemic situation. This module shows the online monitoring results of epidemic situation for users. (iii) Epidemic hot spots. According to the online monitoring results and weighted classification algorithm, the module displays the epidemic related titles of high-risk and high-frequency in a hierarchical rolling cycle, providing users with the latest real-time epidemic's monitoring hot spots. (iv) Spatio-temporal analysis. The module integrates three types of data sources from open-source network, hospital monitoring and Internet users' voluntary submission, and it displays the distribution characteristics and correlation of various types of data in real time by using visualization technologies such as GoogleMap and complex network. (v) Hot spot tracking. This module provides users with interactive display interface of epidemic monitoring. (vi) Network report. This module provides users with information input window and platform, and lays a foundation for further interaction with users. (vii) Epidemic topic. This module customizes disease topic according to users' needs, and monitors diseases in a panoramic way. (viii) Data retrieval. This module provides users with online search function, and makes time analysis and visualization display on the popularity of searched objects on platforms such as Weibo, the correlation of keywords and the social network of bloggers.

3.3.4 Evolution Mechanism and Handling Strategies of Major Terrorist Attacks

(1) **Decision-making interaction and evolution law between deliberate disaster causers and emergency decision makers**

In deliberately-caused disasters, the factors such as action purpose, behavior preference and benefit evaluation of the disaster causer are complex and changeable, so it is difficult for emergency decision-makers to judge the attack ability and target selection of the disaster causers. At the same time, the semi-open defense strategy of decision-makers makes it difficult for the disaster causers to confirm the resource allocation scheme and the possibility of successful attack. Shen Shifei's research group of Tsinghua University applied the game theory to the study of deliberately-caused disaster scenarios, and established a benefit matrix model under the conditions of zero-sum game (between the disaster-causers) and matching attack and defense. Besides, the research group discussed the solution of the game under completely rational conditions and the solution ideas under irrational conditions. At the same time, a signal model for the game has been constructed to discuss the influence of asymmetric information on the resource allocation strategy of emergency decision makers. The loss of decision makers with different game times is shown in Fig. 3.71. With the increase of the number of games, the advantages of dynamic information strategy gradually appear, and the expected loss of the defensive side is obviously reduced. Because the cost of threat is an important factor that affects the actions of those who deliberately cause disasters, comprehensive ways such as emergency management and public security education can increase the difficulty of deliberately-caused panic, prevent the threat of deliberately-caused disasters, and effectively reduce the expected loss.

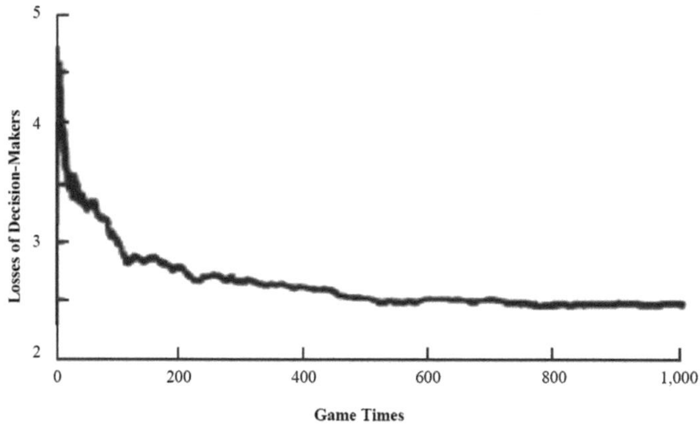

Fig. 3.71 Decision-makers' losses under different game times

(2) **Optimization model of defense configuration under deliberately-caused disaster scenarios**

Shen Shifei's research group established a static game model of defense resource allocation, and used the two-person game structure of basic defense model to discuss the most reasonable and effective emergency resources' allocation scheme, especially the overall defense structure, the correlation degree among disaster-bearing bodies, and the resources' allocation scheme of multiple disaster-bearing bodies. The research results of the impact of total resources on the expected loss show that it is necessary for emergency decision-makers to determine the principle of defense resources' allocation according to the type of intentional disaster causers. Compared with the government, it is a better scheme to allocate defense resources among disaster-bearing bodies evenly when the power of deliberate disaster-causers is weak. For example, distributing police forces more evenly to crack down on small-scale sabotage activities is rational; in view of the strong and deliberate disaster-causers such as terrorist organizations, it is necessary to determine the key defense objectives, identify key facilities and allocate resources in a targeted manner to achieve the best disaster prevention effect. In addition, because the correlation of disaster-bearing bodies will lead to regional risk's sharing, the expected loss according to the system will gradually increase with the increase of correlation coefficient. Auxiliary defense measures such as regional traffic control, individual travel restrictions, semi-closed isolation of key units and buildings, and traffic restriction around sensitive areas can effectively reduce the system's intentional disasters' risk assessment result.

(3) **Adaptive and dynamic decision-making model of "scenario-decision-making-feedback"**

Shen Shifei's research group, based on the research of static resource's allocation and information strategy, built a strategy optimization model based on multi-stage game in view of the multiple confrontations between the two sides under asymmetric information. In this model, the multiple confrontation process between the intentional disaster-causer and emergency decision-makers is described as multiple game stages, and the optimal strategy of each game stage is calculated. Both parties dynamically adjust and optimize the strategy according to their game history. Based on the multi-stage strategy optimization model, the team proposed a dynamic resource allocation method for emergencies deliberately caused, simulated the dynamic resource allocation based on Agent method, and analyzed the optimal defense strategies of emergency decision maker against deliberate disaster-causer with different strategies (such as "belligerence," "caution," "suspicion" attack strategy and other dynamic strategies). The trend of average loss with confrontation times under different attack strategies is shown in Fig. 3.72. The results show that when it is difficult to predict the attack strategy of the intentional disaster-causer, the dynamic defense strategy should be adopted, and the defense strategy should be dynamically adjusted according to the historical results of the confrontations and the gains obtained in the confrontation stage, so as to effectively reduce the loss. This reflects the intelligence of emergency decision-makers who sum up experience and make decisions in the repeated confrontations of asymmetric information.

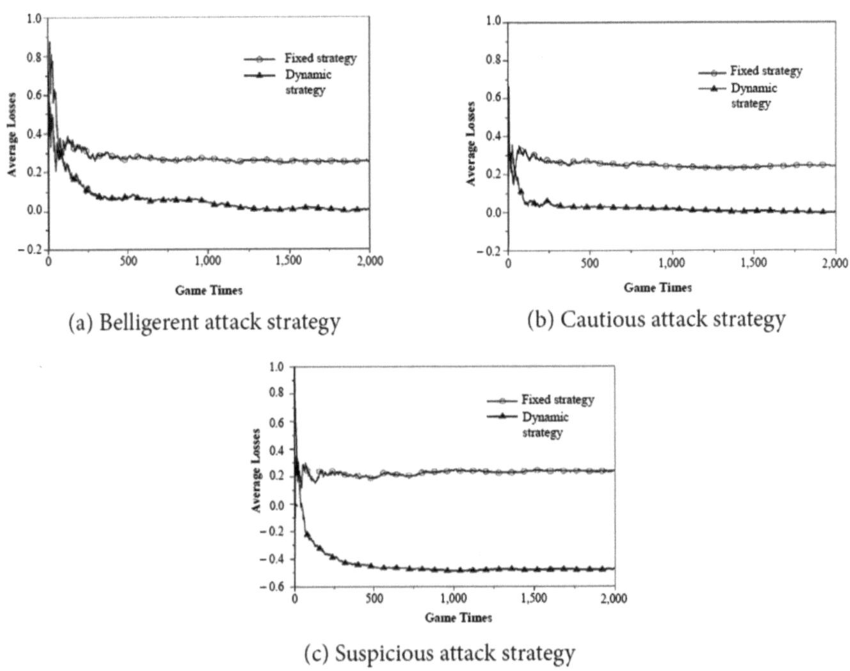

(a) Belligerent attack strategy

(b) Cautious attack strategy

(c) Suspicious attack strategy

Fig. 3.72 Change trends of average loss with game times under different attack strategies

3.3.5 Emergency Evacuation of Large-Scale Crowd and Relevant Intervention Mechanism

(1) Psychological and physiological reaction of panic and spread of panic behavior

Sun Jinhua's research group of University of Science and Technology of China studied the formation and influencing factors of people's cooperative behavior in crisis environment, as well as the influence of personality characteristics on behavior and its neural mechanism. In order to explore the neural basis of human's escape behavior, the simulation platform of fire escape behavior (Fig. 3.73) and the functional magnetic resonance imaging (fMRI) pre-experiment of simulating fire's escape test were developed. After the test started, an obstacle began to burn, the temperature in the room increased, and the protagonist's life value decreased with the increase of temperature. The subject controlled the protagonist with the keyboard to try to escape. The success rate of escape varies with the density of people as shown in Fig. 3.74. The results showed that with the increase of crowd density, the success rate of escape decreased.

Functional magnetic resonance imaging (fMRI) is the most advanced brain function measurement method at present, which can detect the brain activity of human specific behavior without damage, with high temporal and spatial resolution. However, this method has certain technical difficulties, and there is no study on the related problems of fire escape by this method at home and abroad. Sun Jinhua's research group used functional magnetic resonance to measure the brain nerve activity of the subjects when simulating escape behavior. The brain activity during fire escape is shown in Fig. 3.75. The research reveals that under the disaster environment, the physiological and psychological reaction mechanism of panic behavior when people fail to escape, that is, the activity of the front prefrontal lobe of the subjects with high personnel density (low success rate of escape), is related to the processing of information such as emotion and society, and the immediate and positive interference of authoritative information can significantly relieve the tension.

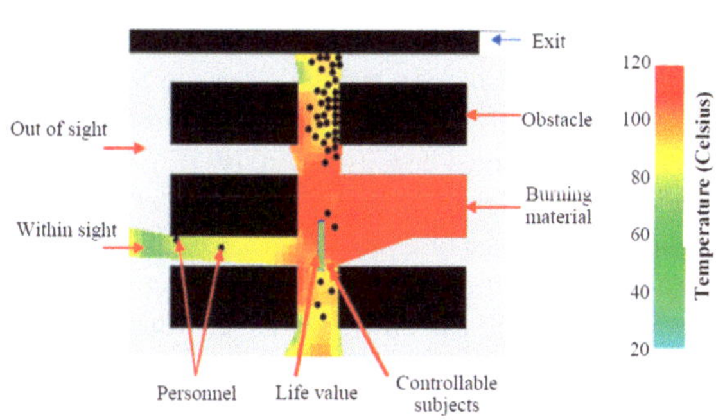

Fig. 3.73 Simulation platform of escape behavior from fire

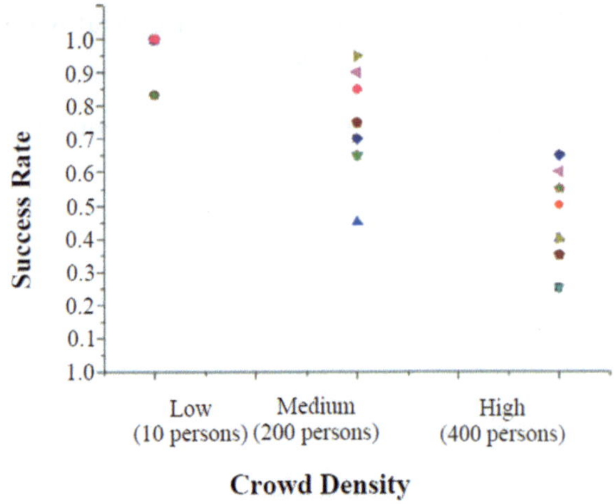

Fig. 3.74 The change of success rate with crowd density

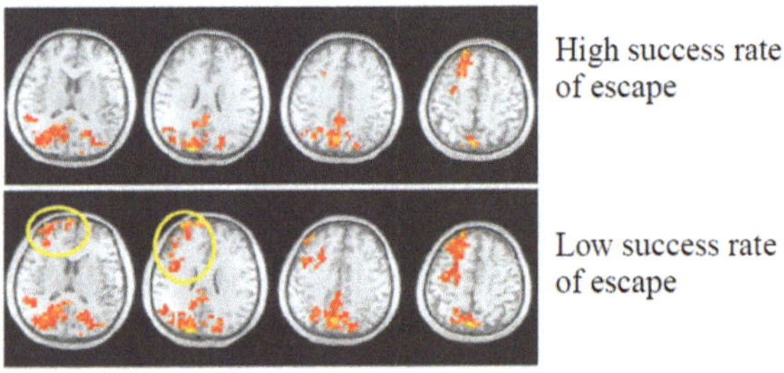

Fig. 3.75 Brain activity of people when they escape from fire

(2) Evacuation risk model considering evacuation guidance and crowd characteristic density

Sun Jinhua's research group analyzed a series of characteristic densities that affect large-scale crowd flow (including people's flowable state and extremely crowded state), and established a quantitative risk model for large-scale crowd's evacuation based on the infinite crowd queuing theory and Hughes' continuous crowd flow model, taking into account crowd flow's intensity, evacuation channel's service capacity and evacuation system's service intensity. This paper analyzes the influence of typical crowd density on evacuation strategy in large-scale evacuation, and puts forward "crowd density risk axis" (Fig. 3.76) to judge the efficiency of evacuation strategy. Effective flow indicates that evacuation efficiency can be improved by setting the number of exits and other human intervention measures. Non-effective flow means that people can't move, and conventional evacuation interventions are ineffective, which can easily lead to accidents in crowded conditions. Critical zone refers to the buffer of evacuation's intervention strategy from effectiveness to failure, and it is the difference between theoretical model and actual simulation. Emergency and powerful intervention measures before or in the buffer zone may avoid failure stage. Based on the numerical assumption of evacuation risk, the three-dimensional

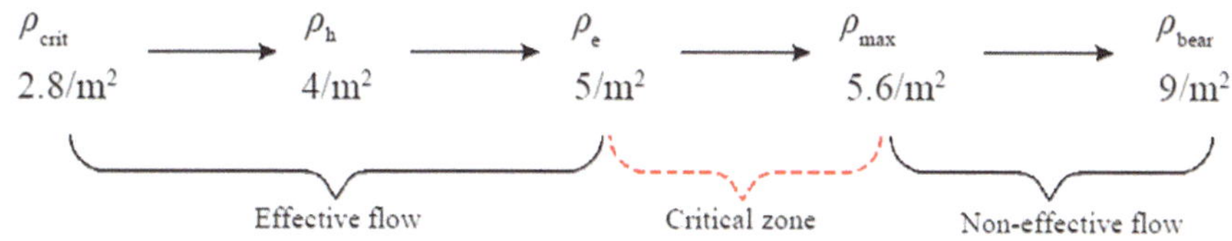

Fig. 3.76 Risk axis of population density

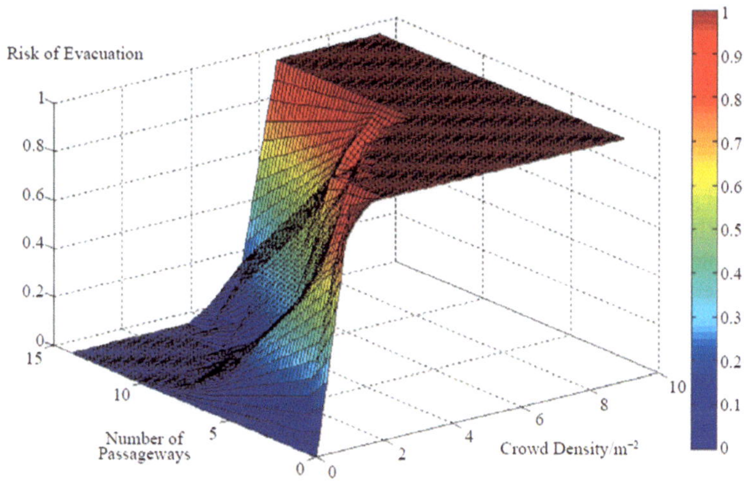

Fig. 3.77 Three-dimensional distribution of evacuation risk for crowds with different density in different evacuation channels

distribution form of evacuation risk matching with the risk axis of population density is obtained (Fig. 3.77). The results show that when the crowd density is certain, the more evacuation routes there are, the less risk there is. When evacuation routes are certain, the higher the crowd density, the greater the risk.

(3) **System of emergency evacuation and guidance**

Zhang Kan's research group of Institute of Psychology, Chinese Academy of Sciences, studied the laws of group activity and evacuation intervention measures under emergencies by means of experiments, investigation and simulation, and made a demonstration in Beijing with remarkable results. According to data from spot measurement of flow of the pedestrian by monitoring cameras and the data of pedestrian path preference obtained by investigation, the statistical model of regional pedestrian flow is built, and the pedestrian flow of the region and each road is obtained in real time, which theoretically solves the problem of expanding the traffic data from point to area. Considering the basic principles such as systematicness, continuity and easy identification of guiding signs, and based on the optimization theory, with the goal of maximizing the role of emergency evacuation broadcast and guiding signs

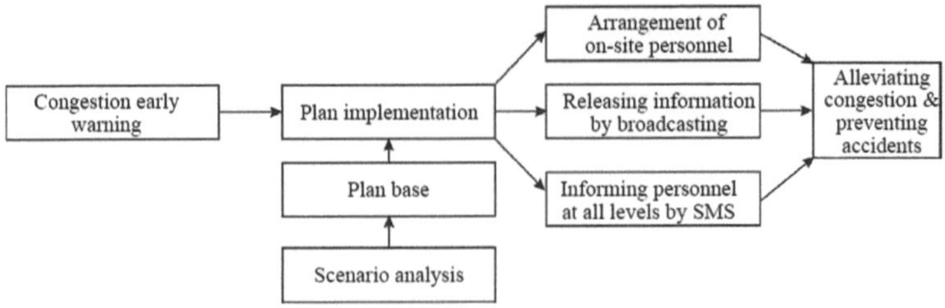

Fig. 3.78 Process of emergency evacuation and guidance

in evacuation guidance, a layout model was established to determine the layout and guiding direction of 60 evacuation and guiding signs. The simulation analysis verified the effectiveness of the evacuation signs. Based on the proposed emergency evacuation and guidance process (Fig. 3.78), an intelligent emergency evacuation system was developed to realize the functions of automatic early warning of congestion, plan implementation, plan configuration, plan release's statistics, and emergency resources' management. It is to realize the intelligentization and informatization of the emergency plan and improving the efficiency of intervention on crowd.

(4) **Security analysis and evaluation of large-scale crowd's evacuation and transportation and evacuation route optimization model**

According to the coupling process of macro-evacuation and transportation organization and micro-individual behavior in subway passengers' evacuation and transportation, Shi Congling's research group of China Academy of Safety Science and Technology has established a series of theoretical prediction models of the evacuation and transportation capacity of major subway traffic nodes (stairs, escalators, gates, passages, ticketing and ticket checking systems, etc.), and studied and put forward a security analysis and evaluation model for emergency evacuation and transportation of subway passengers. The above mentioned models include channel-traffic evacuation and transportation's security analysis model, model of channel saturation, channel's evacuation and transportation capacity's prediction model, public area evacuation and capacity's analysis model, evacuation and transportation network's allocation analysis model, etc., and these can be used to analyze traffic organization, channel's transportation capacity, channel's saturation, evacuation and transportation capacity in public areas and subway network's evacuation capacity in emergency situation.

Zhong Maohua's research group of China Academy of Safety Science and Technology put forward an algorithm in evacuation route optimization based on microscopic model, aiming at the shortest evacuation time. It took into account the crowd distribution, exit location, exit width and other factors and used iterative algorithm to find the optimal exit for each pedestrian to obtain the optimized evacuation route.

Taking the evacuation of a pedestrian commercial district as an example, the evacuation time before and after the optimization of the evacuation route is compared and analyzed by using the classical random walking model without going back but with route deviation. The results show that the function of optimizing the evacuation route is to guide people to evacuate to the exits with wider width and lower density of people, which can evacuate people in commercial streets faster.

Representative Integration Results

Weicheng Fan, Chunchang Shan, Shouyang Wang, Tiemin Liu, Feiyue Wang, Xiaofeng Meng, Lei Wang, Xiaolin Zhou, Wenguo Weng, and Yi Liu

The three representative integration results of the major research plan of "Research on Emergency Management of Unconventional Emergencies" are the new generation of emergency intelligence's computing theory and application for national security, the basic research and application of results of "scenario-response" integrated platform, and the research and application of psychology and behavior of emergency management of unconventional emergencies.

W. Fan (✉)
Tsinghua University, Beijing, China
e-mail: wfan@tsinghua.edu.cn

C. Shan
Counsellor's Office of the State Council, Beijing, China

S. Wang
Academy of Mathematics and Systems Science, Chinese Academy of Sciences, Beijing, China

T. Liu
China Academy of Safety Science and Technology, Beijing, China

F. Wang
Institute of Automation, Chinese Academy of Sciences, Beijing, China

X. Meng
Renmin University of China, Beijing, China

L. Wang · X. Zhou
Peking University, Beijing, China

W. Weng · Y. Liu
Institute of Public Safety Research, Tsinghua University, Beijing, China

W. Fan (ed.), *Unconventional Emergency Management Research*, Reports of China's Basic Research, https://doi.org/10.1007/978-981-99-6798-8_4

1 Theory and Application of New Generation Emergency Intelligence Computing for National Security

1.1 Research Background and Present Situation

Since the 21st century, with the leap-forward development of science and technology, advanced technologies such as the Internet, Internet of Things, cloud computing, big data and artificial intelligence have gradually integrated into the daily life of human society. The iterative updating and large-scale popularization of innovative technologies have unprecedentedly and violently affected and changed the traditional production and lifestyle of human beings, thus shaping a brand-new living form and cognitive system of human society and promoting the living space of human beings to gradually expand from the real physical world to the network virtual world. In the network virtual world interwoven by Internet and mobile network, people's social cognition, attitude's formation, emotion shaping and behavior response to emergencies are increasingly restricted by the indistinguishable event reports provided by news sites, WeChat, Weibo, APP, forums and other platforms, as well as the spread of related information flows. The new world system and life style of the network virtual world have shown an irreversible trend, human's social activities and their organizations have been reshaped with a new look, and the Cyber-Physical-Social Systems (CPSS), which is closely interconnected between people and things, has gradually become a reality. These changes have fundamentally changed the prevention, response, control and management mode of unconventional emergencies, and also brought about the reconstruction and reshaping of national and social security issues. Traditional national, social, economic, biological security issues and other major security issues related to the national economy and people's livelihood have been given a series of new connotations.

In order to adapt to the severe challenges to national security under the new situation, in 2014, General Secretary Xi Jinping put forward the grand strategic thought of "Overall National Security Concept," and said that "the connotation and extension of China's national security are richer than that at any time in history, the space–time field is broader than that at any time in history, and the internal and external factors are more complicated than that at any time in history," and said that "a national security road with Chinese characteristics should be created."

China is a country with frequent disasters, and various unconventional emergencies often occur, such as SARS in 2003, Wenchuan earthquake in 2008, H1N1 pandemic in 2009, violent terrorist incident at Kunming Railway Station in Yunnan in 2014, and the "Occupy Central" incident in Hong Kong in 2014. In terms of the trend, the trigger source, development and evolution, monitoring and prevention, emergency response and management of unconventional emergencies are becoming more and more abnormal, and the Internet and mobile network have gradually become the main positions to propel the evolution of incidents. The dissemination of Internet information related to incidents often involves major national security issues, and these incidents often seriously threaten the overall national security, bring unprecedented severe challenges and also bring great opportunities to emergency management. (i) The overall national security is a complex giant system and all kinds of security problems are interrelated and inseparable, thus emergencies and national security problems coexist and are easy to transform into each other. These problems are open and the definition boundaries are unclear; (ii) The basic characteristics of CPSS are: the deep coupling and strong feedback between real physical space and virtual network space, "human in the loop," the uncertain diffusion of high-frequency real-time information flow and difficult prediction and control of unconventional emergencies. (iii) CPSS brings a series of new changes in human life and interactive organization mode, and the existing emergency management technology with planning, organization, leadership and control as the core encounters with many difficulties; (iv) CPSS brings a series of new methods and technologies. People become information producers, spreaders and sensors in the loop. It is possible to mobilize the whole society to participate in all aspects of emergencies, and management science's innovation is facing great opportunities.

It is a major scientific problem to propel the perception, model and experiment of complex emergency system under CPSS environment and realize the data and intelligence support for real-time decision-making. The major powers in the world have launched fierce competition in this frontier field, and have seized the strategic commanding heights in the network era to defend the national security and sovereignty of the network virtual space.

At present, it has become a basic consensus of developed countries in Europe and America to develop advanced information technology to mine large-scale open-source information from the Internet to ensure national security. European and American countries have put a lot of investment in forward-looking and strategic scientific research. For example, Social Media in Strategic Communication (SMISC), Open Source Indicators (OSI), and Quantative Global Analysis (QGA), respectively carry out research from micro, mesoscopic and macro dimensions. Micro analysis focuses on semantic understanding, behavior modeling and so on; Mesoscopic analysis focuses on network analysis for user interaction and its influence; Macro analysis focuses the visualization of large-scale data, emphasizes the perception and prevention of social risks from the aspects of content understanding, mechanism reasoning and decision-making support. The research results of these projects provide new perspectives, new methods and new technologies driven by big data and characterized by computation for the security work in European and American countries, and develop a number of system platforms applied to national and social security risk perception and prevention and control.

In China, it has become a basic consensus that the rational use of the Internet's open-source information universally is important and urgent for safeguarding national security.

In order to cope with the great changes brought by CPSS to human society and deal with the challenges from developed countries in Europe and America in this strategic frontier field, in 2009, the first major research program of the Management Science Department of the National Natural Science Foundation of China, "Unconventional

Emergency Management Research," through top-level design, clearly proposed to focus on funding a number of projects in this strategic emerging field, absorb a number of top experts and scholars in related fields, build a multidisciplinary national talent team with strong scientific and technological innovation capabilities, and achieve a series of original scientific and technological fruits with completely independent intellectual property rights. In this way, it will provide a solid driving force for ensuring overall national security.

Under the call and support of this major research program (Table 1 for funding), a large number of experts and scholars from dozens of institutions, including Tsinghua University, Institute of Automation, Chinese Academy of Sciences, Peking University, China Academy of Safety Science and Technology, Beijing University of Posts and Telecommunications, National University of Defense Technology, University of Science and Technology of China, Institute of Computing Technology, Network Information Center of Chinese Academy of Sciences, University of Electronic Science and Technology of China, Beijing University of Aeronautics and Astronautics and People's Public Security University of China, have carried out comprehensive exploration in this emerging field, and gradually formed core research teams.

At the beginning of this major research program, foreign research on emergency intelligence's computing under CPSS environment was still at an early stage. Meanwhile, domestic research scale was limited. There was no theoretical system and related technology for emergency computing that could adapt to the new situation of CPSS and meet the overall national security needs, and no systematic solution was applied to relevant national departments. In this strategic field, the scientific and technological guarantee for the overall national security was basically blank.

Therefore, the team of scientists in this major research program, based on the current social development situation and the urgent need for ensuring the overall national security under the new situation, integrates information technology, computing, psychological and behavioral analysis and emergency management theories, establishes the basic framework of emergency intelligence's computing, and carries out a series of scientific and technological innovations in basic theories, core methods, key technologies and actual system, and makes great breakthroughs in heterogeneous super-network models and technologies. At present, the theory and method of emergency intelligence's computing have been closely integrated with national security-related departments. Through cooperation with the Emergency Office of the State Council, the Ministry of National Security, the Ministry of Public Security, the Publicity Department of the Communist Party of China, Cyberspace Administration of China, Xinhua News Agency, China Daily, etc., the theoretical achievements have been widely applied in practice, forming a set of systematic and complete solutions to guarantee national security under the new situation of CPSS.

Emergency intelligence's computing is mainly oriented to the whole life cycle of emergency evolution and development. It integrates incidents' monitoring data in the real physical world with incident-related information flow in the network world, focuses on the development of cutting-edge technologies such as information computing, intelligent control and artificial intelligence for large-scale unstructured data's processing, and reshapes the management and decision-making modes driven

Table 1 Research on emergency management of unconventional emergencies: Funding of major research projects in emergency intelligence's computing

Classification	Project leaders	Project types	Names of the major research program funded project
Representative fruits	Fang Binxing	Major supported project	Social computing method for online emergency perception, early warning and crisis intelligence navigation of unconventional emergencies
	Zeng Dajun	Major supported project	Evolution law of social network structure and its influence on unconventional emergency handling strategy
	Fang Binxing	Cultivation project	Research on the mechanism and related technology of network public opinion in unconventional emergencies
	Qiu Xiaogang	Integrated project	Integrated sublimation platform for dynamic simulation and computational experiments of unconventional emergencies based on parallel emergency management
	Liu Yijun	Cultivation project	System modeling, simulation and analysis of the formation, evolution, guidance and intervention of public opinion in unconventional emergencies
	Shang Mingsheng	Cultivation project	Evolution analysis of social network structure and its application in early warning and control of public opinion and epidemic situation
	Zhang Heping	Major project	Dynamic evaluation model for the whole process of unconventional emergency response
	Cao Zhidong	Cultivation project	Research on public opinion emergence mechanism and artificial public opinion generator of unconventional emergency scenario evolution
Other fruits of innovation	Ding Zhiming	Major supported project	Internet of things technology and system for active perception and emergency command of unconventional emergencies
	Li Jianhui	Major supported project	Cloud service system and key technologies for unconventional emergency management
	Jin Xiaoming	Cultivation project	Research on real-time mining of network visual data flow in unconventional emergencies
	Huang Lihua	Cultivation project	Research on the cognitive model, dissemination rule and early warning mechanism of network information of unconventional emergencies

(continued)

Table 1 (continued)

Classification	Project leaders	Project types	Names of the major research program funded project
	Du Junping	Cultivation project	Research on cross-media data's mining for emergencies based on Agent
	Wang Houfeng	Cultivation project	Research on dynamic detection, extraction and fusion technology of emergency information based on Internet
	Liu Xiao	Cultivation project	Principles and methods of systematic integration of unconventional emergency technology
	Wang Yanzhang	Major supported project	Integration principle and method of support model for evolution analysis, response and decision-making of unconventional emergencies
	Luo Xiangfeng	Cultivation project	Research on Web information's dissemination and evolution mechanism of unconventional emergencies based on complex associative semantic chain network
	Chen Xiaodong	Cultivation project	Research on the analysis method and early warning mechanism of network public opinion of unconventional emergencies

by big data to adapt to the new situation. Thus, it greatly improves the efficiency of departments and risk control ability of the whole country to cope with major security threats, and helps the country to remain invincible in the changeable international and domestic complex environment.

The research on the theory and application of the new generation of emergency intelligence's computing is a clear embodiment of responding to the major needs of the country and actively practicing the overall national security concept. It has important practical and strategic meaning for ensuring the overall national security.

1.2 Important Research Fruits

Emergency intelligence's computing refers to in the whole life cycle of emergencies in CPSS, using computer, automation, artificial intelligence and other information technologies, real-time perception and accurate extraction of all elements of incidents, integrating knowledge in emergency management and security, panoramic and in-depth analysis of incident elements, analysis of relevant mechanisms and principles, and assessment of potential threats. It is to form data and intelligence support for emerging emergency management and decision-making. The core scientific issues include two aspects: (i) How to identify, discover, track and mine specific information that potentially constitutes overall national security from cross-platform, multi-modal, complex and changeable big data of emergencies; (ii) How to effectively integrate knowledge in security field by computing model and analysis based on the specific information of emergencies, and realize the data-based decision-making for overall national security issues.

Emergency intelligence's computing includes the following five aspects: (i) Using large-scale open-source information such as the Internet and mobile Internet to solve national and social security problems based on emergency objects; (ii) Computing and analyzing the specific information of emergencies, and forming relevant judgment and evaluation of the overall national security issues; (iii) Based on the judgment and evaluation of emergencies, the data-based decision-making should integrate with knowledge in the security field; (iv) Adapting to the change and development of overall national security issues, and updating the theory and method of emergency computing and analysis in time; (v) Shaping the actual needs of ensuring national security and driving theoretical development. Emergency intelligence's computing is suitable for solving many types of major security problems, such as new types of terrorism, social group incidents, major activities' security guarantee, public opinion's attack and relevant defense, public health crisis, alien species invasion, destruction to ecological environment, economic and financial crisis, serious natural disasters, energy and resources security, food security, major incidents and disasters, etc.

At present, with the support of the National Natural Science Foundation of China's major research program, a number of innovative research achievements have been made in emergency intelligence's computing: a set of emergency intelligence's computing theories and key technologies with independent intellectual property rights have been constructed, which has promoted the development of interdisciplinary and innovative emergency management theories, trained a number of high-level comprehensive talents in interdisciplinary fields related to emergency intelligence's computing, improved the efficiency and technological level of relevant national departments, and enhanced the overall security guarantee ability of China. The core research fruits of emergency intelligence's computing are as follows: the new generation of emergency intelligence's computing theory, method, key technology and systematic overall solution are put forward to meet the development needs of the new situation; Based on the large-scale open-source information of Internet and mobile network, a set of key technology chains of big data's acquisition, storage, transmission, computing, analysis, understanding and visualization for efficient intelligence computing have been constructed, and the system platform and programming environment for emergency intelligence's computing have been built up. Through the integration of interdisciplinary knowledge and methods, it provides technical support and scientific decision-making means for the real-time monitoring, forecasting, early warning, prevention and control of unconventional emergencies in China. The relevant research fruits have been widely applied in relevant national departments, and have improved the decision-making level and actual handling ability of relevant national functional departments in forecasting, early warning, emergency response, emergency management and emergency control, and have provided new theory and method, core technologies and actual handling platform as support for the overall national security concept and strategy.

The core research fruits of emergency intelligence computing cover three aspects: perception and acquisition, modeling and analysis, management and decision-making. In these three aspects, this major research project has made great breakthroughs, among which the most typical achievement is the new method of heterogeneous super-network model and its application in national security departments' practice. The research group explained and explored the five scientific mechanisms behind the complex super-network phenomenon: heterogeneity mechanism, super-network mechanism, node influence mechanism, community clustering mechanism and crisis information's release mechanism, and the research group put forward a mathematical modeling method, which has been scientifically verified in practice. The following content focuses on the representative research fruit: heterogeneous super-network model, and other research fruits will be briefly summarized.

1.2.1 Heterogeneous Super-Network Model: Mechanism and Modeling

In the information and physical society systems, once an emergency occurs, all kinds of incident elements such as people, organizations, institutions, information and resources related to the incident will emerge rapidly. High-frequency information flows online and offline in real time, and the incident elements are intricately distributed in every corner of the information and physical society systems. Although these elements are highly dispersed, there must be some correlation, and all kinds of elements will adaptively and dynamically change along different interactive channels. Therefore, how to depict and describe the dissemination, influence and function of incident elements or incident-related information flow in the information and physical society systems is the key to understand the evolution law of emergencies and gain insight into trends of incidents.

Traditional emergency management methods are mainly based on statistical analysis of incident elements, and belong to typical linear thinking. In reality, many elements of emergency management are often tightly and organically coupled to propel the evolution of incidents during the interaction. Therefore, emergency management thinking needs to rise from linear thinking to a higher level-network thinking, so that emergency management and decision-making have a holistic, overall and systematic consideration. Heterogeneous super-network model is an emergency management method extracted from network thinking, which is of great significance for deep understanding of the complex system of emergencies and a key to propel scientific decision-making of emergency management.

Heterogeneous super-network model fully integrates the basic characteristics and evolution mechanism of emergencies in the information and physical society system, and scientific research is carried out from the following six aspects. (i) Exploring the heterogeneous mechanism of network combination and connection, and establishing a formal mathematical model. (ii) Understanding the phenomenon and mechanism of super-network composed of different elements, and establishing a formal mathematical model. (iii) Combining heterogeneity and super-network, exploring the internal mechanism of network nodes' role and influence formation in different structure networks, and establishing a formal mathematical model. (iv) Exploring the node settlement composed of heterogeneous nodes, exploring the cooperation mechanism of social clusters and establishing a formal mathematical model. (v) Exploring the internal mechanism of how crisis information shows different modes due to different networks from a macro perspective, and establishing a formal mathematical model. (vi) Using theoretical models and technical tools, the variability, controllability and predictability of complex networks are analyzed.

(1) **Heterogeneity mechanism and modeling**

Exploring and mining complex social systems with network thinking is a research hotspot in computer science, social science, management science and other related fields, and there has been a series of achievements. Traditional research methods usually assume that the nodes and connections in social networks are homogeneous, while the existence, interaction and association of things in the real world are often diversified. For example, in the social network composed of Weibo and WeChat, different users have great differences in participation degree, interest and preference, number of fans and influence. Therefore, exploring and studying heterogeneous individuals and self-adaptive interaction behaviors in complex communication systems is of great value for understanding the formation and evolution of emergencies in information and physical society systems.

Emergency management and scientific control of public health incidents require us to have a profound and accurate understanding and mastery of the basic characteristics, evolution rules, change patterns, key influencing factors and the impact of prevention and control measures on the epidemic situation of new infectious diseases in specific time, space and people. This is to make the decision-making method of public health emergency management have sufficient scientific basis, and make the available epidemic prevention and control resources achieve the maximum overall benefit in a limited time. The existing mainstream modeling methods regard the epidemic process of infectious diseases as a simple system, so their basic premise is certainty, homogeneity and complete random mixing, which simplifies the highly complex and dynamic epidemic process of infectious diseases. Therefore, the model construction lacks feedback interaction and iterative evolution, and the simulated communication system often lacks equilibrium point.

Different from the traditional mainstream modeling thought, Zeng Dajun's research group of Institute of Automation, Chinese Academy of Sciences believes that individuals with diseases in the process of spreading infectious diseases are not independent, but interrelated and interactive. The highly dynamic infectious diseases' transmission system in complex environment needs to understand the individual role and evolution process from the whole transmission system, and then understand the complex behaviors and phenomena of the whole dynamic system of infectious disease transmission. Therefore, for the emergency management and prevention and control for new and sudden infectious diseases, the premise of mathematical modeling needs to regard the epidemic process of infectious diseases as a complex system rather than a simple system, and take uncertainty, heterogeneity, nonlinearity, network structure, equilibrium point, dynamic feedback and micro to macro as the guiding principles of modeling. Under this guiding ideology, the core scientific issues of public health emergency management can be summarized as follows: how to understand the complex transmission dynamics system of infectious diseases from the perspective of network thinking and system science, explain the complex phenomena in the epidemic situation, and realize the prediction, early warning and optimal control of the epidemic situation (Fig. 1).

Zeng Dajun's research group explored the heterogeneity mechanism in the complex transmission system of public health incidents, put forward the formal expression of heterogeneous social contact network, and analyzed all kinds of behaviors of transmission dynamics and their influences. Based on the premise of complexity and heterogeneity of the complex transmission system of infectious diseases, a heterogeneous spatio-temporal transmission's dynamics model of epidemic and spread of infectious diseases was established (Fig. 2). In this model, the ontology of complex transmission system is abstracted as atomic system of disease transmission (transmission source). (I) The susceptible person (S) has a certain probability to become a new transmission source (I → S) through transmission channels, and the complex transmission system is recognized and understood by examining the nonlinear interaction among a large number of I → S atomic systems. Atomic

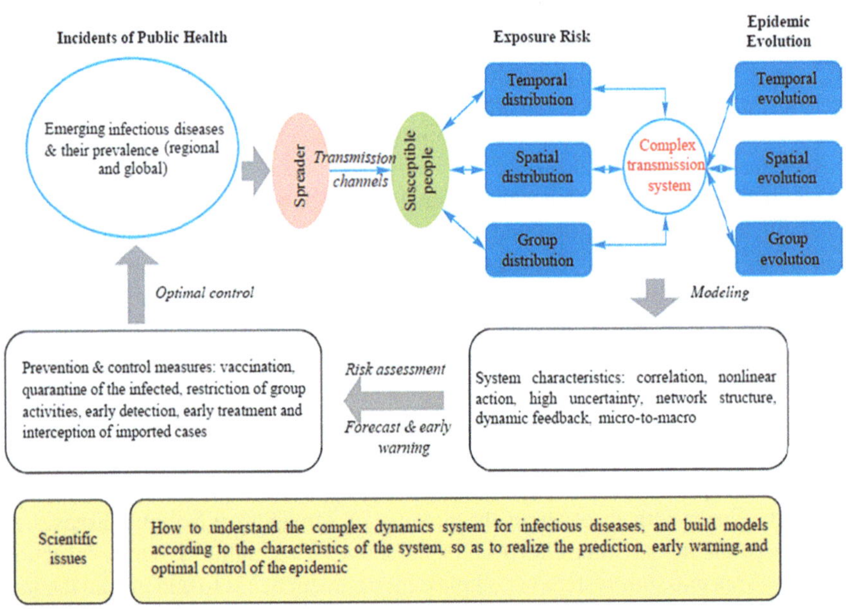

Fig. 1 Modeling thought and core scientific issues of complex system of public health emergency management

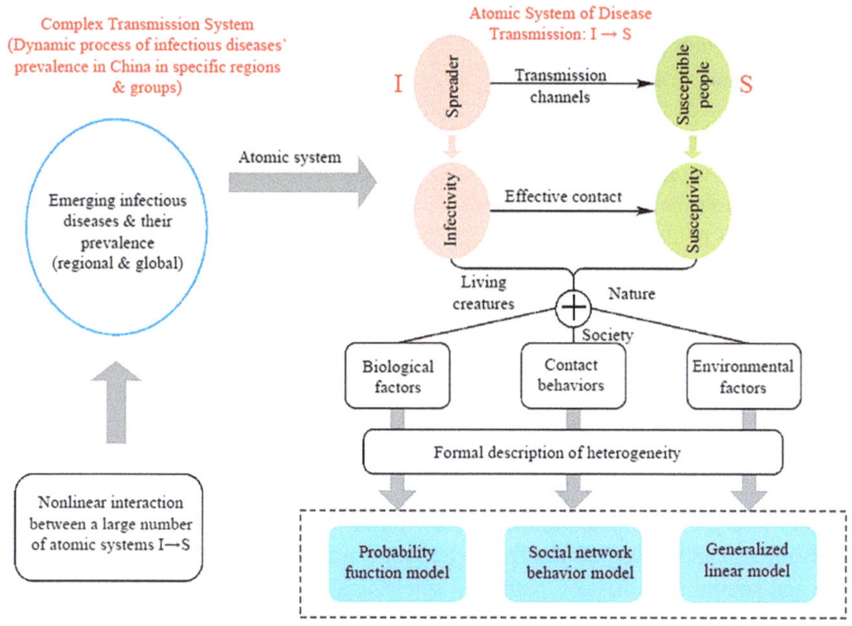

Fig. 2 Basic framework of heterogeneous spatio-temporal transmission dynamics model for epidemic and spread of infectious diseases

system is influenced by biological characteristics of pathogens, contact behavior between individuals and environmental factors, and the composite action of these influencing factors determines the basic dynamic behavior of atomic system. Zeng Dajun's research group classified the influencing factors of the spread and prevalence of new and sudden infectious diseases into three categories-biological factors, social factors and natural factors, and put forward a formal description model for the biological characteristics of diseases (virus infectivity, incubation period, infection period, crowd susceptibility, etc.), crowd activity patterns (close contact network, effective contact rate, crowd spatial mobility, etc.) and environmental effects (average temperature, relative humidity, etc.). Heterogeneous spatio-temporal dynamics model has high granularity, assemblability and expandability, and has a wide range of applications, which can be used for risk assessment, prediction and early warning of infectious disease transmission and epidemic under complex and changeable epidemic scenarios.

In recent years, the heterogeneous spatio-temporal dynamics model has been applied to study the epidemic situation of major infectious diseases that bring serious threats to China. Zeng Dajun's research group explored the Power-law Phenomenon, hierarchical clustering phenomenon and super-transmission phenomenon in complex transmission system, and carried out a series of studies in the aspects of epidemic risks' assessment, assessment of vaccine effect, epidemic control, epidemic situation's deduction, etc., quantitatively evaluated the intervention strategies such as combination and optimization under heterogeneous conditions, and obtained a number of valuable research fruits.

Severe acute respiratory syndrome (SARS) was the first global infectious disease epidemic faced by mankind in the 21st century, and it was also the most serious public health crisis faced by our country, with far-reaching influence. SARS first broke out among people in Guangzhou, and then spread to Beijing, Singapore, Canada and other dozens of countries and regions around the world. Among them, Beijing was the most serious SARS epidemic area. Therefore, the SARS epidemic situation in Guangzhou and Beijing can be used as the best research object to know and understand the characteristics, laws and emergency management of public health crisis, and such research is of great value. Zeng Dajun's research group systematically studied the characteristics and temporal and spatial evolution rules of SARS transmission and epidemic in Beijing and Guangzhou by using the heterogeneous spatio-temporal dynamics model and combining with the first-hand data resources monitored during SARS transmission and epidemic in Guangzhou and Beijing, and summarized the epidemic characteristic parameters such as the infection rate, latency, infection period, regeneration number and population immunity rate of SARS transmission in these two cities. Relevant research revealed the spatio-temporal pattern of SARS spread, high-risk area and spatial transmission path, and scientifically evaluated the implementation effect of control measures such as vaccination, early detection and isolation, and restriction of crowd activities under assumed conditions, and obtained a number of valuable research conclusions. This study not only scientifically evaluated the gains and losses of SARS epidemic's prevention and control in Beijing and Guangzhou, but also provided technical reference for emergency management and prevention and control strategies for SARS and similar major infectious diseases's spreading among people in Beijing, Guangzhou and similar megacities. The heterogeneous contact network and its temporal and spatial characteristics of SARS's spread in Beijing in 2003 are shown in Fig. 3. The research shows that the spatial spread of SARS has a strong regularity, which approximately obeys lognormal distribution, and the spatial distance of the maximum spread risk is 6–8 km.

When new infectious diseases spread, schools become the most dangerous areas because there are a number of students in a limited space. In 2009, more than 70% of the outbreak of influenza A in Beijing occurred in various schools (primary schools, middle schools and universities). Therefore, the emergency management and control after the outbreak of infectious diseases in schools is very important, and it is the key to the overall epidemic control in cities. During the H1N1 epidemic in 2009, a university town around Beijing encountered the largest epidemic of influenza A in China. The whole process of control and emergency response took 20 days. There were 586 influenza-like cases, 226 of which were confirmed by detection and about 13,000 people were isolated. The disease control department of military conducted a comprehensive and systematic investigation on the epidemic situation and obtained valuable first-hand data. Zeng Dajun's research team cooperated with the disease control department of military, carefully tracked the whole process of each case from exposure, getting infected, isolation, getting cured and getting rid of isolation, traced the disseminator, infected persons, occurrence time and place of each infected case,

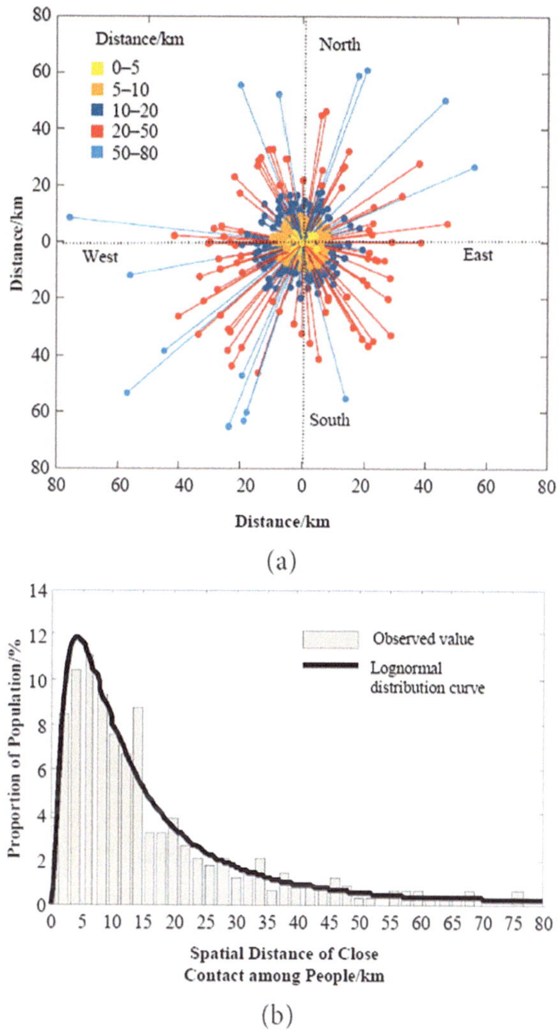

Fig. 3 Heterogeneous contact network and its temporal and spatial characteristics of SARS spread in Beijing in 2003

and completely sketched the whole transmission chain. At the same time, the school outbreak was reconstructed by using the heterogeneous spatio-temporal communication dynamics model. The model well simulated and predicted the distribution law of influenza A epidemic in time, space and population (Fig. 4), and based on this, carried out computing and experimental analysis, quantitatively evaluated the effect of emergency prevention and control measures such as isolating dormitory buildings and prohibition of rooms' mutual visit, and summarized the experience and lessons, and it provided technical reference for emergency management and decision-making for public health crisis in schools. The research shows that: (i) the spread in schools is wide and deep, and it develops rapidly, which needs timely intervention; (ii) The

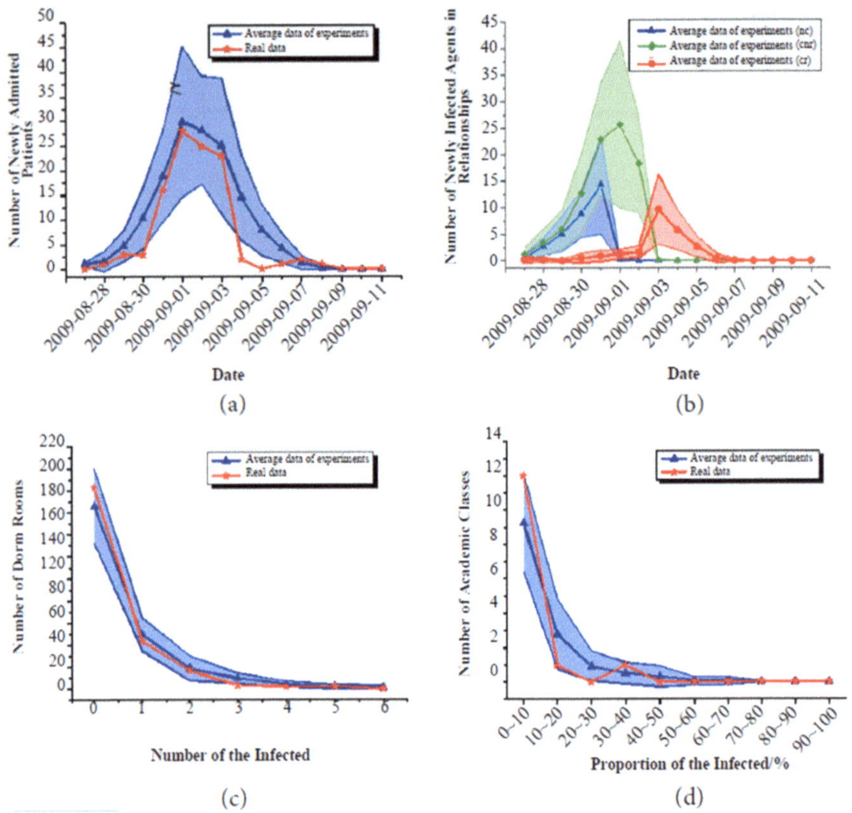

Fig. 4 Reconstruction of the transmission chain of H1N1 in schools and evaluation of comprehensive intervention strategies

risk of transmission is controlled by the mixture of social relations and spatial relations of students, and the high-risk population causing epidemic spread is freshmen; (iii) Three emergency measures adopted in this epidemic (the first measure is to isolate dormitory buildings and prohibit people from frequent entering and leaving, the second measure is to prohibit students' mutual visit in dormitory, and the third measure is to isolate all students who are suspected or confirmed cases in dormitory buildings), and it can substantially reduce the transmission risk, with remarkable effect, but with a certain time lag; (iv) The effect of the second emergency response measure is limited. Although it effectively prevents the cross-spread of infectious diseases among different dormitories, it aggravates the risk of cross-spread among roommates in the same dormitory. The third emergency response measure has the best effect, but it has high requirements for the resources that can be mobilized and used in emergency situations.

In the new outbreak of infectious diseases, the number of second-generation infected people caused by individual disseminators is often uneven, that is, not completely random and not normally distributed. The number of second-generation infected people caused by some individual communicators will far exceed the average level (great difference), this is called super-transmission phenomenon. Super-transmission was first observed in the Spanish influenza pandemic in 1918, but it has been ignored until the SARS global pandemic in 2003. In 2005, some scholars wrote an article in *Nature* about super-transmission and its influence. Since then, the phenomenon of super-transmission has been widely reported in serious public health crises such as H1N1 pandemic and Ebola virus epidemic. Super-transmission phenomenon means that a few individuals can influence or even control the epidemic trend, and it has important research value. How to predict and discover the occurrence probability of super-transmission phenomenon, and how to quickly identify and control super-transmission individuals with disease in advance is an important scientific proposition. At present, there are many reports about this phenomenon, but the scientific explanation at the mechanism level is still blank. Zeng Dajun's research group built a heterogeneous spatio-temporal dynamics model, and made a systematic study on the largest known outbreak of influenza A in schools in China in 2009, focusing on the scientific demonstration and explanation of the super-transmission phenomenon in this epidemic at the mechanism level. Zeng Dajun's research group regarded the activity frequency of disease-spreading individuals, the aggregation degree of susceptible population and the duration of time when disease-spreading individuals are infected as three types of influencing factors, built up the probability distribution function of these three types of influencing factors, and integrated them into the heterogeneous spatio-temporal transmission dynamics model, thus carrying out simulation, computing and analysis under different outbreak scenarios, and deducing the occurrence probability of super-transmission phenomenon under different integrated conditions. According to the "20/80" law, if 20% of individuals cause 80% infection in the population in simulation environment, it is considered that there is super-transmission. Experiments show that the super-transmission phenomenon may be caused by the nonlinear combination of three factors: high-frequency activity of disaster-spreading individuals, high-density aggregation of susceptible individuals and long-term free transmission of disease-spreading individuals. This finding provides a logical and self-consistent scientific explanation for the mechanism of super-transmission phenomenon.

In the complex system of self-organization, the phenomenon of power law distribution often appears, and the research on it has extensive and far-reaching significance. The transmission and spreading of infectious diseases in specific populations also have power-law distribution phenomenon, which was first seen in Nature in 1996. The author investigated measles outbreaks in a relatively closed island with 25,000 people during recent hundred years, and found that the number of infected people (outbreak scale) caused by each outbreak of measles and its frequency obey power-law distribution. There have been a lot of reports about the existence of power law distribution in epidemic situation, but the internal mechanism of this phenomenon is still unclear. In 2009, A (H1N1) spread widely in Beijing, and a number of outbreak data were obtained through monitoring by Beijing CDC. Zeng Dajun's research group analyzed 207 identified group epidemics of H1N1 in Beijing before October 18, 2009 (especially in schools). It is found that in relatively isolated and densely populated areas (such as schools, communities, companies, etc.), there is a strong regularity in the outbreak scale of H1N1, which obeys the Law of Power and has fixed scale. Zeng Dajun's research team built a dynamic model of heterogeneous spatio-temporal spread, realized the probability function's expression of disease parameters such as incubation period and infection period, and built the interactive contact between people in epidemic areas into a multi-level hierarchical contact network, thus simulating the heterogeneous contact behavior of people in different regions in reality; Using a large number of repeated computing and experiments, it is found that the population caused by the spread of infectious diseases under the multi-level hierarchical contact network obeys the Power Law Distribution. The research shows that the epidemic of infectious diseases does not spread out in a wave-like equilibrium, but develops in a leapfrog way. First, it spreads rapidly in a dense subgroup with strong connectivity, quickly infects most individuals in the subgroup, and then spreads out with a certain probability through weak connectivity (with a certain time delay), thus detonating other subgroups with strong connectivity. This study reveals that the main reason of power-law distribution in the spread and prevalence of infectious diseases may be the multi-level hierarchical network contact pattern of people in epidemic areas. Therefore, it is the key of public health emergency management by controlling the weak connection in the crowd contact network to prevent the epidemic from spreading.

(2) **Super-network mechanism and modeling**

The network public opinion of emergencies has become an important factor related to national security, and its role in social management can't be ignored. Public opinion incidents in network society are similar to emergencies in real society. It is necessary to clarify six elements: when, where, who, what, why and how (namely 5W1H), so as to determine the occurrence and development process and dynamic evolution mechanism of network public opinion. There is a disconnect between mechanism cognition and guidance in existing studies: the mechanism mostly focuses on the physical expression of public opinion on various complex networks, and is not closely related to actual incidents; The guidance is mainly qualitative description, lacking the basis of quantification and evaluation.

According to the social combustion theory, the social shock wave theory and the public opinion mechanism of social behavior entropy, the research group of Liu Yijun from Institute of Strategic Consulting, Chinese Academy of Sciences, used the "5W1H" public opinion comprehensive integration research method to build a public opinion super-network model consisting of public opinion subjects, environmental information, psychology of subject and released opinions. By modeling the behavior of public opinion subjects, building the evolution model of public opinion under complex networks, judging the public opinion situation, the formation, evolution, guidance and intervention of public opinion are systematically modeled and simulated, and qualitative consultation and quantitative simulation are combined to study the guiding strategy of super-network public opinion, which explores the social complex problem of network public opinion from a new perspective of super-network. By establishing social, psychological, environmental and viewpoint sub-networks, a super-network model describing public opinion is established, a logical and self-consistent interaction mechanism with multi-layer network is designed, and a "super-link prediction" algorithm and a "super-link ranking" algorithm are proposed. Liu Yijun's research group combined the super-network model with simulation method, computed and analyzed the formation and evolution process of public opinion in unconventional emergencies, studied the mechanism and strategy of public opinion's guidance and intervention in unconventional emergencies, discussed the "mechanism of public opinion," and solved the problem of "controlling public opinion."

The super-network model of network public opinion is shown in Fig. 5, which shows the super-network modeling framework and the relationship and driving mechanism among the subnets of each layer. Super-network consists of four subnets: (i) Environmental network. It represents the process of information's dissemination, and each piece of information represents a new environmental node. The formation and evolution of the super-network model of network public opinion is based on the introduction of new environmental information, and the environmental network is the external driving force for the evolution of other subnets; (ii) Social network. It represents the interaction relationship between individuals, that is, the mutual reply between netizens; (iii) Psychological network. It is the internal driving force to social subnet. When an individual receives new information, he will choose whether to receive the influence of the new environment according to his own psychology (that is, to receive the viewpoint spread by new information); (iv) Keyword network. By observing the opinion network formed by netizens, we can detect the social effect of public opinion (cluster behavior caused by micro-individuals), and then propel the emergence of new environmental information.

Liu Yijun's research group calculated and evaluated three social intervention strategies of public opinion by using the super-network model. (i) Isolation strategy means that one (or several) negative super-edges in the super-network are isolated, and the super-network structure is intervened under certain rules, so that the node attributes change on the basis of network structure's changes. It will make the negative super-edge attributes in the super-network change into positive super-edges; (ii) Embedding strategy refers to embedding one (or several) positive super-edges into the super-network, and intervening the super-network structure under certain rules, so that the attributes of negative super-edges in the super-network change into positive super-edges; (iii) Reconstruction strategy refers to reconstructing one (or several) negative super-edges in the super-network, and intervening the super-network structure under certain rules, so that the attributes of other negative super-edges in the super-network change into positive super-edges.

To solve the prediction problem of super-network model, Liu Yijun's research group proposed a super-link prediction algorithm based on super-network (Fig. 6). Firstly, the connectivity matrix is abstracted from the constructed public opinion super-network model. Then, the adjacency matrix, the number of super triangles in the super network model and the similarity between any two super edges (the

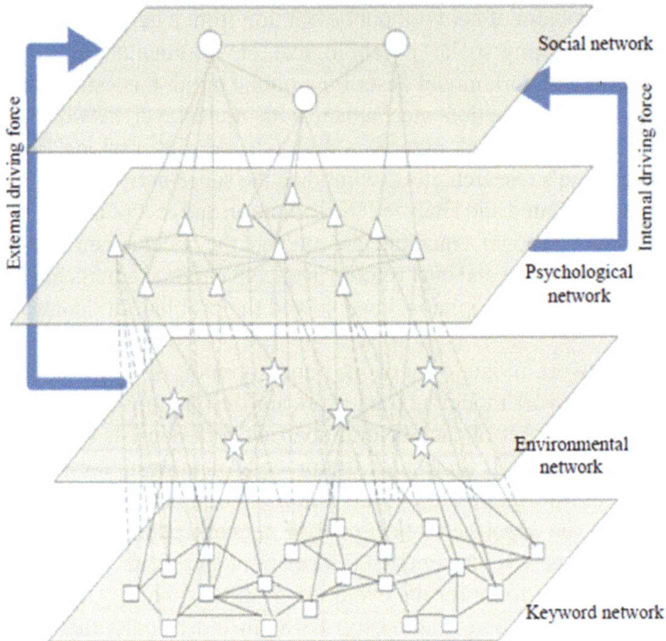

Fig. 5 Super-network model of internet public opinion

number of super triangles contained together) are computed. Finally, the prediction algorithm and evaluation method of super-link in subnet layer in public opinion super-network are proposed, and the super-edge prediction algorithm is extended to the whole network.

(3) **Mechanism and modeling of node influence**

Unconventional emergencies can easily lead to rumors and panic. There are many ways to spread rumors, including the Internet, mobile phone short messages and traditional face-to-face and mouth-to-mouth methods. Among many ways of transmission, the method of network is the fastest and the most harmful. For example, the national salt grabbing incident caused by nuclear radiation in Japan and the network rumors in the "7/23" Ningbo-Wenzhou serious railway traffic accident have reduced the government's credibility; Internet rumors caused riots in several major cities such as London. Therefore, it is of great significance to clarify the incidents and effectively guide the development of online public opinion.

It is believed that opinion leaders play an important role in spreading rumors or dispelling rumors. Therefore, quantitatively computing the influence of nodes, discovering and mining opinion leaders, and evaluating their influence have become the primary issues. Therefore, Shang Mingsheng's research group of University of Electronic Science and Technology of China proposed a LeaderRank method for mining opinion leaders in social networks, which is obviously better than the method

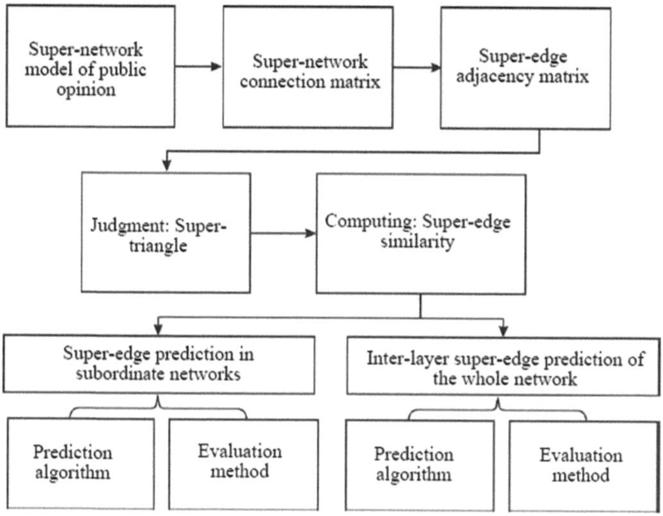

Fig. 6 Super-link prediction algorithm based on super-network

that only considers the degree of nodes. Compared with PageRank algorithm, LeaderRank algorithm also has the advantages of high accuracy, good robustness and no parameters. Comparison of the effects of LeaderRank and PageRank in speeding up message transmission and enhancing robustness against attacks is shown in Fig. 7.

In emergencies, the process of dispelling rumors corresponds to the spread of rumors, which is one of the important means to control public opinion. Therefore, it is of great significance to intentionally train opinion leaders to dispel rumors. Shang Mingsheng's research group found that in typical social networks, the leadership of opinion leaders has scale-free characteristics (Fig. 8). This shows that a few influential opinion leaders control the spread of online public opinion, thus controlling opinion leaders can control public opinion to a great extent. Shang Mingsheng's research group further theoretically analyzed the formation mechanism of this scale-free feature, and found that the formation of opinion leaders follows the mechanism of "the more interested, the more voice," that is, people with broad interests and better judgment will become the leaders of the social network opinion market. This conclusion has important reference significance for shaping opinion leaders of online media.

Shang Mingsheng's research group also proposed a semi-local centrality's recognition method based on neighborhood information, which is used to identify the most influential nodes in undirected networks, and its computing speed and effect are both good. Based on a large number of practical complex networks, compared with the identification methods based on Degree Centrality, Betweenness Centrality and Closeness Centrality, the semi-local centrality identification method based on domain information can quickly and effectively identify the most influential nodes in complex networks.

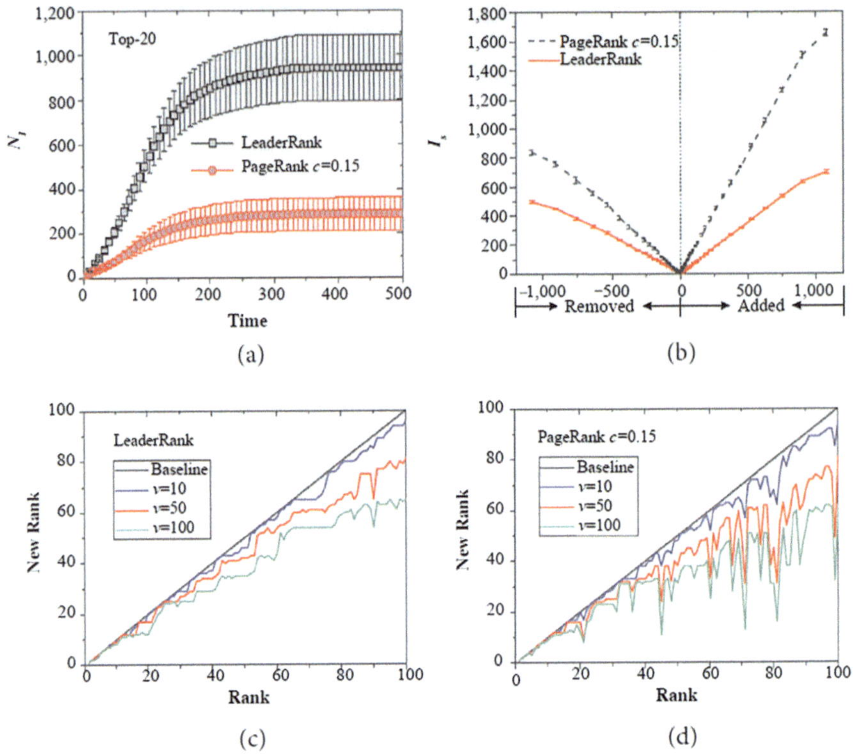

Fig. 7 Comparison of the effects of leaderrank and pagerank in enhancing message dissemination, robustness and anti-attack

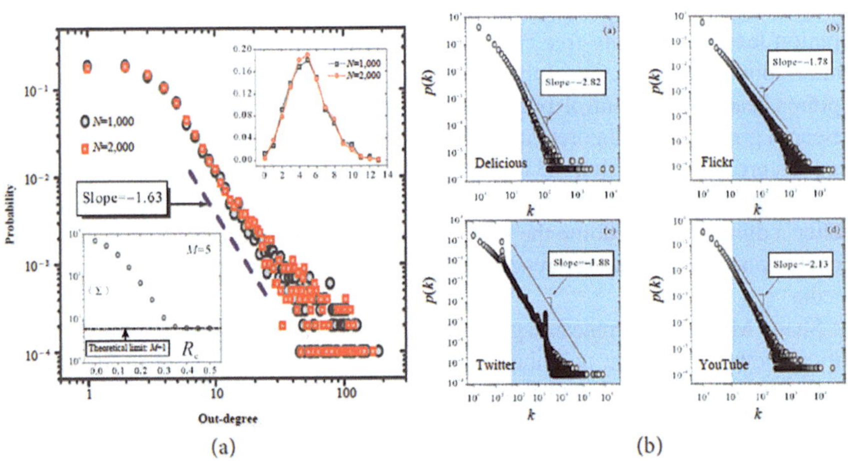

Fig. 8 The scale-free phenomenon of opinion leaders' leadership in social network

(4) **Community aggregation mechanism and modeling**

Community in complex network is a special subset, which has the characteristics that nodes in the same community are closely interconnected, and nodes in different communities are less connected. In practice, complex social network changes dynamically, and new nodes may join, old nodes quit, new connections are generated or old connections disappear at different times, and the connection relationship between nodes may also change. Therefore, the main purpose of discovering community is to mine static community structure from complex networks, and to study the birth, structural change and extinction of communities through dynamic evolution of communities.

The research group of Fang Binxing from Beijing University of Posts and Telecommunications explored the mechanism that nodes aggregate to form a community, and proposed a community division method based on the characteristic threshold of label propagation algorithm (LPA) of infectious diseases' transmission model. Through semi-supervised machine learning method, the community was divided by using the intelligent exchange of network node labels and the process of community integration. In order to improve the running speed of LPA algorithm, convergence speed and community division accuracy, especially the division accuracy of overlapping communities, we should start from the network connection matrix of label information's spread, the maximum non-zero eigenvalue of the matrix is combined with the threshold value of network label information's spread, which significantly improves the operation efficiency. The effectiveness of the new algorithm is verified by LFR Benchmark simulation and test network, random network and real social network data.

The research shows that compared with the typical LPA, the time complexity of the new algorithm proposed by Fang Binxing research group is greatly reduced; Compared with the typical CORPA algorithm based on LPA model, the new algorithm has higher accuracy in dividing overlapping communities, especially when the overlapping communities are obvious. The division accuracy of the new algorithm is close to the high-precision algorithms GA, N-cut and A-cut, and obviously superior to the classical algorithms such as GN, FastGN and CPM.

Fang Binxing's research group also proposed a method to automatically discover the most influential set—the CTMC-ICM influence spread model based on the SpreadRank node influence's measuring, that is, to find the most influential node set from the complex network. Based on the community structure, the node role is redefined. Firstly, new two-dimensional PageRank measurement methods-Inner Rank and OutterRank are proposed to describe the influence of nodes inside and outside the community. Then, a new method of node role's definition is proposed. According to two kinds of metrics, nodes are divided into four roles according to community structure: core node, bridge node, extremely important node and ordinary node. At last, the influence analysis is carried out based on SpreadRank, and a new spread model CTMC-ICM is established by introducing continuous-time Markov chain (CTMC) and combining with independent cascade model (ICM). Comparison of spread range between SpreadRank and PageRank is shown in Fig. 9. Experiments

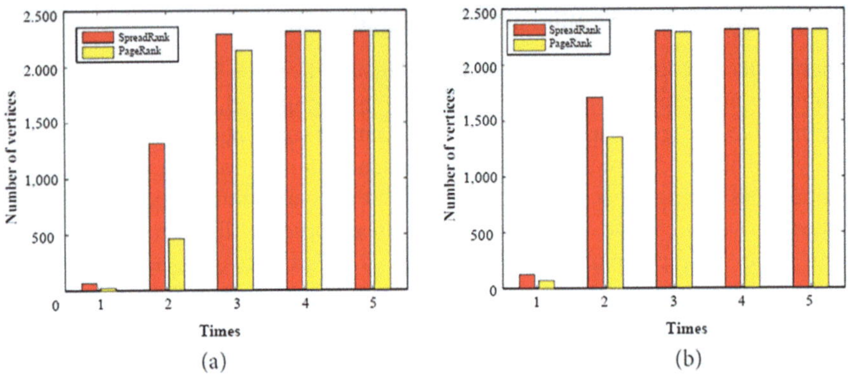

Fig. 9 Comparison of spreading range between SpreadRank and PageRank

show that, compared with the distance-based measurement method, SpreadRank method is more efficient, which can extract the influential node set from the network and maximize the network information spread range by activating the node set.

(5) Mechanism and modeling of crisis information releasing

With Wenchuan earthquake as the background, Zhang Heping's research group of University of Science and Technology of China revealed the public's demand rules for disaster information, entertainment information, education information and life service information after large-scale natural disasters, and explained the release mechanism of crisis information on social platforms in Weibo, that is, the evolution mode of emergencies is highly correlated with the public's risk perception. According to the duration of crisis information release, focusing on three typical modes of centralized release, continuous release and fluctuating release, a crisis information's spread model is established by using Logistic Model, and the official information release strategy based on Weibo platform under different crisis scenarios has been put forward. The research shows that the spread curve of crisis information in Weibo is S-shaped. At the initial moment, the number of individuals releasing information and the duration of information release will have a great impact on the final coverage of information spread. The transmission efficiency of crisis information in weak relation network is higher than that in strong relation network, and the transmission range is positively related to the credibility of information and the coefficient of transmission tendency.

(6) Variability, controllability and predictability of contact and spread

In order to understand the predictability of infection source's spread in complex networks, Shang Mingsheng's research group of University of Electronic Science and Technology of China studied the variability, controllability and predictability of contact and spread, and compared and analyzed the variability of spread of different layers of nodes regarded as initial infection sources. Experiments show that, from a global perspective, network spread has low variability and good predictability, but for local communities, it has great variability in the early stage of spread. The farther the spread source is from the node, the worse the predictability is, which indicates that the accurate prediction of spread can not be realized. After introducing random factors, namely contact mode based on network structure, the spread process shows unpredictability; The real social network not only has a more complex network structure, but also is affected by the extremely changeable virus infection, human dynamics, multi-information coupling and other factors, which increases the variability of the complex network spread system and makes the spread process unpredictable.

In complex networks, the self-adaptive behavior of nodes will change the structure of the contact network and the spread process, which in turn will promote people's self-adaptive behavior and lead to changes in the network structure, forming a feedback loop with mutual influence between network topology and dynamics. This kind of network is called self-adaptive network. In order to deeply understand the role of self-adaptive behavior of nodes in complex networks in incident's evolution, Shang Mingsheng's research group studied how the self-adaptive behavior of people affects the development of epidemic situation in disease transmission, and computed and analyzed the influence of different isolation strategies on epidemic spread from the aspects of two basic disease control measures, namely immunization and isolation, and then put forward a control strategy based on community effect. Experiments show that the earlier the control starts, the better the effect will be. When the community is strong, the effect of control is the best; Compared with the immunization strategy based on community structure, the isolation strategy based on community structure has better effect.

1.2.2 Perception and Acquisition of Large-Scale Open-Source Information

Obtaining large-scale open-source information from the Internet is an important basis for emergency intelligence computing, and it is a necessary prerequisite to perceive and obtain multi-source heterogeneous information related to emergencies, with high efficiency and high accuracy. However, in the information and physical society system, the physical-cyber space's coupling interaction, the network and content tend to be socialized, the netizen group shows the characteristics of self-organization, high complexity, virtual and real interaction, and the information shows the characteristics of dynamic change, deep hiding, different source formats, etc. The emergency intelligence's computing under the new situation is facing severe challenges.

Zeng Dajun's research group of Institute of Automation, Chinese Academy of Sciences and Fang Binxing's research group of Beijing University of Posts and Telecommunications systematically studied key technologies such as network information's acquisition and filtering, specific field information's classification, distributed inverted index, network information flow's mining, incident's semantic extraction, atomic event evolution pattern's mining, event information's traceability, etc., and made substantial breakthroughs. They developed a self-adaptive emotional crawler technology with independent intellectual property rights. And this technology can collect, process and integrate large-scale Internet's open-source data in real time, proactively and adaptively.

1.2.3 Modeling and Analysis of Incidents' Big Data

Extracting valuable knowledge and information from the acquired incident-related large-scale data is the major research direction of emergency intelligence's computing. Modeling and analyzing big data, and forming scientific judgment on netizens' emotions, behavioral motives and intentions are the key challenges for emergency intelligence's computing.

In this major research project, the team of scientists has made substantial progress in network incidents' news discovery based on Dirichlet process, automatic recognition technology for incident entity, computing model with association mode of terrorism-related information, network construction of complex association semantic chain, deep semantic analysis technology, incident tracking oriented to semantic evolution, Weibo emotion analysis technology with multi-dimensional vectors, big data analysis and modeling technology, etc., and independently developed a number of core algorithms. Typical achievements include: in emotion computing, a multi-granularity emotion computing method is proposed, which not only has high performance, but also does not need to manually mark training examples In the aspect of behavior computing, the traditional association rule's mining technology is developed, and the learning framework and algorithm of behavior rules based on knowledge are put forward, and it can realize the automatic learning and generation of individual or organization behavior rules from webpage data, recognize behavioral goals and intentions and infer their future behavior patterns in combination with observed individual and organization behaviors. In the aspect of semantic analysis, the memory activation theory of cognitive psychology is used to strengthen the semantic representation of the text, activate the potential hidden information in the text and add it to the original text. This text representation method based on cognitive activation theory can extract the implicit information of text that accords with human intuition and effectively assist the task of text mining.

1.2.4 Management and Decision-Making for Actual Combat in the Field of Security

Zeng Dajun's research group has made substantial progress in key technical fields such as behavior analysis of hostile terrorist organizations, behavior rules based on knowledge, early warning of abnormal security incidents based on spatio-temporal scanning statistics, multi-sequence emergencies' discovery based on Hidden Markov Model, and interactive visual cloud services. As for the hidden groups on the Internet, the technologies of cross-platform identity mapping and hidden groups' discovery are proposed. Cross-platform identity mapping is realized by extracting behavior contents such as identity characteristics, texts' topic characteristics and users' writing style of cross-platform users, and the association between physical space entities and cyberspace entities is established. Based on potential feature space, hidden groups are identified and discovered. Zeng Dajun's research group fully integrated the knowledge in the field of national security, and put forward a big data-driven emergency intelligence's analysis and situational decision-making model for actual needs, and developed an integrated system solution.

1.3 Application and Popularization

The team of scientists in this major research project put forward the theory and technology of emergency intelligence's computing which is suitable for the new situation of information and physical society system, and developed a systematic overall solution to meet the important needs of national security. The research results have strong practical value. Through in-depth cooperation with government departments, facing the actual needs of business at strategic level, an emergency intelligence's computing software system has been developed, which has overcome a series of difficulties such as business modeling, data integration, computing platform's construction, analysis engine and interactive visualization, etc. It has been applied in actual combat and showed good effect in the discovery and analysis of terrorism-related clues, the study and judgment of behavior and intention of non-governmental organizations involved in China, the analysis and prediction of social security situation in Hong Kong, Macao and Taiwan, and the epidemic control of acute infectious diseases.

1.3.1 Application in Social Security

Zeng Dajun's research group fully investigated the major needs of relevant national departments in actual combat, and based on the theory and method of emergency intelligence's computing, the research group developed three practical application systems: "Internet public opinion management and control system," "open-source intelligence's analysis system" and "knowledge encyclopedia and intelligence map system," which realized the integration, modeling and computing analysis of public data, semi-public data and closed-source data, and effectively overcame the difficulties to government departments in actual combat. The system solution put forward in this research have become a model for ensuring national and social security by innovative scientific and technological means.

In order to meet the needs of large-scale open-source data's collection, storage and fast computing, Zeng Dajun's research group has built two platforms: "Smart Cloud Computing Platform" and "Internet Big Data Cloud Platform." At present, the two platforms can monitor 17 types of data such as news data, Weibo data, WeChat data, Twitter data, Facebook data and Youtube video in real time, scan nearly 100,000 multilingual website data, and automatically collect various forms of Internet open-source big data such as news, forums, online discussions, blogs, Weibo and Q & A for the whole day, covering 133 countries and regions on six continents. Covering more than 75,000 news websites and about 170,000 channels, it collects and updates more than 100 million pieces of Chinese and English data every day, and has stored more than 2 billion pieces of public and semi-public data. The two platforms also integrate a variety of departments' data, which has become an important data support for relevant departments in China.

Relying on the above three systems and two platforms, the research results of Zeng Dajun's research group in Internet's content security, open-source intelligence analysis, and innovation and practice of anti-terrorism law are widely used in social security departments of the state and the military, including the National Computer Network and Information Security Center, Beijing Public Security Bureau, Shanghai Public Security Bureau, Publicity Department of the Communist Party of China, State Council Information Office, Xinhua News Agency, China Daily, and national ministerial-level and military intelligence business units. The research results have achieved a series of innovative technological achievements that have been evaluated by departments as great contributions in the field of intelligence computing. The actual application effect of related fruits are outstanding, and many mature practical tactics and supporting systems have been promoted throughout the country, making outstanding contributions to national and social security. Zeng Dajun's research team won the second prize of the Science and Technology Progress Award of the Ministry of Public Security in 2015 in the theoretical research and system application of the public security department, and the research team participated in the formulation of the standards' construction of the national public security's public opinion system; In the field of theoretical research and system application of the media propaganda department, the research group won the first prize of the 2015 Wang Xuan News Science and Technology Award, and was repeatedly instructed by the central leadership; Theoretical research and system application in the field of social security won the first prize of Science and Technology Award of China Automation Society in 2013.

1.3.2 Application in Public Health

Zeng Dajun's research group is based on the current construction situation of monitoring, early warning and control of major infectious diseases in the country and the army, and through long-term and continuous in-depth cooperation with major disease control departments of the country and the army, such as PLA Academy of Military Medical Sciences, National Center for Disease Control and Prevention, Beijing Center for Disease Control and Prevention, and China Center for Animal Disease Prevention and Control. The transmission characteristics and temporal and spatial evolution laws of infectious diseases such as SARS, influenza A, hand-foot-mouth disease, seasonal influenza, avian influenza and dengue fever, which pose a great threat to China since the 21st century, are systematically studied. Early warning and transmission dynamics models are put forward, the potential risks and impacts of the epidemic are judged, and the effect of large-scale vaccine delivery in epidemic control and the number of sufficient vaccine reserves are evaluated. These measures provide scientific decision-making basis and technical reference for national and military disease prevention and control departments and public health emergency decision-making departments.

The research fruits are mainly applied to the monitoring and early warning of infectious diseases in Beijing and the on-site disposal of major infectious diseases in the military. Beijing mainly uses it to improve existing monitoring and early warning system of major infectious diseases, enhance the computing and analysis ability of epidemic monitoring data, and improve the timeliness and sensitivity of monitoring and early warning of infectious diseases in the capital. This achievement won the third prize of Beijing Science and Technology Award and the third prize of China Preventive Medicine Science and Technology Award. The military mainly uses it to compute and analyze the digital operation mode and intelligent epidemic data of the military epidemic's handling, which significantly improves the work efficiency and emergency decision-making ability of the military epidemic's handling. The epidemic control and analysis system led by Zeng Dajun's research group has been popularized and used in 18 military disease control units, which has effectively improved the work efficiency and emergency decision-making ability of military epidemic's handling. In the theoretical research and systematic application in the field of public health, the research group won one first prize (2015), three second prizes (2011, 2012 and 2017) and one third prize (2012) of Beijing Science and Technology Progress Award.

1.4 International Comparison and Influence

The team of scientists in this major research project proactively carried out international academic exchanges and cooperation, and invited many famous international scholars in this field to exchange visits, including Nabil R. Adam of Rutgers University, Katia Sycara of Carnegie Mellon University, Kwok L. Tsui of Georgia University of Technology and so on. These scholars spoke highly of China's research work in this field.

Zeng Dajun's research group has organized and hosted many influential academic conferences in this field, including the international conference on intelligence and security informatics, the top conference in this discipline, which expanded China's international influence in this field. The achievements published by the scientist team in this major research project in theory, algorithm and system have also had an important impact internationally. Many papers were published in Management Information Systems Quarterly, Informs Journal on Computing, Scientific Reports, High-level academic journals and magazines such as *Journal of Management Information Systems, ACM Transactions on Information Systems, IEEE Intelligent Systems* and

Decision Support Systems, Scholars from Harvard University, Cambridge University, Yale University, Carnegie Mellon University, University of Michigan, University of California, San Diego and University of California, Davis, as well as NASA, IBMWatson, HP Labs, Yahoo! Well-known experts from Labs and other institutions have cited and made positive comments on the papers published by the research group.

Professor Michael Gordon, a famous scholar in the field of management information system at Ross School of Business, University of Michigan, systematically reviewed and commented on the mainstream research fruits analyzed on social media in the world, and clearly proposed that the research fruits of Zeng Dajun's research group were fully absorbed in the CUP framework designed by him, thus supplementing and perfecting the latest social media analysis framework. Professor Gard Weiss, a multi-agent modeling expert from Maastricht University in the Netherlands, positively commented on the method proposed by Zeng Dajun's research group to simulate agent belief based on Bayesian learning expression and update mechanism, and thought that this method can make multi-agents have stronger learning ability and high value. An article published in the Bulletin of the World Health Organization pointed out that the positive deviation cumulative sum algorithm put forward by Zeng Dajun's research group is consistent with the mainstream conclusions in the world.

2 Basic Research and Achievements' Application of "Scenario-Response" Overall Integration Platform

2.1 Research Background and Present Situation

Unconventional emergencies have many characteristics, such as poor predictability, strong destructiveness, high complexity, difficulty in forecasting and early warning, etc., and have become one of the hot and frontier issues in the emergency field at home and abroad. From an academic point of view, related research can be summarized into two categories: (i) Theoretical and methodological research aimed at providing decision support, including realizing accurate perception of emergency situation, making reasonable and effective analysis and accurate deduction, making scientific decision based on perception-analysis-deduction, and realizing the organization and management mode and mechanism necessary for efficient decision-making, etc. (ii) Facing the interdisciplinary and interdisciplinary characteristics of unconventional emergency response, aiming at the specific needs of emergency response, it is necessary to realize the effective integration of multidisciplinary theories and methods, and realize the efficient integration of emergency fruits in different regions, different disciplines, different scales and granularities and different expression ways with the support of existing technical means, so as to provide strong support for emergency decision-making. In view of the deficiencies in the research on related theories and methods, organizational mechanisms and decision-making's support platforms, it is necessary to deeply analyze the basic internal evolution law of unconventional emergencies, and provide important basic support for improving the government's emergency management ability.

The research project on basic scientific problems of unconventional emergency management and "scenario-response" total integration platform aims at the project of "scenario-psychology-decision-making emergency management's integration platform for unconventional emergencies" in "research on unconventional emergency management," focusing on the core scientific goals and overall integration of major research programs. In basic scientific research, first of all, real-time situation's perception and analysis of physical and social space is needed to obtain the overall description of the current situation. Then, considering the influence of psychological factors, the situation deduction, comprehensive judgment and decision-making can be carried out. Finally, considering the characteristics of China's national conditions, reasonable emergency management system and process can be formulated to achieve efficient emergency management of unconventional emergencies. In the research of open overall integration platform, both research of scientific issues and actual emergency decision-making need complex interaction and collaboration of multi-disciplines, multi-departments and multi-links, cooperation of various scientific methods and technical means, and integration of research fruits in different regions and disciplines.

2.2 *Major Research*

2.2.1 The Evolution of Individual and Group's Psychology and Behavior in Emergencies

Han Shihui's research group of Peking University studied the neurophysiological mechanism of human psychology and behavior in emergency, established the characteristic database of individual and group psychology and behavior indicators, studied the measurement and standardization of psychological behavior, and built the evolution model of individual and group psychology and behavior in emergency. In the aspect of neural mechanism, the object-based attention theory was studied by subliminal paradigm (Striving to reduce top-down attention control) The cognitive and neural mechanisms of death information's processing were studied by using incident-related potentials and functional magnetic resonance imaging (fMRI). The measurement theory and method of interpersonal susceptibility are established. Based on the theory of uncertainty management, the research group studied the individual's psychological and behavioral performance under emergencies and conducted a nationwide survey of social security. The neurophysiological mechanism of human psychology and behavior in emergency was discussed.

Combining functional magnetic resonance imaging (fMRI), transcranial magnetic stimulation (TMS) and psychophysics experiments, Fang Fang's research group of

Peking University found that perceptual learning effect can migrate from learned visual tasks to unlearned visual tasks, which leads to the remodeling of brain function of normal adults' visual system.

Human beings often experience the threat of death and physical and mental injury in emergency situations; Non-affected groups, emergency management personnel and rescue workers will have information and emotional contact with the affected groups, have empathy experience, and carry out pro-social behaviors such as donation; Different people will selectively perceive and process information, which can be caused by differences in genes and neurophysiological mechanisms, as well as cultural values. The actual work should be mainly guided by the construction of "one system and two bases," which refers to psychological and geographic information system, psychological knowledge base and tool base for psychological behavior measurement.

In the information construction of psychological geography system, Wang Lei's research group of Peking University has completed the collection and operation of basic data and the geographical distribution research based on many psychological indicators, including openness, extroversion, emotional stability, agreeableness and responsibility.

In the psychological knowledge base of unconventional emergency management, Wang Lei's research group reviewed all the literature abstracts of 18 top journals in applied psychology from 2000 to 2016, and screened according to the relationship between literature content and unconventional emergency management, combining the frontier theory of psychology with the practice of unconventional emergency management (Fig. 10). Adjusting and intervening the negative emotions of people who have experienced unexpected incidents after the disaster can lead them to adopt the method of cognitive re-evaluation and reduce the negative emotions by re-evaluating the incidents they have experienced. In view of the catastrophic experience of natural disasters, the disaster victims can be guided to use the following cognitive re-evaluation methods: (i) The method of separation and re-evaluation can be adopted in the early stage to achieve the effect of reducing negative emotions quickly; (ii) In the later period, positive re-evaluation can be adopted to gain positive subjective experience.

With regard to the construction of tool bank for psychological behavior measurement, Wang Lei's research group compiled a tool bank for measuring people's psychological and behavioral indicators in emergencies, combining the progress of psychological theories and methods, as well as the empirical research and practice results at home and abroad. Eighteen kinds of psychological and behavioral indicators have been revised, collected and sorted out, including interpersonal susceptibility, psychological resilience, individual traditional (sexual) values, "neuroticism" in Big Five personality, sense of fairness, self-efficacy, social support or social support system, social skills and mood, etc.

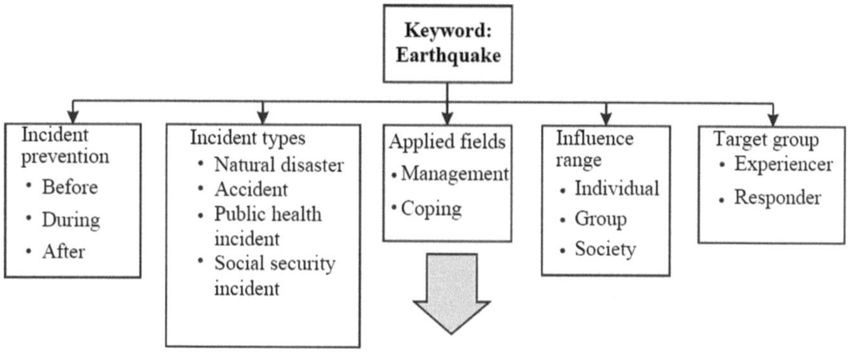

Fig. 10 Psychological knowledge base of unconventional emergency management

2.2.2 Scenario Analysis Method for Comprehensive Integration of Multi-granularity Information

Based on on-site monitoring data and collected data, through data analysis and integration, the situation assessment and comprehensive judgment of emergencies are carried out, and finally the early warning and emergency handling against emergencies are realized.

Big data's mining technology provides technical support for the research of emergencies' transmission in information space, and also provides an opportunity for the research of emergency decision-making supported by big data. Emergencies can easily lead to discussion and wide spread in various media. Through real-time monitoring and analysis of media data, we can realize the perception of public opinion situation in emergencies, and finally provide support for emergency management and decision-making.

Big data's mining technology also provides technical support for the research of emergencies' transmission in social space. By monitoring and analyzing the media data in real time, we can realize the perception of public opinion situation in emergencies, explore the social sensor network and social contact network, discover the opinion leaders on social networks in real time, and finally provide support for emergency management and decision-making.

The research contents of physical-information-society ternary spatial data integration and real-time sensation of scenarios include four aspects: (i) Real-time acquisition of multi-source heterogeneous data (aiming at multivariate data); (ii) Integration technology based on Internet of Things and construction of Internet of Things model of scenario sensation for emergencies (aiming at physical spatial data); (iii) Construction of social sensor network with real-time proactive perception (aiming at social spatial data); (iv) To realize the construction of "one system and two databases" of psychological data's integration, the investigation of national happiness and the study of psychological vulnerability (aiming at psychological spatial data).

In terms of the integrated technology and system of the Internet of Things, the comprehensive integration based on the Internet of Things is realized. In the aspect of prototype system's development, Guo Danhuai's research group of the Network Information Center of Chinese Academy of Sciences constructed the "1-3-5" framework of "one platform, three types of subjects and five layers of interfaces" (Fig. 11); The data injector of Internet of Things data and the simulated regulator of space–time carrier's parameters before and after disasters were developed, and the data of group behavior characteristic were obtained to support platform's integration. In order to protect related technologies, patent protection has been applied for key parts. In addition, the research fruits won two international academic awards, which verified the advancement of the research fruits.

On the basis of integration of physical-information-social spatial information, scenario analysis of multi-granularity information's integration and spread aims to identify the key elements when an emergency occurs through different granularity information, and comprehensively analyze the initial process of the incident, thus

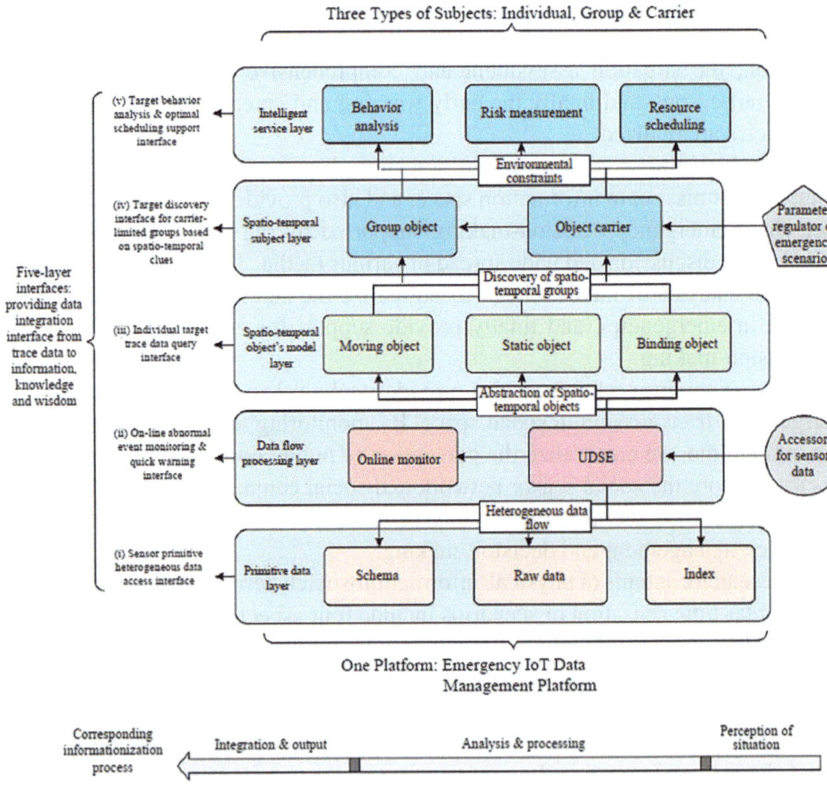

Fig. 11 "1-3-5" framework

laying a foundation for scenario deduction, decision-making and analysis. Multi-granularity information refers to macro, meso and micro information. Through the comprehensive integration of multi-granularity information, the emergency management's decision-making can be improved (Fig. 4.12).

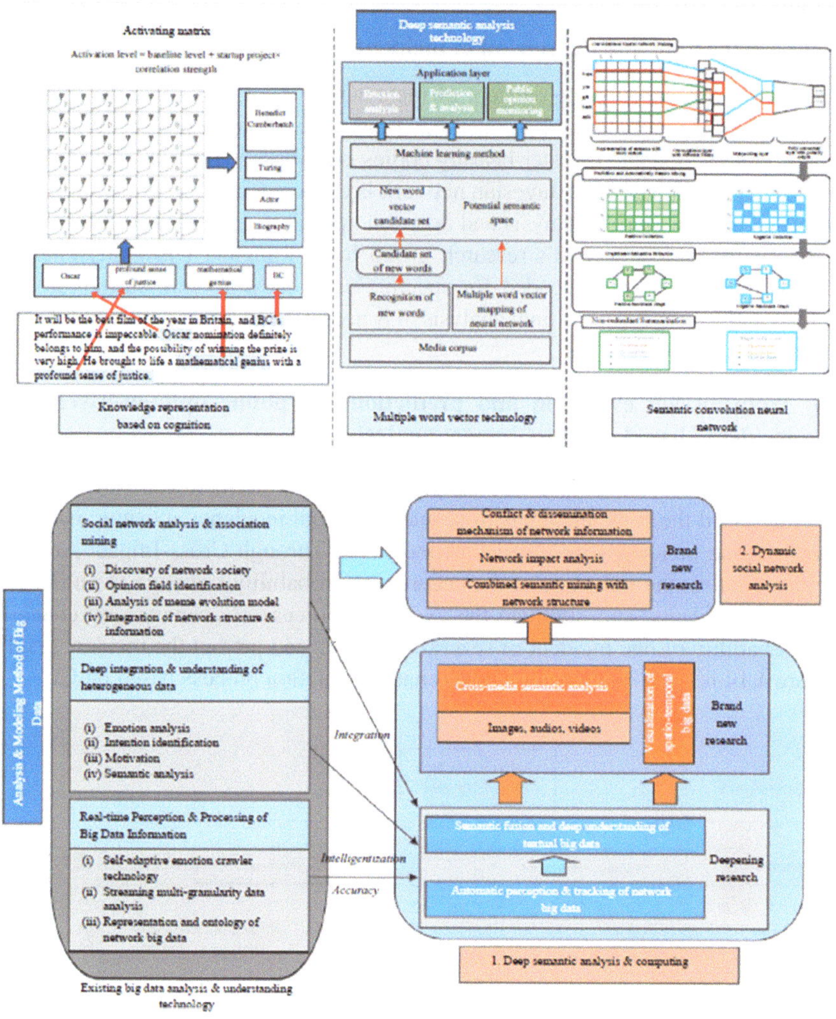

Fig. 12 Comprehensive integration of multi-granularity information

2.2.3 Scenario Deduction Method of "Incident Chain-Public Opinion's Transmission-Psychological Behavior" Coupling Feedback

The scenario deduction method of "incident chain-public opinion's transmission-psychological behavior" coupling feedback is the core of "scenario-coping" decision-making theory.

(1) **Construction and expression of incident chain and its mechanism**

In terms of mechanism, the concept of meta-action is used to study the modeling theory of incident chain, and a generalized theoretical framework including emergency, meta-action and disaster-bearing bodies is proposed. The analysis method of emergency chain based on Bayesian network is explored, and the application of the proposed method in risk analysis and decision optimization is studied. In the emergency platform, Zhang Hui's research group from Tsinghua University analyzed various models and data needed for specific emergency. To link various disaster models and data to form the final analysis, it is necessary to adopt the idea of disaster chain's integration (Fig. 13).

(2) **Transmission evolution and prediction of "public opinion-psychology-incident" based on cross-impact analysis**

Zhang Hui's research group used cross-impact analysis (CIA) to generate scenarios, and expressed the basic elements of scenarios (initial incidents, dynamic incidents, and outcome incidents) by a series of incidents. Through Cross-Impact Analysis, we can confirm the influence of the occurrence probability's change of an incident on other incidents, and simulate the evolution process of scenarios by changing the probability of one incident or several incidents, and predict the incident results. Figure 4.14 is exactly an example of scenario deduction process based on CIA.

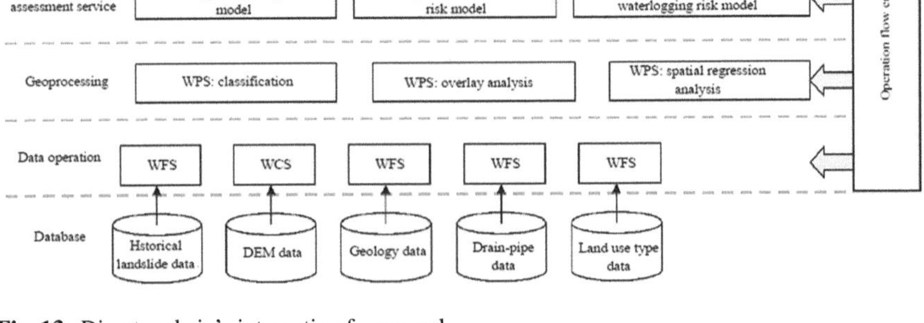

Fig. 13 Disaster chain's integration framework

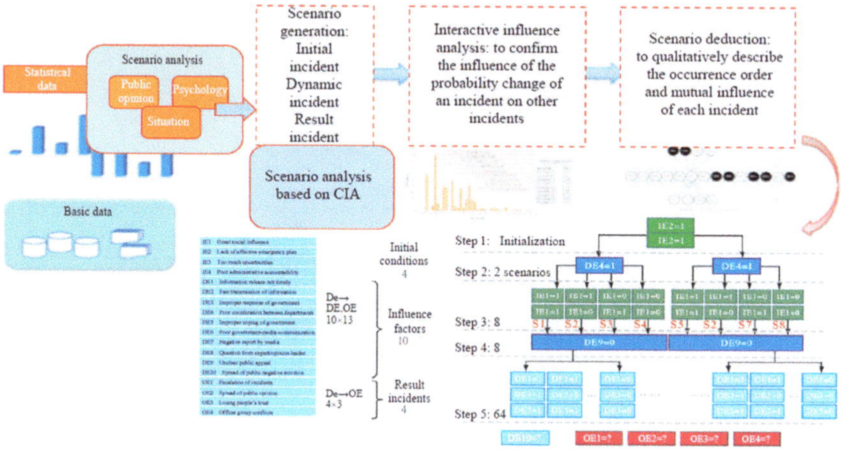

Fig. 14 Example of scenario deduction process based on CIA

Emergency decision-making research based on "scenario analysis-task ontology-model combination-data interaction-solution" decomposes complex emergencies into sub-scenarios, and makes the sub-scenarios form a task network by planning engine. Based on the task requirements, the corresponding models and data are used for reasoning, and the solution for incidents is obtained. The scenario deduction method of "incident chain-public opinion's transmission-psychological behavior" coupling feedback is shown in Fig. 15.

2.2.4 Innovation of Emergency Management Mode Based on China's National Conditions

The research on multi-agent emergency decision-making method based on China's national conditions is mainly carried out from five aspects: top-level design, bottom-level innovation, decision-making mode, organizational system and emergency response process's optimization.

(1) **Risk assessment and risk management model**

The evolution process of physical and social risks is presented dynamically, and the key nodes and paths of risks' occurrence and spread are determined. Xue Lan's research group from Tsinghua University has realized the transition from "One Case, Three Systems" to a comprehensive system through intervention in advance, handling during the incident and introducing risk transmission mechanism, formed a "risk government affairs" model with all risks and throughout the whole process, revealed the risks' information transmission, physical transmission, policy transmission mechanism, and the evolution mechanism of national security risks, proposed the national resilience's enhancement mechanism, and established the comprehensive situation's analysis method and risk pre-judgment system for national security.

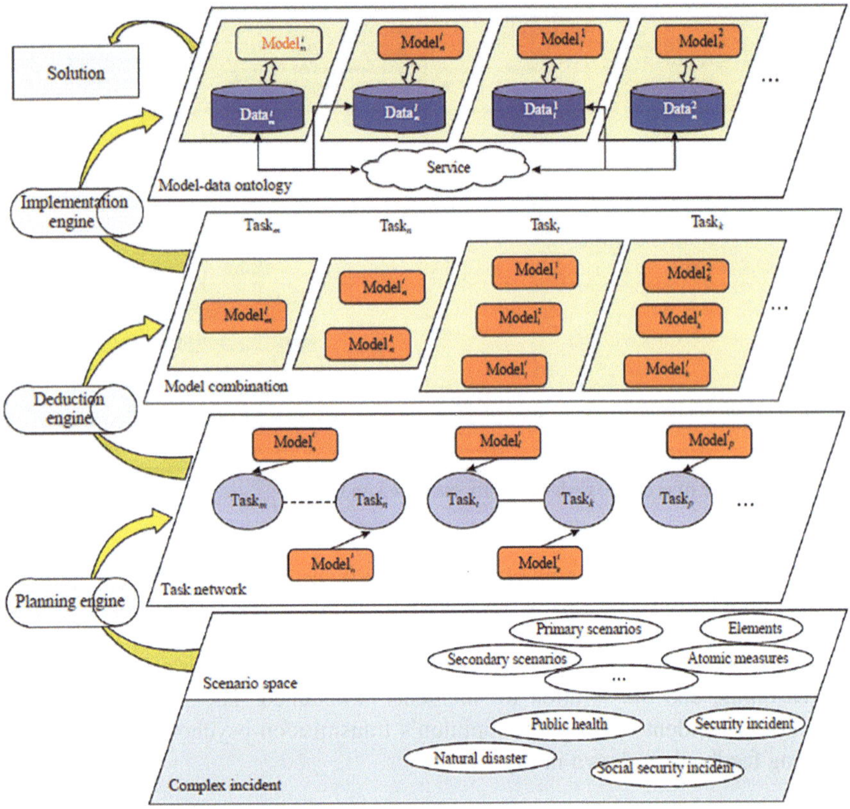

Fig. 15 Scenario deduction method for coupling feedback of "incident chain-public opinion's transmission-psychological behavior"

At present, China has formed an all-hazard approach that considers factors such as risk identification, spatio-temporal analysis, cyberspace, public opinion information, social situation, international image, etc., an analysis method that establishes the correlation of security incidents, an evaluation theory and method of multi-hazard coupling risks, and technologies and methods of overall risky factors' monitoring and whole process prevention and control against risks.

(2) **Framework and model innovation of China's unconventional emergency management system**

Based on system analysis and "structure–function" method, aiming at the shortcomings of China's existing unconventional emergency management organization system (Fig. 16), Xue Lan's research group outlined the ideal structure of the national emergency management system (including components, structural features, operational functions, etc.), which provided an analysis framework for subsequent research in case comparison, and also provided a basic platform for top-level design and model

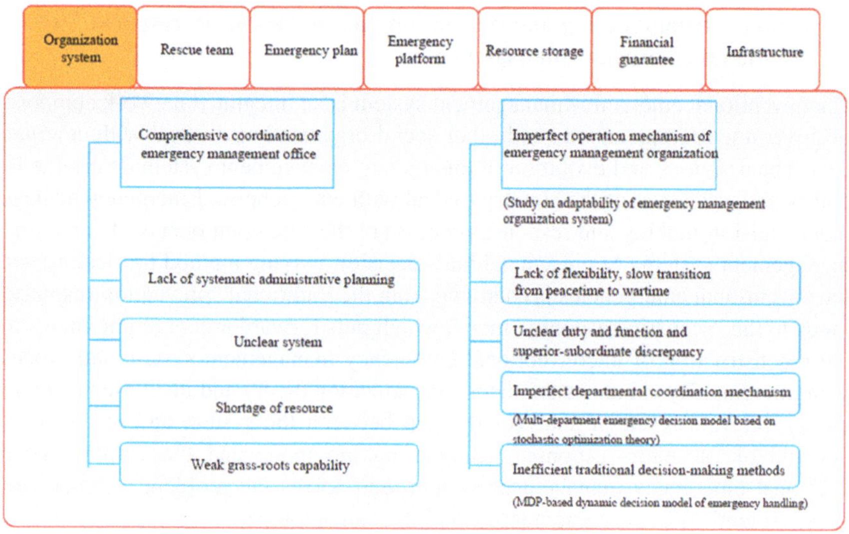

Fig. 16 Shortcomings of China's existing unconventional emergency management organization system

reconstruction research of emergency management system; Through the comparison of historical cases, regional cases and departmental cases, the research group has summed up the model innovation of China's emergency management system under the situation of "institutional change."

According to the framework of national emergency management system, from the perspective of organizational change and system analysis, combined with the theories of collaborative governance and social capital, and based on the existing emergency management work, relevant research studies the functional orientation and overall planning of unconventional emergency management system from the perspectives of coping subject, management process, management environment and management culture, and analyzes the optimization path of organizational structure. On the one hand, starting from the current situation of emergency management system with "One Case, Three Systems" as the main content, relevant research studies the organizational structure model of the first-level government (including the departments within the same level government in the same region) and the relationship of emergency management's powers and responsibilities between vertical and horizontal governments at different levels; On the other hand, based on the integration model of "government-enterprise-people," relevant research discusses the benign interaction between inside and outside the government system, and establishes an organic cooperation model between the government and non-governmental organizations (including representative international organizations), the public and other countries and regions.

(3) **Process reengineering and realization path of "scenario-response" unconventional emergency management**

Unconventional emergency management system is an integrated network composed of government organizations and other social organizations to deal with emergencies. The activities and evolution of emergency management system determine the ability and efficiency of the country to deal with emergencies. Emergency management decision-making and response process are the core components of emergency management system. At present, China's decision-making method for dealing with unconventional emergencies is changing from the traditional "prevention-response" mode to the "scenario-response" mode, which puts forward urgent requirements for the reconstruction of unconventional emergency management system and process reengineering. By using system theory, organization theory and institutional change theory, relevant research analyzes the gap between the system and mechanism's demand of "scenario-response" decision-making mode and the current unconventional emergency management system, and solves the problem of emergency management's process reengineering and its realization path.

(4) **Panoramic safety management and multi-agent collaborative decision making**

Under the trend of cross-regional ecological environment security systems' mutual influence and polarization of management system, Zhang Hui's research group of Tsinghua University has established a pre-judgment mechanism and an advanced response system for China's national security against external threats. The method of multi-agent, multi-objective and multi-task collaborative decision-making and scenario deduction is established, which reveals the cognitive-behavioral mechanism of government, public and other participants in the public security's governance system, and forms an emergency management's decision-making mode of real-time scenarios, mode switching and social cooperation with the participation of government and public. The research group realized the emergency management and decision-making method which considers the internal environment and external factors. The panoramic security management and collaborative decision-making mode is shown in Fig. 17.

(5) **Innovation of China's emergency management system (taking typhoon as an example)**

In view of the shortcomings of the existing process and system, Zhang Hui's research group took typhoon disaster as an example to demonstrate the research of emergency response system and relevant process's reconstruction based on China's national conditions. Through the analysis of typhoon disasters, a typhoon emergency management system in line with China's national conditions has been formed (Fig. 18).

Zhang Hui's research group carried out a special study on Typhoon Haiyan and Typhoon Hagueby from the perspective of physics, information and social management. The research group analyzed typhoon from the aspect of disaster chain of

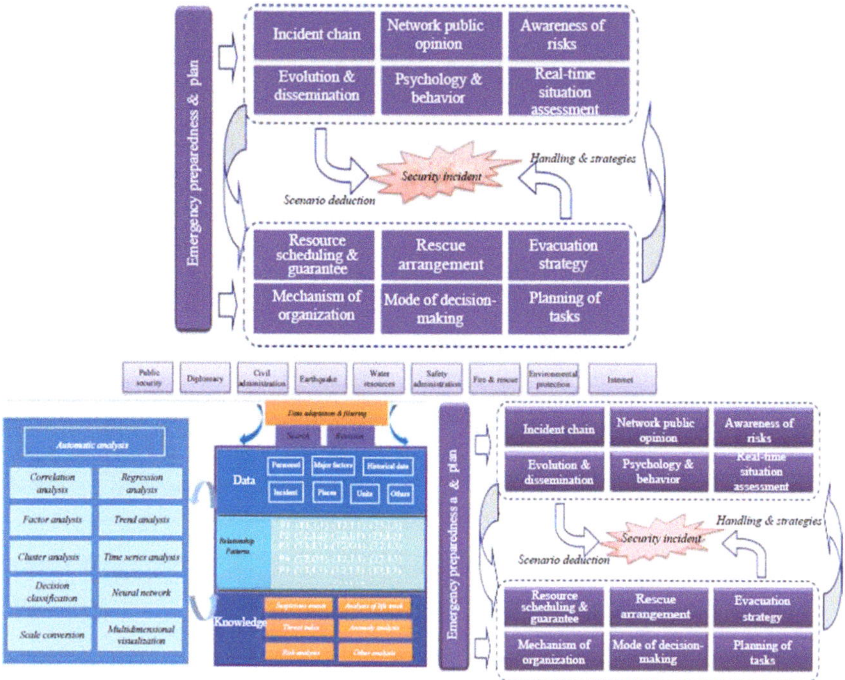

Fig. 17 Panoramic security management and collaborative decision-making mode

physical space, it is necessary to accurately evaluate the current situation and track the path before typhoon landing; After the typhoon landed, the affected areas and losses should be evaluated as soon as possible, and the corresponding emergency measures and information should be released. The research group analyzed public opinion information from cyberspace, and obtain information to assist decision-making through data analysis and processing in information space. In addition, from the perspective of social space, the research group analyzed people's reaction and evacuation under typhoon disaster and the corresponding policies and strategies.

This case adopts the dynamic three-dimensional spatial analysis method (Fig. 19 shows the analysis framework). After an emergency occurs, through obtaining and analyzing the incident-related data, we can understand the public's attitude, opinions and feelings towards the incident, various policies, regulations and responses of relevant departments, and make corresponding management policies in a targeted manner.

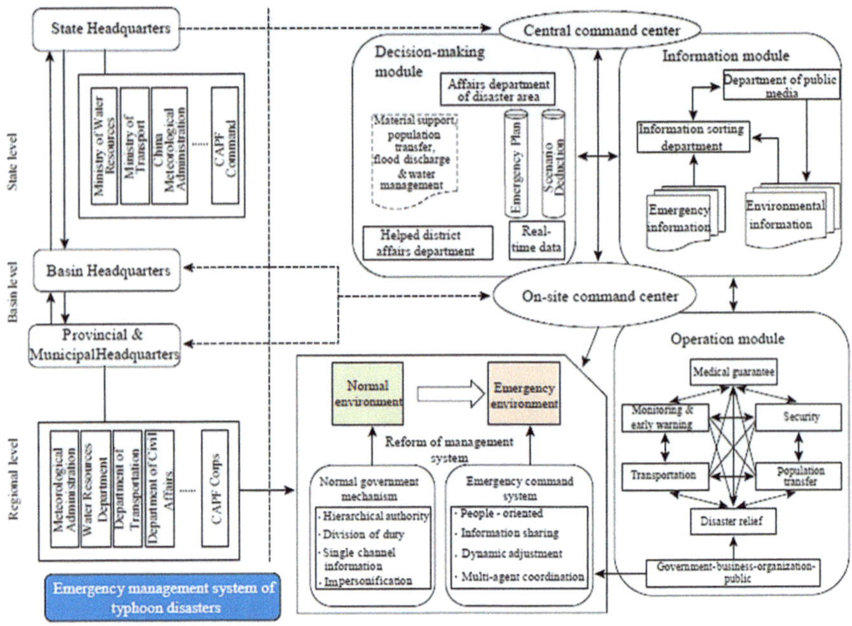

Fig. 18 Emergency management system (taking typhoon as an example)

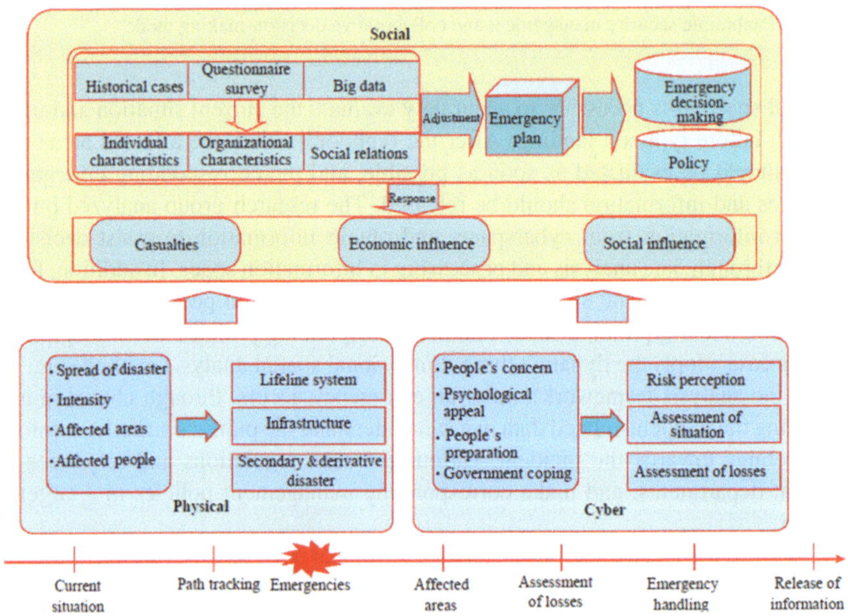

Fig. 19 3D spatial analysis framework (taking typhoon as an example)

2.3 *Application and Popularization*

With regard to the demand of "scenario-response" emergency decision-making, Zhang Hui's research group in Tsinghua University has realized interdisciplinary integrated application demonstration. On the basis of comprehensive integration of "data, model, case and psychology," and based on the task requirement mode's integration framework, three platforms, namely, the basic platform based on emergency platform system, the basic platform based on emergency plan system and the dynamic simulation system for incidents, were constructed, and the open integrated platform of emergency management was integrated, which has been widely applied in the construction of emergency system for national production security.

2.3.1 The Distributed Sharing Platform

Zhang Hui's research group has formed an open and distributed resource sharing platform's integration framework with a three-layer structure of physics, virtual resources and knowledge application (Fig. 20). The physical layer integrates and analyzes data from different sources, including physical spatial data, cyberspace data, social spatial data and psychological behavior data; The research group reflected the data from physical layer's integrated analysis to virtual resource layer for management, optimization and control, and finally realized application in knowledge and application layer.

We should select large-scale traffic examples for evacuation simulation. Large-scale traffic evacuation is one of the important contents of emergency management, which is particularly important in the rapid evaluation and reasonable prediction of the whole traffic scenarios. The topic of large-scale traffic evacuation was jointly completed by Tsinghua University, Huazhong University of Science and Technology, University of Science and Technology of China, Tongji University and National University of Defense Technology. The evacuation platform is divided into five functional modules, namely, emergency plan's formulation module (Tongji University), pedestrian evacuation simulation module (University of Science and Technology of China), road traffic simulation module (Tsinghua University), material allocation's planning module and optimal route's selection module (Huazhong University of Science and Technology). The underlying data are mainly provided by the Network Center of Chinese Academy of Sciences.

2.3.2 Response to Public Health Events (Ebola Outbreak)

Taking the Ebola incident as an example, Zhang Hui's research group predicted the possibility of Ebola outbreak in China and its possible influence scope. This application integration shows the management and decision-making scheme based on physical-social-psychological spatial data's integration (Fig. 21 shows response

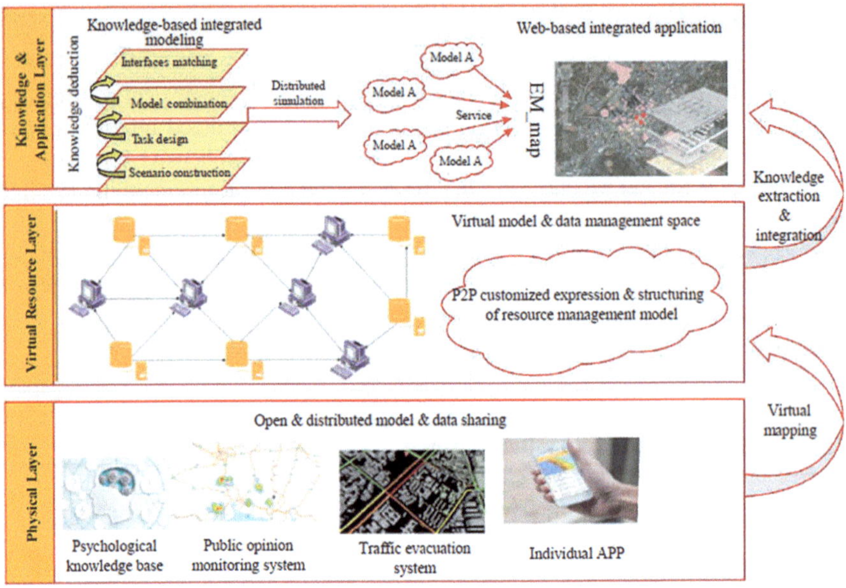

Fig. 20 The integration framework of open and distributed resource sharing platform[1]

to Ebola outbreak). Besides, Framework of Response Measures for Ebola Outbreak Based on Vision-Scenario-Task-Ability was proposed. (Fig. 22).

Zhang Hui's research group used the model parameters of Ebola case development in West Africa to simulate the epidemic outbreak in Beijing without vaccine control. At the same time, the influence of randomly vaccinating a certain proportion of people and discovering cases and isolating Ebola cases one day in advance on Ebola epidemic situation was deduced.

2.4 *International Comparison and Influence*

Professor Zhang Hui from Tsinghua University took the lead in organizing the formulation of the international standard for emergency response capability assessment (ISO 22325). It was unanimously approved and was the first standard adopted by China in the field of security. Professor Zhang Hui, as the international leader of the construction of information system and data standards in the Emergency Operations Center (EOC) of the World Health Organization, WHO), has applied the research fruits to WHO, enhanced the international public health's emergency response capability, and got the trust of WHO. It promoted the cooperation between WHO and the research group, which was beneficial for the Chinese team to continue to play a

[1] EM map is an emergency map developed by Guo Danhuai's research group of Network Information Center of Chinese Academy of Sciences.

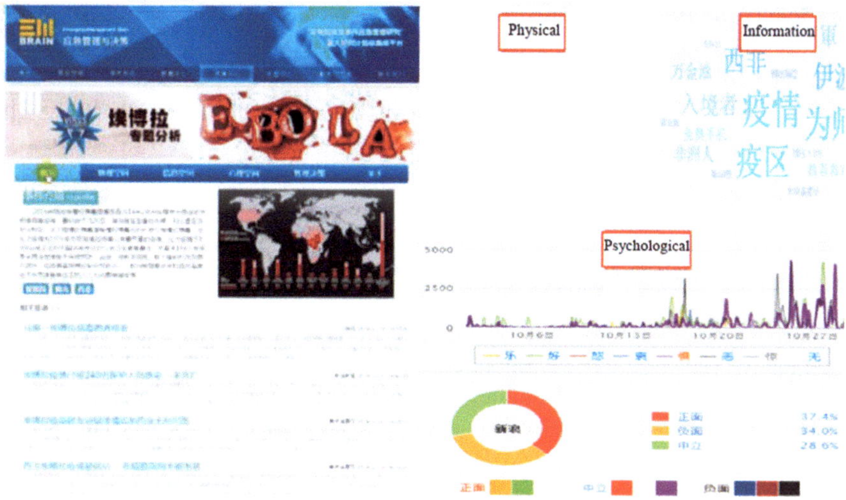

(a) Ebola thematic analysis of the overall integration platform

(b) Response to Ebola outbreak based on ternary space analysis

Fig. 21 Response to Ebola epidemic

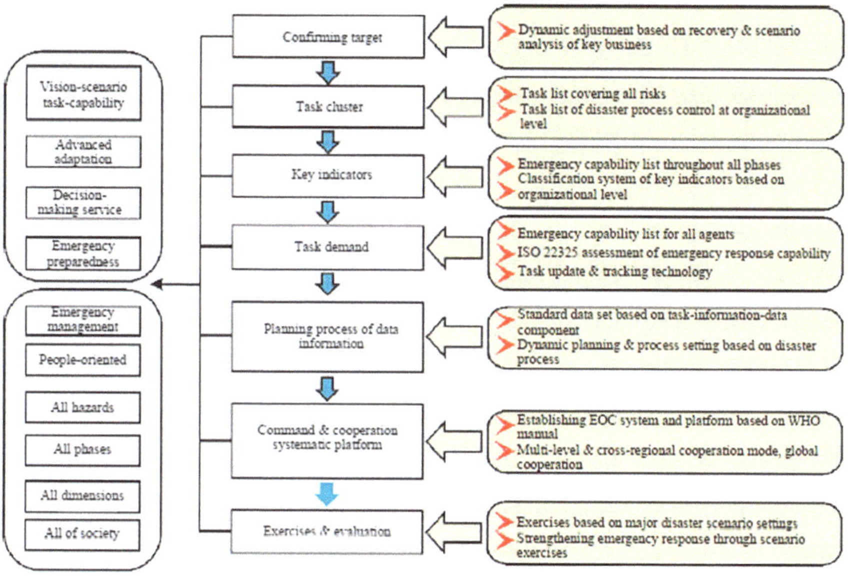

Fig. 22 Framework of response measures for Ebola outbreak based on vision-scenario-task-ability

key role in the follow-up work of compiling the health emergency framework and manual, and to play a leading role in the world. He also propelled the establishment of public health emergency platform 2.0 and the cooperation center of the Health Planning Commission, thus improving China's ability to deal with public health emergencies. The construction of the health emergency platform has received important instructions from Premier Li Keqiang and Vice Premier Liu Yandong.

2.4.1 Construction of WHO EOC Network

In response to Ebola's problems, WHO began to lead the international public health emergency. The fight against Ebola virus in West Africa is the first joint rescue operation led by the United Nations.

The Joint Rescue Action Group put forward the thinking of urban security planning based on the perspective of risk society, emphasizing the importance of regional cooperation. By studying the emergency command centers in the United States, Britain, Australia and other countries, this paper summarizes the effective factors of emergency actions in these countries, provides suggestions for the further construction of China's national public health's emergency response center, and points out that we should learn from the advantages of other countries in terms of objectives, organizational structure standards, human resources' strengthening and information system's construction. Taking Ebola virus' prevention and control as an opportunity, WHO put forward that countries should cooperate with each other to cope with emergencies, and devote themselves to rapid and accurate assessment of working ability under dynamic scenarios. In 2012, WHO established the Emergency Operations Centre Network (EOC-NET) to coordinate the action information and strategic resources for global public health incidents and emergencies, propelled global cooperation in disaster response. It was to ensure the best practices and standards related to public health emergency response centers, and provide support for the capacity building of public health emergency response centers in various member countries. At present, EOC-NET mainly includes three aspects: joint prevention and control mechanism between countries and regions, research and framework design of communication and information management systems; Information sharing, EOC information system's construction and evaluation; For resource sharing, from 2016 to 2018, three major regions (Africa, Asia and Europe) were selected, with 4–5 countries of each region. Public health or national emergency response centers were configured according to the framework guidelines, and scenarios were set up to conduct cross-regional emergency response drills, and finally such method was introduced to the whole world.

In 2014, the members of the research group defeated the United States and other powerful countries through global competition, and participated in and led the compiling of the guiding framework and manual for EOC construction. After more than one year's efforts, it was published after discussion at two WHO EOC-NET conferences (the first conference was attended by public health EOC leaders from more than 50 countries and regions, and the second conference was attended by public health EOC leaders from various organizations within the United Nations and many countries). At present, the system of the World Health Organization and its six regions is quite perfect, but relatively closed. The focus about the problem is to connect the EOC of each member country through the network, while maintaining the independence and confidentiality of the EOC of each country. EOC fruits are shown in Fig. 23.

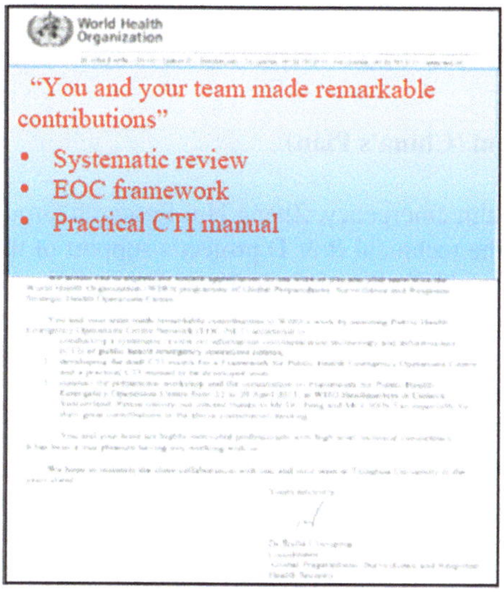

(a) WHO's thanks letter to the leader of the program

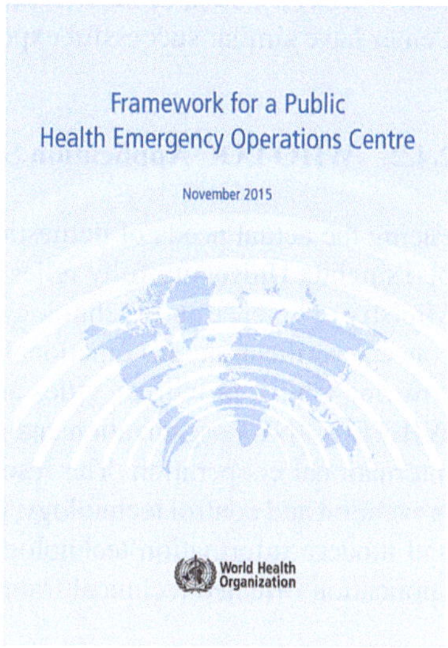

(b) WHO EOC-NET framework

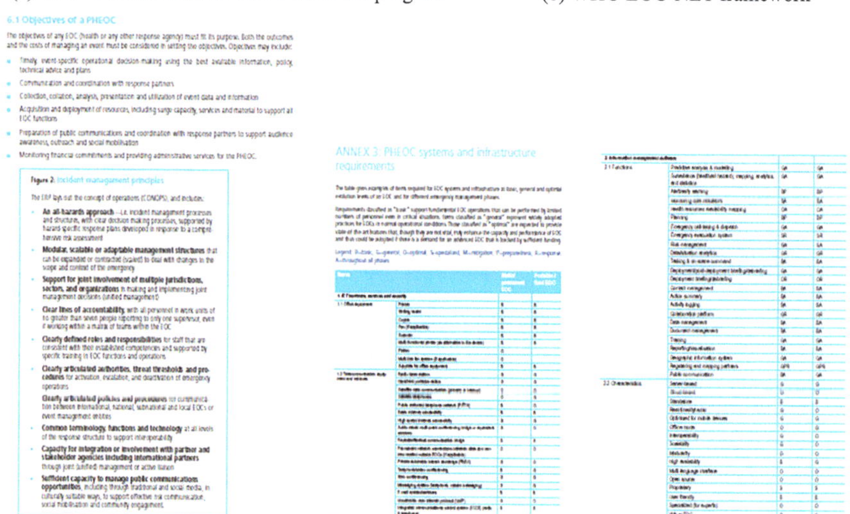

(c) Part of EOC achievements and main contents

Fig. 4.23 EOC's fruits examples

WHO is propelling strategic cooperation with the Tsinghua University research group to discuss the establishment of an international cooperation center. WHO believes that the formation of such a cooperation mechanism will help the Chinese team to continue to play a key role in the follow-up work of the health emergency framework and manual, and play a leading role in the world. WHO is drafting specific plans. In the strategic cooperation, domestic universities and research institutions can follow the example of some American universities and establish cooperation centers with WHO. At China, China Disease Control Center and Guangdong Disease Control Center have similar successful experiences.

2.4.2 WHO EOC Application System (China's Plan)

Facing the actual needs of domestic health emergency, Zhang Hui Research Group of Tsinghua University fully relies on the technical R & D project's support of the Ministry of Science and Technology to quickly build a technical prototype system for concept verification and functional testing, and build a health emergency platform covering China, provinces, cities and counties. The research group connected with WHO EOC-NET organization and introduced these fruits to other countries through international cooperation. The research group made full use of the current disease prevention and control technology, public security's emergency response technology and modern information technology, and has built a technologically advanced and application-oriented technical framework (Fig. 24).

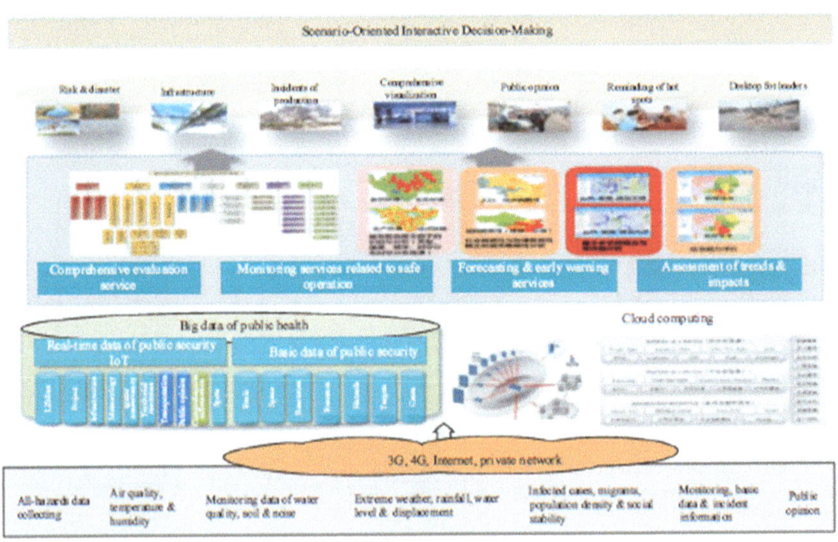

Fig. 24 Technical framework of emergency platform system in health field

The national health emergency platform includes the national public health emergency command center platform, the emergency operation center platform of the Chinese Center for Disease Control and Prevention, the mobile vehicle information platform of the national health emergency team, and the individual soldier information platform, and has many functions such as emergency duty, comprehensive monitoring, risk assessment, early warning and response, auxiliary decision-making, command and dispatch, on-site management, emergency resource pool's management, emergency evaluation, online training, simulation exercise, etc. The national public health emergency command center platform is the core and center of the national health emergency platform, and the carrier of implementing daily emergency management and realizing emergency command decision. The WHO EOC application system (China scheme) is shown in Fig. 25.

3 Research on Psychology and Behavior of Unconventional Emergency Management and Application of Relevant Fruits

3.1 *Research Background and Present Situation*

Unconventional emergencies not only cause loss of people's lives and property, but also lead to abnormal psychological state and behavior pattern of individuals/groups, which makes people in a state of intense psychological stress. Various irrational behaviors may even couple with the influence of incidents and cause secondary disasters.

The difficulty in dealing with unconventional emergencies lies in the unique regularity of its impact on people. The traditional "prediction-response" method is not suitable for dealing with unconventional emergencies. Therefore, it is necessary to build a cross-disciplinary platform for information integration and decision-making, and determine the key elements of information collection, decision-making and execution at each key node, including the influence of individual psychological cognition and emotion, group psychology and behavior rules, etc.

3.1.1 Individual-Group Level

Under unconventional emergencies, public stress response has unique regularity, which makes new requirements for targeted emergency management and psychological intervention. Generally speaking, the individual psychological law is not clear, and the application of psychological intervention is not effective. Therefore, studying the characteristics of stress response and its influencing mechanism can provide scientific explanations for various behavioral problems after the incident,

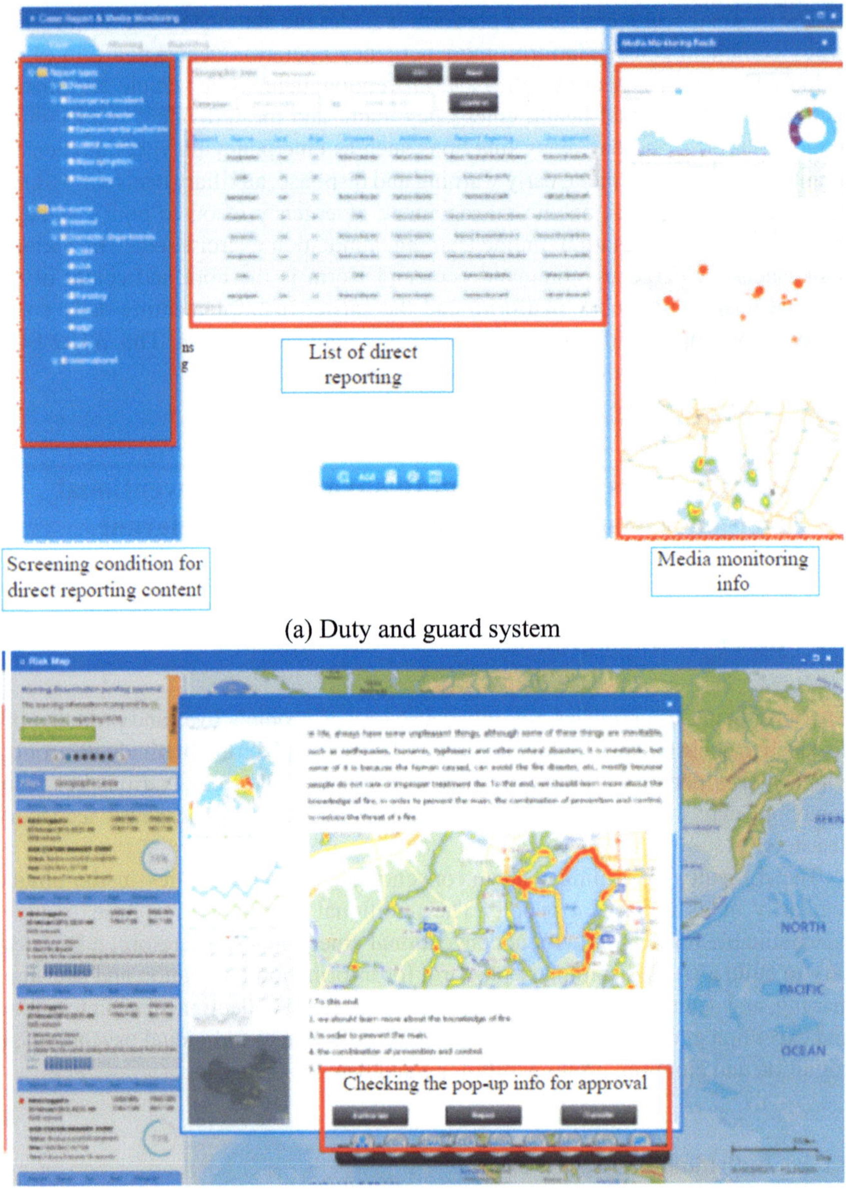

(a) Duty and guard system

(b) Risk cockpit

Fig. 25 WHO's EOC application system (China's Plan)

(c) Coordination system—incident handling and coordination

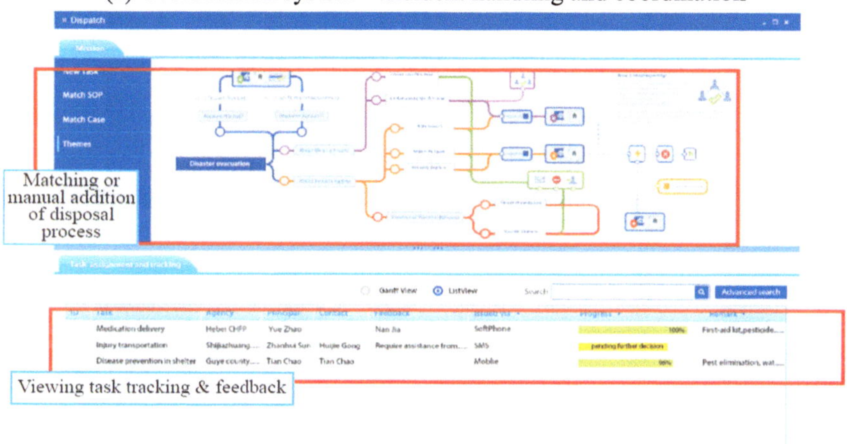

(d) Coordination system—distribution of task process

Fig. 25 (continued)

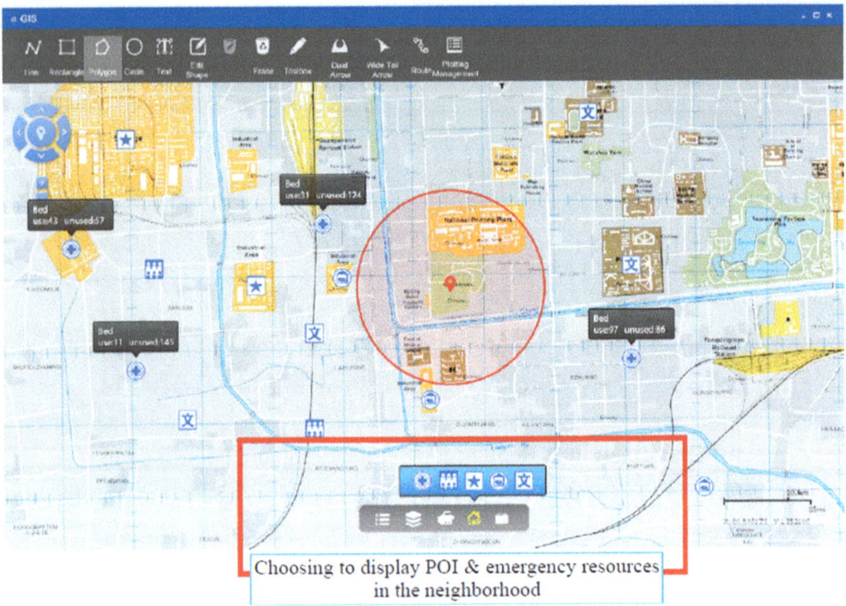

(e) Coordination system—GIS

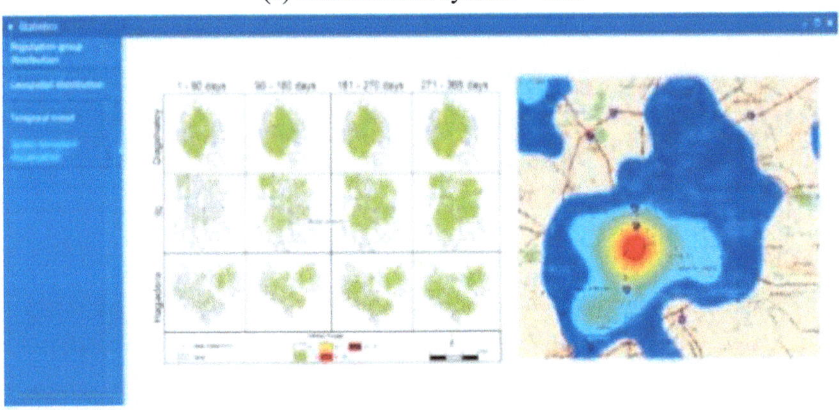

(f) Analysis and judgment system

Fig. 25 (continued)

and provide psychological basis for making emergency management and psychological intervention policies. Through laboratory research, on-site simulation and other means, we can grasp the law of individual psychological process in emergencies, and formulate corresponding pre-disaster prevention, disaster handling and post-disaster counseling strategies for different affected subjects.

According to the distance from the disaster center, the psychological and behavioral laws of the affected people are different, but they may seriously affect social stability. For the directly affected people, the series of psychological and behavioral studies in this major research plan (hereinafter referred to as psychological and behavioral research) pay attention to the law and principle of individual psychological and behavioral response in a short time after major stress; For potential victims, psychological and behavioral research focuses on individual cognition and evaluation of potential risks and coping methods; For the bystanders, psychological and behavioral research focuses on their attitudes and behavioral responses to the affected groups, as well as derived social incidents. The application of various experimental means to simulate the cognitive, emotional and behavioral characteristics of individuals/groups and their brain mechanisms under stress has practical significance for emergency intervention and selection of special talents.

In the aspect of psychological intervention, many researches have focused on psychological reconstruction after disasters, and focus on changing the negative impact of disasters on individuals through intervention, but this has great limitations in practical application. Psychological and behavioral research hypothesis: (i) To cope with crisis, a lot of individual psychological resources need to be consumed. In crisis, to avoid exhaustion of body and mind, individuals need to actively mobilize psychological and physiological resources to create a virtuous circle. (ii) The negative effects of disasters on individuals include both primary injuries and secondary injuries caused by abnormal behaviors of groups in specific situations. Therefore, psychological intervention should be carried out effectively from the macro level by analyzing the psychological and behavioral factors of the evolution law of secondary injuries.

Therefore, the research discusses the psychological and behavioral laws of individuals and groups in crisis situations, that is, under the conditions of stress and exhaustion, discusses the psychological and behavioral changes of individuals and groups, studies the specific influence path of psychological system on physiological system, and puts forward the theoretical conception of physical-mental interaction model; Through the compound research of disaster distance and stress response mode, a more scientific and systematic portrait of the affected subjects is established, and corresponding prevention, handling and counseling strategies are formulated for different affected subjects.

3.1.2 Social Level

In terms of the complex features of multi-subject, multi-target, multi-level and multi-type, Academician Fan Weicheng put forward the security triangle theory, which divides the public security system into three major components: emergency, disaster-bearing bodies and emergency management, and links the three factors through material, energy and information to form an organic whole. However, it is rare to study the integration and analysis of data in all aspects of the system from the perspective of system integration methodology. Taking social psychology and psychometrics as the main subjects, Tong Hengqing's research group of Wuhan University of Technology examined the measurement rules and intervention evaluation's analysis for group negative psychology in emergencies, and realized the monitoring and early warning of scenario response and data drive. Through the observation, experiment and theoretical innovation and comprehensive integration of related multi-disciplines, it has brought a profound scientific understanding of the objective laws of the core link of unconventional emergency management-monitoring, early warning, response and decision-making.

(i) Correlation analysis between public attitude and social instability and relevant monitoring and early warning. The causes of emergencies are extremely complex, which are often caused by various factors such as society, politics, economy, culture and so on, and these external factors that cause emergencies play a role through people's psychological mechanism. Social psychology research shows that for the same social living environment and public policy, different individuals will inevitably hold different social attitudes because of their different gains and losses. Because attitudes guide people's social behaviors, only by understanding public attitudes can we accurately predict their behaviors and identify social unstable factors.

(ii) Theoretical empirical analysis, monitoring and early warning of group psychological instability and relevant evolution models. Tong Hengqing's research group focused on serious group incidents and group reaction's processes that rapidly formed and spread by various media, studied self-organization phenomena such as dissipation, synergy and mutation, as well as the laws of complexity, and paid attention to the description and measurement results of model's evolution and mutation.

(iii) The measurement law and intervention's evaluation and analysis of negative psychology of groups in emergencies. Wang Lei's research group of Peking University, Xie Xiaofei's research group of Peking University, Zhang Kan's research group of Institute of Psychology of Chinese Academy of Sciences, and Zhou Xinyue's research group of Sun Yat-sen University are dedicated to collecting and analyzing the survey statistics of several typical psychological negative reactions (such as fear, anxiety, depression, compulsive behavior, etc.) of incident participants (including managers, rescuers and the public) in the whole process of unconventional emergencies (before, during and after incidents), combined with the theoretical basis of psychology and mathematics.

(iv) Integration method of resilience model for dealing with unconventional emergencies. The existing research has not realized the integration of data from all aspects of the system from the perspective of system integration's methodology, especially the system integration's methodology of big data in objective environment and big data of subjective psychological behavior obtained through questionnaire and survey, which is the key issue that needs to be discussed in studies.

Based on Academician Fan Weicheng's security triangle theory, Shi Kan's research group of Chinese Academy of Sciences comprehensively considered danger, vulnerability and resilience from three aspects: emergency, disaster-bearing bodies and emergency management. Based on the psychological model of resilience of responders and teams in crisis, the team put forward the concept of psychological and behavioral coupling model, and was committed to providing a multi-measure analysis approach for the analysis of massive psychological and behavioral big data. In the system integration platform, the psychological behavior data obtained by mining and analysis was combined with the normal model of psychological measurement and the causal relationship model obtained by psychology and behavior experiment, which realized the system integration of subjective and objective big data, and the team contributed to the improvement of the decision-making function of the whole integration platform.

3.2 *Major Research Fruits*

Psychological and behavioral research starts from three perspectives: biology-psychology, individual-group and decision-maker-executor, and has obtained a series of achievements in psychological and behavioral aspects of emergency management, and it has improved the basic research level and management level of emergency management in China.

3.2.1 Biology-Psychology Perspective

Psychological and behavioral research uses qualitative analysis and quantitative analysis, on-site investigation and laboratory research to explore the characteristics and internal mechanism of psychological and behavioral responses of people who are directly affected by unconventional emergencies, rescue workers and peripheral people from both individual and group aspects. In this way, it provides a new perspective for the study of emergency response.

(i) Xu Yan's research group of Beijing Normal University constructed the psychological energy theory through literature review, interview and questionnaire survey, and developed the Psychological Energy Questionnaire, which has good reliability and validity indexes. This questionnaire is used to explore the relationship between psychological energy and problem behaviors after the disaster. The results show that there is a significant negative correlation between psychological energy and problem behavior.

(ii) Xu Yan's research group explored the internal mechanism of psychological energy's loss and its aftereffect by combining brain imaging technology with behavioral experiments. In terms of mechanism, a series of new concepts such as resilience, buffer protection factor, supplementary protection factor and social justice recovery have been put forward. After the disaster, negative emotional immersion consumes the individual's psychological energy, which shows a high level of activation of Stroop task brain area; After the regret emotion is induced, the individual's psychological energy level is low, which is characterized by poor persistence in boring tasks, insensitive Stroop task response, poor performance in mental arithmetic task and high fatigue level. In terms of the aftereffect, it is found that mental energy's loss weakens the negative deviation effect of emotional information's processing and memory, and reduces the ability and willingness of individuals to process complex information.

(iii) By combining qualitative and quantitative methods, Xu Yan's research group explored the protective factors of psychological energy. Qualitative research shows that discovery of benefit, life meaning's seeking and experience can be used as protective factors for psychological energy. Discovery of benefit can effectively buffer the aftereffect of regret; The cognition for daily positive events can positively predict the level of individual's experience, while negative events can give people negative prediction. Meaning seekers are more sensitive to negative information and tend to interpret fuzzy events negatively, while meaning experiencers are more sensitive to positive information and tend to interpret fuzzy events positively.

(iv) Starting from the real cases, Xie Xiaofei's research group of Peking University analyzed the psychological and behavioral reactions of peripheral people after unconventional emergencies by means of questionnaire survey and laboratory research. It is found that when others are unhappy, the peripheral people may not only show schadenfreude, but also have different degrees of intergroup willingness to help others.

Through laboratory and on-site simulation of emergencies, the research group made full use of daily events, focusing on people's psychological characteristics, reaction mechanism and behavior characteristics in emergency situations. At the same time, the research group paid attention to the public opinion after the incident, especially the evolution characteristics and mechanism of public opinion in new media. Combining mutation theory and emotion control theory, the research group propelled the construction of mutation theory from conventional incidents to unconventional incidents, that is, when an individual is in an emergency situation, the prefrontal lobe can't control the emotional response of the amygdala and makes the individual turn to irrational behavior; At the group level, some factors of control variables can propel the subjects of group incidents to change from nonviolence to violent confrontation. Taking densely populated scenic spots as example, the research group fully analyzed the rules of crowd behavior and risk characteristics, and put forward a series of intervention measures, including the optimization of evacuation signs and the formulation of standards in congestion prevention and control. Risk analysis was carried out for key areas, emergency evacuation plans were prepared, and an intelligent control system for crowd evacuation was developed and put into practical use.

3.2.2 Individual-Group Perspective

On the basis of three levels (social and cultural level, group level and individual level), four links (genetic and environmental factors, occurrence and action mechanism, laws and evolution results, prevention and intervention means) and five types of methods (brain imaging, gene, behavior experiment, network analysis and information processing), this research explored the methods to strengthen the infrastructure and management level of emergency management in China from the individual-group perspective.

(1) **Physical and mental interaction effectiveness and risk communication's obstacles in crisis situations**

In crisis situations, psychological effects produced by altruistic behavior affect individual physiological perception, and self-motivation effect of altruistic behavior can act on physiological system, which shows that the effectiveness of individual coping behavior in crisis depends more on psychological efficacy. Altruistic behavior does not completely mean the loss of resources to the helper, but also will bring positive effects to physiology. The related research of Xie Xiaofei's research group of Peking University supports the theoretical conception of physical and mental interaction efficiency model, and the experimental evidence supports the positive influence of psychological resource's mobilization on physiological system in crisis, which is of great value for survival in crisis.

According to the results, Xie Xiaofei's research group put forward the Lingering Fragrance Effect, which is a typical result from the interaction between physiological system and psychological system, that is, altruism not only brings long-term rewards (kinship altruism and reciprocal altruism) to helpers, but also brings immediate positive feedback. The results show that altruism can be used as an effective way to reduce the weight-bearing feeling of the body, and bring altruists a relaxed and positive physiological experience. The characteristics of crisis situations show that the individual's physiological resources are extremely limited, and the psychological system shows stronger proactive characteristic. Altruistic behavior in daily and crisis situations does not completely mean resource depletion for helpers, but can bring positive effects to physiological level, which is of great value for crisis survival.

Altruism has a positive effect on expression perception, which can increase people's sensitivity to positive expressions and enhance people's evaluation of expression potency. The positive effect of altruism not only has a positive effect on a specific expression, but also promotes people's perception of the whole expression of others and their feeling of life. In addition, compared with non-altruistic behavior, altruistic behavior can relieve pain. The research extends the warmth effect of altruistic behavior to pain perception, which provides empirical evidence for understanding the evolutionary causes between altruistic behavior and pain perception, and shows that the effect of relieving pain through altruistic behavior is conducive to forming a long-term healthy lifestyle.

There are genetic and cultural factors behind the relationship between body and mind. There is a reciprocal relationship that "increases resources" between positive emotion and altruistic behavior, and a reciprocal relationship that "prevents resource loss" between fatigue and altruistic behavior. Altruistic behavior is an important investment behavior for individuals to maintain resources. The reciprocal model of "physical feeling and mental feeling" and "interpersonal altruistic behavior" (Fig. 26) can be regulated by time and DRD4 gene. At the same time, Xie Xiaofei's research group focused on the topic of moral evaluation in crisis, revealing that the individual's evaluation on negative behaviors will be influenced by the characteristics of the situation and the components of incident, which proves that moral evaluation is an important cognization source for the formation of emotions and attitudes in crisis. It is also found that low self-control will increase altruistic behavior and help people understand the motivation behind altruistic/selfish behavior. It shows that whether human nature is altruistic or selfish can't be generalized, and it should be discussed in different situations. Taking the subway congestion crisis as an example, Xie Xiaofei's research group discussed people's psychological state (such as mood and attitude towards life) and behavioral reaction (such as aggressive tendency and withdrawal behavior) under different crowded situations, and explained the motivation of the above altruistic/selfish behaviors.

Expectation difference in crisis will aggravate risk of communication barriers, and psychological cognitive intervention can alleviate the vicious circle of barriers to group emergency response. For the information receiver, the individual's information needs are strong and constantly changing. If the information publisher lacks communication experience, the individual's needs are difficult to be fully and timely understood, and efficient risk communication is difficult to achieve; For the information publisher, due to the limitation of external decision-making conditions and internal cognitive ability, the published information can not be processed effectively and rationally by the receiver, and it is difficult to meet the publisher's expectations. The special characteristics of risk make risk communication very difficult. If there is expectation difference between the two sides during risk communication, factors such as distrust will be caused, which will worsen the barriers of risk communication

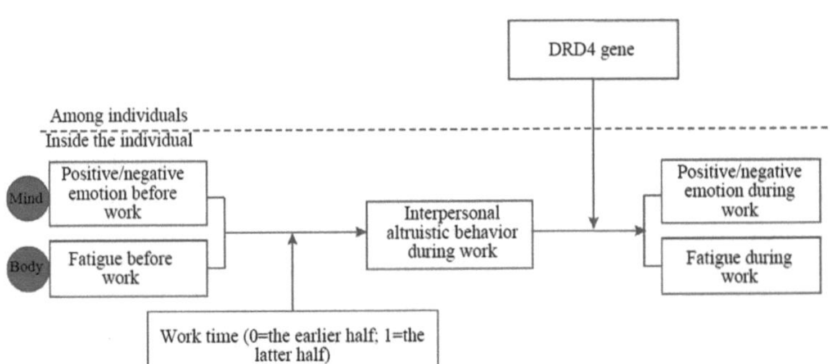

Fig. 26 Reciprocal model of physical and mental feelings and interpersonal altruistic behavior

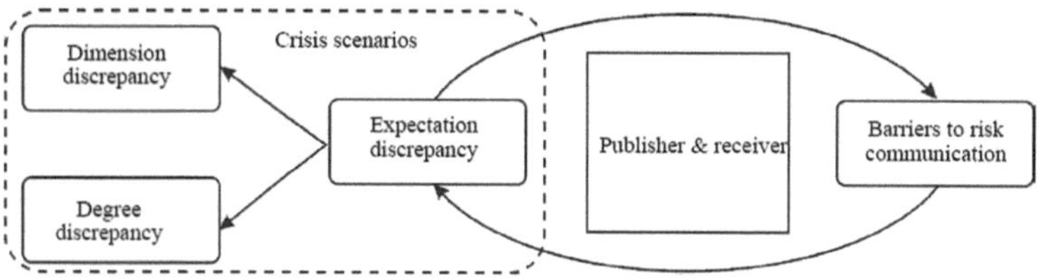

Fig. 27 The circular effect of expectation difference and risk communication and the amplification effect of crisis scenario

and form a vicious circle that are difficult to solve. The circular effect of expectation difference and risk communication and the amplification effect of crisis are shown in Fig. 27. The concrete path of expectation difference's effect is embodied in two aspects: the expectations of all parties in communication are not in the same dimension, and the expectations of all parties in the same dimension are in different degrees. The main factors causing different concerns include the role in crisis, the characteristics of information perception and knowledge, etc. Trust and psychological rights are the main factors that cause different expectations in the same dimension. Breaking the mindset and changing the individual's thinking mode is an effective way to eliminate the difference in expectations.

(2) **Losing the sense of control will change people's cognitive modes**

The research group of Zhou Xinyue from Sun Yat-sen University systematically proved that the loss of control feeling will lead to the disappearance of the differences between the eastern and western ways of thinking. When the sense of control is deprived for a short time, both eastern and western people tend to adopt analytical thinking; When the sense of control is deprived for a long time, East Asians will adopt a holistic way of thinking again. In addition, inducing East Asians to adopt analytical thinking will increase their sense of control. These findings not only show that losing control will lead to the change of cognitive model, but also show that cultural differences in crisis scenarios will not hinder the dialogue and communication between the two sides, which is an extension and supplement to Richard Nisbett's view that East Asians are not good at logical thinking. The related fruits were published in the *Journal of Personality and Social Psychology*. Long- and short-term threats that change individual thinking mode are shown in Fig. 28.

3.2.3 Perspectives of Decision-Maker and Executor

Psychological and behavioral research tries to propel decision-makers to make scientific decisions, so that executors have the ability and quality to cope with challenges and implement decision-makers' decisions. In order to achieve this goal, the research is carried out from five aspects: (i) In the aspect of monitoring and early warning of social conditions, the social unrest index of China from 1998 to 2012 is built

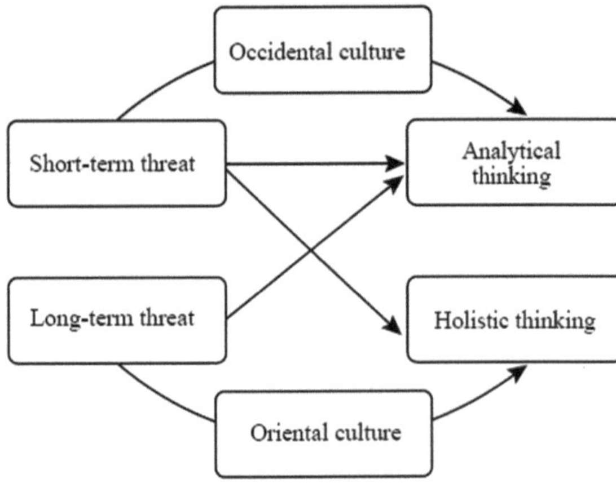

Fig. 28 Long- and short-term threats change individual thinking mode

up for the first time; (ii) In the pre-disaster simulation and training for the whole people, the research deeply used plasticity of cerebral cortex to improve individuals' disaster coping ability; (iii) In the aspect of post-disaster behavior intervention to individuals, biological hormones are used to help post-disaster individuals reconstruct their beliefs and reshape their mentality; (iv) In terms of the guidance for decision-making, the sudden emergence of cultural neuroscience provides a new perspective for dealing with cultural differences in cognitive processes; (v) In the aspect of executors, in order to better select and train executors, it is necessary to improve the theory of psychological and behavioral coupling, build a model of resilience, and optimize the selection methods and training system of executors.

According to the triangle theoretical model of public security, the research of psychology and behavior puts forward the theory of psychological and behavioral coupling. Emergency incidents correspond to risks, disaster-bearing bodies correspond to vulnerabilities, and emergency management corresponds to resilience. In terms of the response to sudden emergencies causing public crisis, the structure model and action mechanism of resilience of crisis responders and teams are revealed by combining qualitative research with empirical research. Through interviews on incidents and affected groups and large-scale data surveys, the five-factor structure model of resilience of crisis rescue workers (including rational coping, strong personality, positive emotions, self-efficacy and belief) and the four-factor structure model of resilience of crisis rescue teams (including group effectiveness, common belief, emotional support and team resilience) were obtained. On this basis, a questionnaire for measuring individual resilience and team resilience was developed, and the reliability and validity met the measurement standards. Through process analysis methods such as risk identification, vulnerability analysis and resilience assessment, the integrated model of objective environmental indicators and subjective psychological indicators is constructed, and the resilience model and related assessment questionnaires are improved. In terms of system integration, the idea of psychological and behavioral coupling model is put forward, which combines the psychological and behavioral data obtained from mining and analysis with the norm of psychological measurement and the causal relationship model obtained from psychological and behavioral experiments, so that the analysis function of the integrated system can be enhanced. And the following contents have been done: Provide information technology for process analysis, data mining and emergency management, design the scheme, compile the "evaluation system of rescue workers' psychological and behavioral model," and design a set of IT programming scheme; Completing the "mental security's training system" (demonstration of integration) for crisis response of coal mining enterprises, and incorporated the fruits into the training and screening standards of Chinese rescue workers.

3.3 Application and Popularization

The application results can be summarized as "four banks in one," that is, psychological and behavioral index bank, psychological-geographic information bank, knowledge bank, case bank and prevention and response system. These application results inject scientific connotation into the national emergency platform system and provide important reference information for emergency decision-making.

(1) **Psychological and behavioral index bank**

Combining the psychological theory, methodological progress and empirical research and practice results at home and abroad, Wang Lei's research group of Peking University compiled a tool library for measuring people's psychological and behavioral indicators under emergencies, and revised and collected 18 categories of psychological and behavioral indicators.

(2) **Psychological-geographic information bank**

A large number of studies have confirmed the important influence of natural and social environment on human cognitive patterns and psychological characteristics. According to incomplete statistics, there are 15 related articles published in Nature and Science in recent ten years. Among them, some studies have expounded that differences in geography will lead to differences in personality traits; Other researchers compared 49 cultures and found that the national personality in the form of prejudice helps to maintain national unity. It has been proved that GIS can improve the management level of crisis events.

Based on the collection and operation of basic data, Wang Lei's research group completed the study of geographical distribution based on multiple psychological indicators, combined the psychological characteristics of people in different regions with the geographical characteristics of the region, and obtained the typical level of psychological characteristics based on geographical integration and classification. Psychological indicators include openness, extroversion, emotional stability, agreeableness and sense of responsibility. The psychological and geographic information system (psychological GIS) constructed by this research group is the first in unconventional emergency management, which will provide more comprehensive psychological and geographic information for decision makers and enable managers to adopt more scientific and effective methods to deal with unconventional emergencies.

(3) **Knowledge bank**

Wang Lei's research group has collected more than 2,000 important psychological and management literatures related to emergency management, involving psychological assistance for unconventional incidents, emergency team management, emergency rescue procedures, post-disaster reconstruction methods and many other aspects. The team chose information according to the relationship between literature content and emergency management of unconventional emergencies, combined psychology theory and emergency management practice, and provided theoretical support. For example, through multiple retrievals, we can know the theories, knowledge, methods and technologies related to emergency management in time, so that decision makers can make scientific analysis and prediction based on previous theoretical and empirical studies.

(4) **Case bank**

The case bank compiled by Wang Lei's research group includes theoretical bank and case bank. Theory bank includes 74 core theories related to unconventional emergencies; The case bank includes five categories of cases: public security, medical and health care, food safety, natural disasters and man-made panic, involves the psychological and behavioral analysis of the crowd in three stages: before, during and after the case, and removed the gap between theoretical research and intervention practice. Taking Kunming Railway Station Terrorist Attack as an example, the monitoring and prediction of information dissemination in this area before the incident can provide reference for later monitoring and prediction. The psychological process of the affected individuals and the fermentation process of group behavior in the incident are the typical expression of emergency norm theory and irrational infection theory in psychology. By grasping relevant theories and summarizing practical cases, we can effectively predict and intervene emergencies.

For the first time, Wang Lei's research group tried to sort out the change rules of cognition, emotion and behavior at different time and event nodes from the perspective of emergency management process of unconventional emergencies. This paper analyzes and evaluates the response experience from the perspective of the response process of various typical unconventional events in the early stage and the present, and provides reference for emergency management platforms at all levels to help them make correct decisions and response measures.

(5) **Prevention mechanism and response system**

The research on the theory and method of emergency management risk is incomplete and unsystematic. In view of unconventional emergencies, it needs theoretical breakthrough, technological innovation and system reform to change from passive emergency response to risk prevention and response in advance and upgrade from traditional risk theory to new risk theory.

Liu Xia's research group of Shanghai Jiao Tong University established an open source database of international terrorism in China. Docking the data of 3S technology (namely geographic information system/GIS, global positioning system/GPS, satellite remote sensing system/RS) in real time, and constructing a quantitative description of the vulnerability of public facilities "community-service-infrastructure" for terrorist attacks; Using multicollinearity method, Cronbach reliability analysis method and principal component analysis method, the structural factors and functional factors are extracted, and the structure–function analysis framework of grid "region-population-infrastructure" resilience is established. Related research has developed the risk theory system, and established the methods of risk quantitative description, risk control and risk preparation, which has methodological value.

3.4 International Comparison and Influence

With the concentrated support of this major research program, through the research with relatively short time, concentrated teams, strong financial support and clear objectives, the research on psychology and behavior of emergency management has made quite gratifying achievements, with a large output of academic achievements and a batch of high-quality achievements. Papers published in nature human behavior, PNAS, annual review of psychology, brain and human behavior, organizational behavior and human decision processes, Journal of behavioral decision making, personality and social psychology bulletin, biological psychology and other top international academic journals. At the same time, a team cluster involved in emergency management psychology and behavior research has been formed, and new theories, knowledge and tools have been put forward and developed, which has been recognized and actively paid attention by international counterparts, initially establishing China's influence in this field, making China's emergency management psychology and behavior research rank among the top in the world, and providing a scientific basis for building a safe society, a healthy society and a resilient society.

The research fruits in this field have also been interviewed and reported by many news media, such as The New York Times, The New Yorker, The Wall Street Journal, Atlantic Monthly and so on. In addition, the research team was invited to write chapters for books such as *handbook of personal security, encyclopedia of industrial and organizational psychology, encyclopedia of personality and individual differences.* This indicates that Chinese scholars have been recognized by their international counterparts in related fields.

(1) **Risk decision, vulnerability analysis and management**

The research on the methods of risks' quantitative description, control and preparation has a certain international influence, and relevant research groups were invited to give special reports in well-known universities such as the University of South Carolina and strategic research institutions such as RAND Corporation. In addition, Liu Xia's research group of Shanghai Jiao Tong University was invited to the Institute of Natural Disasters and Vulnerability of the University of South Carolina, the Terrorism Research and Terrorism Early Warning Center of the US Department of Homeland Security at the University of Maryland, and the Disaster Research Center of the University of Delaware.

(2) **Crisis management and post-disaster reconstruction**

Zhang Kan's research group of Institute of Psychology, Chinese Academy of Sciences actively carried out international academic exchanges and cooperation, and invited many famous international scholars in this field (including many disaster and crisis management experts from Southeast Asia, Africa and other countries and regions) to exchange ideas, and also attracted many professionals with practical experience who are active in the fields of humanitarian assistance, post-war recovery and development. International first-class experts and scholars spoke highly of the research work of the research team.

(3) **Study on the interaction effect between body and mind**

Outstanding fruits have been published in top journals in the field of management and decision-making (such as *organizational behavior and human decision processes, journal of behavioral decision making, personality and social psychology bulletin*), and the fruits have been widely recognized and paid attention by domestic and foreign counterparts. For example, in 2011 and 2013, Professor Wang Lei was invited by the Asian Social Psychology Conference to give a special lecture; In 2014, Professor Xie Xiaofei was invited by the Ninth International Congress of Surveying (Spain) and the International Congress of Applied Psychology (France) to give a special lecture.

Prospect

Weicheng Fan, Chunchang Shan, Shouyang Wang, Tiemin Liu, Feiyue Wang, Xiaofeng Meng, Lei Wang, Xiaolin Zhou, Wenguo Weng, and Yi Liu

1 Strategic Needs in the Field of Public Security in China

At present, China has frequent public security incidents. The complexity of public security problems intensifies, the potential risks and new hidden dangers increase, and the difficulty of emergencies' prevention, control, disposal and rescue continues to increase. The task of maintaining public security is important and arduous. Although the overall level of the four core technologies of China's public security, namely risk assessment and prevention, monitoring, forecasting and early warning, emergency response and rescue, and comprehensive support, has already entered the international advanced groups, surpassing the average level of developing countries, and the overall gap between China and the international leading level is narrowing, but it

W. Fan (✉)
Tsinghua University, Beijing, China
e-mail: wfan@tsinghua.edu.cn

C. Shan
Counsellor's Office of the State Council, Beijing, China

S. Wang
Academy of Mathematics and Systems Science, Chinese Academy of Sciences, Beijing, China

T. Liu
China Academy of Safety Science and Technology, Beijing, China

F. Wang
Institute of Automation, Chinese Academy of Sciences, Beijing, China

X. Meng
Renmin University of China, Beijing, China

L. Wang · X. Zhou
Peking University, Beijing, China

W. Weng · Y. Liu
Institute of Public Safety Research, Tsinghua University, Beijing, China

W. Fan (ed.), *Unconventional Emergency Management Research*, Reports of China's Basic Research, https://doi.org/10.1007/978-981-99-6798-8_5

still relatively lags behind about ten years. Compared with the international leading level, China's public security technology level has formed the basic pattern of "partially advanced, overall backward," and most parts of China's relevant technologies are behind the advanced level. Compared with leading countries, China's ability to transform basic research fruits into superior technologies is weak, and in technological competition China is still at a disadvantage position; The modern scientific and technological innovation system that leads and supports the national public security's governance system and governance capacity needs to be improved.

The destruction scale and complexity of all kinds of emergencies in the world are becoming more and more serious, and disasters are developing from single disasters to integrated disasters. Its development has brought great challenges to the risks' assessment, prevention, monitoring, forecasting and early warning, emergency response and rescue and comprehensive support under emergency situation. Due to the coupling of multiple disaster-causing factors, secondary and derivative disasters, complex disaster-causing process and mechanism, and the constraints of many factors, poor process reproducibility and difficult quantification of indicators, China's risk assessment still stays at the qualitative or semi-quantitative level, which is difficult to meet the future needs of quantitative, comprehensive, systematic, cross-industry and cross-field risk assessment. In order to realize the risk management goals of "all incidents can be avoided" and "all risks can be controlled," and to reduce the system risks to the minimum, it is necessary to carry out qualitative, semi-quantitative and quantitative risk assessment methods. It includes the following aspects: risk assessment methods based on index system, historical disasters' probability statistics and real scenarios' simulation or construction, carrying out comprehensive and systematic risk assessment based on multiple disasters to realize comprehensive assessment of potential risks in specific areas, carrying out multi-disaster coupling and disaster process' simulation and scenario construction based on different scales and multi-physical fields, and realizing the whole chain risks' assessment and management crossing different fields and industries.

Facing national strategic deployment, especially the proposal and promotion of "the belt and road initiative" and "xiong'an new area," the problems of social security management, emergency resources' sharing and coordination guarantee, Internet of Things infrastructure's destruction, network attack and privacy protection, and urban lifeline's monitoring and maintenance are becoming more and more prominent. At the same time, the risks of terrorism, epidemic spread, biological invasion, complicated transportation and logistics and public opinion spread between regions and countries require public security field's emergency management to be more rapid and efficient. The trend of China's social and economic development shows that cities are developing towards internationalization, virtualization, flattening, informatization and networking, and this trend can propel the change and development of emergency management mode. Information ubiquity, prominent individual role, social structure's change and other factors constantly propel emergency management to change from one-way management to two-way interaction, from offline management to online and offline integration, and from government supervision to individual and social collaborative governance.

Future public security should be prepared more comprehensively, predicted more accurately, responded more scientifically and recovered more quickly. At the same time, the development of public security science and technology should pay more attention to prevention, response and resilience, and propel the development of public security guarantee towards controllable risk, intelligent prediction, efficient response and integrated guarantee.

(1) **Emphasis on identification and evaluation of conventional and unknown risks**

The identification and evaluation of conventional and unknown risks are the premise and foundation for realizing the development concept and idea of proactive management and advanced implementation of public security's risk prevention. In order to achieve this goal, we should start from two aspects. (i) Risk identification and assessment is an important premise and foundation for realizing proactive management of public security with prevention first and key nodes moving forward. As for the known and predictable risks, we should make a comprehensive analysis from three aspects: emergency, disaster-bearing body and coping ability, promote the quantification and standardization of risk assessment, and reduce the danger and potential disaster-causing degree of known disasters or risks to a minimum. (ii) As for the revolutionary development of new materials, new processes and new technologies in the future, we should build risk prediction tools based on technologies such as Internet of Things, big data, cloud computing, artificial intelligence, simulation and scenario deduction, introduce risk assessment before industrial planning, process design, policy release and technology's application, and consider the unknown and fuzzy risks generated by their collision and integration with human beings, environment and society. At the same time, we need to carry out pre-risk assessment and preventive management with subversive ideas to ensure that new industries, new materials, new policies and new technologies will not bring unbearable or unknown risks and disasters to mankind. We also need to predict the unknown risks and study and pay attention to the previous incidents before the disaster, so as to build the comprehensive control ability of conventional risks and unknown risks.

(2) **Changes from monitoring and early warning to the integration of proactive perception, intelligent prediction and early warning**

Informatization, networking and intelligence have become the main characteristics of the development of international public security's monitoring, forecasting and early warning technology. The rapid development of integrated information technologies such as remote sensing satellite, network information technology, mobile communication and big data analysis provides new technical methods for monitoring and forecasting natural disasters such as meteorology, floods and droughts, earthquakes and tsunamis, as well as social security issues such as public health, hazardous chemicals, transportation, anti-terrorism and public opinion. The focus of monitoring and forecasting has started to extend from the occurrence stage of security incidents to the whole process of gestation and recovery afterwards, and it should be a normal state to carry out systematic monitoring and early warning in cooperation with many industries such as transportation, water conservancy, communication, health and public opinion. The connotation and related fields of public security monitoring and early warning are constantly expanding, and the related methods are constantly enriched. From the perspective of monitoring and early warning methods, with the rapid development of modern information technology, intelligent and networked monitoring and early warning technology and information release technology have been continuously developed and improved. The monitoring and early warning technology of public security will develop toward the emergency linkage of automatic monitoring, identification and early warning with the support of information technology; In terms of the development trend of engineering science and technology, the important issues of public security's monitoring and early warning in recent years are the application of public security's monitoring data, the development of monitoring, evaluation and early warning mode (identification and early warning of emergencies' causing factors) and the construction of early warning and emergency linkage mode. The improvement of data analysis and computing technology propels the integration of public security's monitoring, forecasting and early warning to proactive perception, intelligent forecasting and intelligent early warning. Early warning and emergency linkage system based on big data analysis and large-scale rapid calculation has become an international trend.

(3) **Development of emergency platform and emergency equipment towards intelligence and automation**

Through continuous efforts during the Eleventh Five-Year Plan, Twelfth Five-Year Plan and Thirteenth Five-Year Plan, China has gradually established a national emergency platform with the emergency platform of the State Council as the center and provincial emergency platforms and emergency platforms of other departments as the structure. It integrates sensing, monitoring, communication, transmission, forecasting, coordination, command and dispatch, realizes decision-making based on data, risk and forecasting, and provides a powerful guarantee for emergency management of governments at all levels and large enterprises. In order to meet the great demands of artificial intelligence, Internet of Everything, ubiquitous information and the implementation of national strategy for public security, we need to build a new public security emergency platform integrating intelligent hardware, platform, software and think tank, and establish automatic, intelligent and integrated large-scale or special emergency equipment for emergency rescue. We need to establish a multi-level, multi-dimensional, high cohesion, low coupling, self-adaptive, self-learning, safe and reliable system of platform software. We need to establish a scientific, professional, systematic and forward-looking public security think tank to scientifically predict, warn and predict the medium and long-term situation. This is to conduct a pilot analysis of possible crises and impacts, and to provide strategies for the overall situation, direction and victory of public security.

(4) **Intelligent and emerging technologies propel the systematic and synergistic development of public security and emergency management**

With the progress of national key strategies, especially the "the belt and road initiative" initiative, the internationalization and regionalization of security issues have become increasingly obvious. First of all, foreign security threats have become more serious due to interconnection, such as international terrorism and political separatist forces, which have become severe challenges to China's national security. Secondly, terrorism is intertwined with ethnic, religious and territorial disputes, and it makes national defense, border situation's control and disasters' prevention encounter various risks. In view of this, new concepts such as common security and cooperative security have increasingly become the main theme of today's security concept worldwide.

With the development of new technologies such as artificial intelligence, intelligent terminals in the future will create a panoramic digital life with the Internet of Everything, man–machine integration, and realize seamless interaction and comprehensive emotional sensation among people, machines and cloud. The focus of public security's needs has changed from establishing and having access to voice communication to obtaining various forms of information and services in real time; Information flow and unlimited bandwidth will achieve a high degree of coordination among government, society and the public, and guide public security to innovation, coordinated development, integration and open governance. By integrating collaborative organizations and platforms, public security's daily information monitoring, analysis and early warning platform, information release and crisis handling's command platform, platform of key areas' management and control based on Internet of Things, platform of recovery and development evaluation, and public security's education, training and public participation platform, we can break the traditional single-line and vertical contact mode of the original public security's participants, thus forming a brand-new mode of networked public security service and finally realizing the systematicness and synergy of public security.

(5) **Handling of emergencies has developed from simple mode to diversification, coordination and high efficiency**

With the development of globalization and regional cooperation, the trend of internationalization and regionalization of public security has become increasingly obvious. Facing the national key strategic deployment, especially the proposal and development of the "the belt and road initiative," the trend of internationalization and regionalization of security has become increasingly obvious. Thus, we need to establish a three-dimensional guarantee mechanism for the coupling of national security and public security along the "the belt and road initiative," coordinate multiple forces to promote the deep cooperation of countries along the "the belt and road initiative" in the information field, form a community with common interests, responsibilities and destiny, and realize the deep sharing of resources. We also need to study and establish an information strategy database, an emergency resource database, and a fast emergency response and rescue platform in the field of national security and public security for countries along the "the belt and road initiative." With the deployment of the "the belt and road initiative" initiative as a breakthrough and demonstration, systematic risks' identification, analysis and early warning technologies for major national strategies will be formed, and an implementation mode with technology and policy's coupling and optimization will be formed to ensure the efficient implementation of national strategies.

The development of new technologies, such as artificial intelligence, the Internet of Everything, man–machine integration, information ubiquity and unlimited bandwidth provide effective technical methods to achieve high degree of collaboration among governments, society, public, communities, cities, countries and regions. The development of emergency response technologies will propel the formation of a networked emergency response system with high efficiency. The acquisition, mining, identification and analysis of multi-source and heterogeneous public security's big data and the coupling experiment and simulation of multi-disasters in cities and towns will effectively propel the rapid improvement of China's comprehensive public security's guarantee system and maximize the security of people's lives and property.

(6) **Security and resilience cities become development trends based on the concept of life cycle**

Based on the triangular framework of public security science and technology (emergency, disaster-bearing bodies and emergency management) and the overall view of urban complex giant system, the whole life cycle concept is adopted to manage and control urban hazardous chemicals, important energy sources and key facilities, so as to ensure the sustained, stable and safe operation of urban hazardous chemicals, important energy sources and key facilities, realize the related diagnosis, automatic response, independent protection, proactive service and rapid recovery in the field of public security, and comprehensively enhance the resilience and security maintaining capability of cities. We should apply the Internet of Things and Internet technologies to deeply integrate the public security's emergency management process (risk assessment and prevention, monitoring and early warning, emergency response and rescue, and comprehensive security) with the whole life cycle management (production, import, storage, transportation, sales, use and abandonment) to ensure the continuous, stable and safe operation of urban hazardous chemicals, important energy sources and key facilities. We should realize the accurate monitoring, prediction and early warning of the potential disaster-causing factors and operating conditions of urban hazardous chemicals, important energy sources and key facilities. We should ensure the rapid monitoring, inspection and rescue of dangerous elements at the incident site, as well as the post-incident investigation and incident recurrence. We should widely adopt technologies such as Internet of Things, large-scale computing, artificial intelligence, data mining and virtual reality, integrate multi-industry resources, and realize diversified, intelligent and integrated information acquisition, information sharing, joint operation monitoring, joint prevention and control, intelligent decision-making, efficient response, accurate post-incident inspection and incident reconstruction and other measures to comprehensively enhance the integrated security maintaining capability in the field of public security.

(7) **Public safety and emergency management change towards deep integration, mutual sharing and universal application for multiple industries**

With the rapid development of network information technology, "Internet plus" propels China to break the "isolated information island" and "data chimney," and accelerates the deep integration of Internet and various industries. With the rapid advancement of informatization and internationalization, new technologies such as Internet of Things, big data and cloud computing have propelled the development of science and technology in the field of public security, but they have also caused new risks and hidden dangers, and brought new challenges to public security science. In this context, it is difficult to meet the security needs under the situation of cross-border integration. By widely adopting technologies such as Internet of Things, cloud computing, artificial intelligence, data mining and virtual reality, and integrating the resources of various industries, we can realize the real-time acquisition and sharing of diversified, intelligent and integrated information on life, transportation, health, medical care and disasters. We can realize emergency linkage and prevention and control in traffic security, hazardous chemicals, water security and power grid security, and finally realize the comprehensive integration of industry resources such as transportation, power, water conservancy, communication, hazardous chemicals, lifeline and public opinion, the sharing of industry information and the deep application of industry insurance.

2 Assumptions and Suggestions for In-Depth Research

2.1 Assumption of In-Depth Study in the Field of Public Security's Emergency Management

In China's public security field, it is necessary to strengthen the construction of security resilience from individual to national level, adhere to the problem orientation and demand orientation, take "Xiong'an New Area" and "The Belt and Road Initiative" as application fields, and strive to solve the emergency problems influencing the development of megacities, and solve the problems of collaborative emergency management among countries and regions. China should focus on transforming the "unified-diversified-systematic" emergency management model into an intelligent and all-round public security net, and promote emergency management from incident-driven to data-driven, from object-centered to human-centered, from subjective decision-making to decision-making with emergency computing's support, and from passive emergency management to proactive response.

China should innovate the connotation of emergency management's system and mode, change the trigger mechanism of emergency management from the traditional emergency-driven to (big) data-driven and push the risk assessment to a new stage from the occurrence of incidents. China should pay attention to the complexity of emergency management, regard the grass-roots level as the main battlefield of public security, and change the core objects of emergency management from the occurrence and development process of objective incidents to individuals, highlighting the role of individuals in emergency management. China should adapt to the virtual characteristics of the network information age, and make use of Internet, cloud computing, Internet of Things and other technologies to transform emergency decision-making from government decision-making to qualitative and quantitative decision-making based on data support and computing support. At the same time, China should change the traditional comprehensive guarantee of public security from focusing on post-disaster relief to focusing on the active emergency guarantee mechanism combining prevention, disaster relief and rescue.

Since the 21st century, the new machine learning technology has made rapid progress, and its computing ability in dealing with specific fields and tasks has been gradually enhanced. In the future, we will need to build resilient cities and a proactive and intelligent public security net, so that it can basically maintain the same functions, systems and identities when it encounters the impact and pressure from future social, economic, technological systems and infrastructure. At the same time, we need to develop a set of theories, methods and systems of intelligent emergency computing that support resilient societies.

A resilient city supported by intelligent emergency computing system needs to be based on a public security net that connects all things and integrates individuals, facilities and information. It also requires us to design a proactive perception mechanism based on edge computing, a disasters' early warning mechanism based on data assimilation, an infrastructure's design mechanism based on observation, an information collection mechanism based on the Internet of Everything, and a structural invulnerability mechanism based on software. Improving the disasters' perception, prediction and anti-attack ability of infrastructure and strengthening the individual's proactive disaster bearing ability can strengthen the resilience of the subjects. Using the invulnerability of facilities and equipment connection, the security of information flow itself, the diversity of connection brought by the Internet of Everything and the proactive connection brought by edge computing, we can realize the city's connection resilience. Ensuring the functional continuity of laws, regulations and organizational structure can improve the robustness of public security network's structure, achieve structural resilience, and build intelligent cities with resilience of subject, resilience of connection and resilience of structure.

We need to develop an intelligent emergency computing system and establish a proactive intelligent public security net. We should combine cutting-edge technologies such as Internet of Things, the fifth generation mobile communication technology, big data and edge computing, and adopt software to define public security networks. At the same time, we should make emergency computing have the comprehensive application ability for multiple disasters and deep integration of multiple fields, so as to realize the online and offline two-way monitoring and management of the perfect combination of infrastructure and data and software services. We should also build an integrated platform for public security's comprehensive guarantee from two dimensions: physical information society and network information society, and realize individual protection, ensure the security of key facilities, key places, communities and mega-city circle from bottom to top.

The development of intelligent emergency computing system should achieve the following three goals:

(1) **Realizing the information ubiquity and the resilient society's structure in the form of the Internet of Everything**

Unlimited data, unlimited storage and unlimited bandwidth will be the trend. Its specific features include the following aspects: Internet of Everything and human–machine interaction, highly integrated and personalized data acquisition-processing-analysis, highly interactive and collaborative government-institution-public, and flat decision-making structure and process. These contents ensure the strength of a resilient society in connection and structure.

(2) **Resilient society's deduction to realize the deep integration of "data-calculation-reasoning"**

The future deduction theory and method based on the deep integration of "data-computing-reasoning" can realize the monitoring, forecasting and early warning of key places and facilities, important energy and reserve resources, major projects throughout the life cycles; It can predict the impact of the new technological revolution and impact of global pattern's changes on the public security in the future; It can propel the construction of the future public security's response system, and ensure the robustness of a resilient society by means of deduction and technology.

(3) **Realizing a prediction-oriented resilient society that can actively assess unknown risks**

Based on the technologies of Internet of Everything, Edge Computing, Data Assimilation, Software Defined Society, etc., the risk prediction tools can analyze whether there are potential risks that are unbearable or unrecognized risks hidden in new technologies, new materials, new industries, new policies, and risk prediction tools can also help predict unknown risks and study disasters' omen events, and achieve the goal of proactive prediction for a resilient society.

Under the support of resilient urban emergencies' theory system, intelligent emergency computing system and proactive and intelligent public security net, we should improve the emergency management infrastructure of the Internet of Everything at individual and community levels; We should establish an international emergency standard system based on the three-dimensional network of public security; We should break through the emergency management mode of "One Case, Three Systems" and establish an emergency management system for megacities; We should design a collaborative emergency management and sharing mode that crosses regions, borders and departments, build an integrated platform for public security's comprehensive guarantee, and realize proactive public security's guarantee based on deep integration of multi-industry resources and integrated application of multi-field technologies.

2.2 Suggestions on Future Basic Research in the Field of Public Security

(1) **Comprehensive perception, efficient prediction, intelligent decision-making, and active security technology in the context of the Internet of Everything**

In the future, cities will develop in the direction of real-time perception of cities, integrated space, dynamic expression, scientific decision-making, intelligent management and safe operation. The network formed by the security infrastructure in the city will fully realize the unification and integration of the three-dimensional space of people, things and information in the city, and realize the overall perception of urban things. On the basis of comprehensive perception, we should study and comprehensively utilize modern information technologies such as the Internet, Internet of Things, big data, cloud computing, and spatial geographic information; We should study the optimal layout of sensor network for infrastructure and monitoring of engineering building, and study the data's collection model and optimization method of social public opinion information; We should study the data mining for real-time and long-period monitoring, the early detection of abnormal signs and the prediction and early warning technology for emergencies; We should combine information transmission and social interaction data to conduct deduction research on cross-disciplinary emergencies; We should study the early detection method of abnormal signs for urban operation safety, and build a "smart security city" model based on monitoring data, integrating physical model, information transmission and social interaction; We should study the whole cycle and whole chain risk assessment and prevention system, all-weather detection system for all kinds of disasters, big data analysis and large-scale rapid computing system for public security, deeply integrated emergency decision-making, rescue and handling system and collaborative linkage system, and we should establish an integrated platform for comprehensive security of public security to realize comprehensive perception, efficient prediction, intelligent decision-making and proactive security response. At the same time, we should realize the "Resilience Project" to increase security and resilience of individuals and the country.

(2) **Theory and method of large-scale fast computing based on big data for complex security problems**

The theory and method of large-scale rapid computing is an important foundation to support risk assessment, prediction and early warning and emergency decision-making of complex security problems. The development of big data, high-performance computing, cloud computing and edge computing provides a good foundation and opportunity for the research of large-scale fast computing for complex security problems. We should focus on the following aspects: Large-scale rapid simulation method for studying the dynamic evolution process of public areas and large-scale rapid analysis technology for disasters' evolution process driven by model and data; studying the multi-parameter and multi-dimensional matching between the simulation result of urban disaster process and the on-site data, and the dynamic correction method of disasters' evolution process; Developing a large-scale rapid simulation method for urban regional disasters' evolution driven by model and data; Studying the theory and method of emergency perception and early warning based on big data; Studying the fast solution method and computing theory of large-scale complex security issues based on high-performance computing; Studying fast 3D modeling method based on large-scale spatio-temporal data with multiple scales; Studying the theory and computing method of human–machine interactive simulation and modeling based on cloud computing and edge computing; Studying the coupling solution and calculation method of multiple physical problems in emergencies; Studying the integrated computing theory and method of physics-information-society integration.

(3) **The key technology system and resilience guarantee platform of resilience city in the field of public security**

With the acceleration of urbanization, big cities, megacities and urban agglomerations appear constantly, and the situation of urban public security is serious, and the fragility of urban buildings, roads, lifeline networks and places where people gather increases. We must build resilient cities as soon as possible and improve the city's ability to resist major emergencies. We need to focus on the following aspects: Studying the comprehensive evaluation technology of resilience city; Studying the collapse mechanism of urban high-rise buildings, large comprehensive facilities, construction scaffolding and other buildings and structures, as well as the security resilience's evaluation and optimization technology of urban buildings and important infrastructure; Studying the security monitoring and disaster prevention system of urban transportation and rails' operation and the security's operation guarantee and risks' management and control technology of urban subway; Studying the security's monitoring technology of important lifeline systems (water supply and drainage, gas, heat supply, bridges, etc.) and the security's monitoring, detection and early warning technologies and equipment of utility tunnel in urban municipal pipe network; In view of the public risks in crowded gathering areas with cities (towns) as the carriers, the behavior database conforming to the characteristics of all kinds of people in China is established, the behavior characteristics of people in crowded areas and the cause deduction mechanism of crowd trampling incidents are studied, and intelligent monitoring, early warning and analysis techniques are studied; Studying the resilience planning technology of urban integrated public security. We need to select typical cities, research and develop urban security's management and resilience guarantee platform, carry out application demonstration, and enhance urban public security's guarantee ability.

(4) **Large simulator in the field of public security for multi-circle coupling, multi-field integration and cross-regional cooperation**

The purposes of this large simulator are as follows: to study the simulation technology that comprehensively considers the interaction and complexity of atmosphere, ocean, biology and solid earth, and to realize the simulation and expression of the interaction coupling mechanism of various circles and physical and biochemical processes in the earth system; to study the risk assessment technology of comprehensive index system, data statistics, scenario evolution and other methods to realize the whole life cycle management and whole chain's risk assessment of hazardous chemicals, key facilities (hazardous chemicals warehouse, urban pipe network, deep-sea pipeline, transportation hub, etc.) and important energy (electricity, oil, gas, hydrogen); to study early warning of risks and evacuation technology of dense crowd, large-scale traffic evacuation's simulation technology and emergency traffic assessment technology, and build a regional evacuation and refuge system; to study the perception, reproduction, simulation and deduction technology of major disaster scenarios, realize the effective superposition of digital information and physical information, and realize the simulation and scenario deduction for complex disaster scenarios and their development and evolution process; to study the deduction technology of public opinion's transmission on Internet and the simulation and deduction technology for special individuals' identification, tracking and tracing process for anti-terrorism, and realize the incident simulation and deduction with online-offline integration; to study the simulation technology of public security issues related to various industries such as power system and water conservancy system, and realize the deduction of multiple incidents and incident chains; to build a large comprehensive simulator for the simulation of earth system and emergency evolution with multi-circle coupling and multi-field integration.

(5) **Resilience theory of public security and evaluation method of emerging comprehensive risks**

Resilience in public security refers to the ability to bear the impact and pressure of unexpected incidents on social, economic, technological systems and infrastructure but still maintain the basic structure and function under the undermining of unexpected incidents. Resilience theory is useful in enhancing the ability to resist disasters and recover after disasters, and improving the ability to respond to emergencies, so that the system, community or society exposed to hazards can resist, absorb and adapt to the impact of disasters and recover after disasters timely and effectively by protecting and restoring important basic structures and functions. It mainly includes studying the connotation, composition and measurement method of resilient city, studying the interaction among urban systems and analyzing the characteristics of dynamic uncertainty, studying the complex self-adaption's evolution mechanism, robustness analysis and control methods of urban systems and emergencies; studying the assessment of needs, optimal allocation and whole-process management methods of emergency resources and emergency capabilities. At the same time, the development of emerging technologies such as artificial intelligence, additive manufacturing, new modes of transportation, and the Internet of Everything not only provide a lot of convenience and advantages for human, but also bring new security risks to public security. Meanwhile, economic globalization, social aging, the intensity and frequency of natural disasters increased significantly, terrorism and other new situations have caused new challenges and made new requirements for future comprehensive risks' assessment in public security field. For example, it is urgent to research risk assessment methods based on technologies such as Internet of Things, big data, cloud computing, artificial intelligence, simulation, and scenario deduction, study the reasoning method of complex multi-case analysis of public security, study the methods of emergency knowledge's management, deduction and evaluation based on historical cases, study the mechanism and risk assessment method of regional multi-disaster coupling and secondary derivation and research risks' perception and prevention and control methods based on big data.

(6) **Theory and method of emergency management, prevention and control for whole lifecycle and safe operation and integrated prevention and control**

Whole life-cycle security is one of the important concepts of public security. Large cities, chemical parks, important powers, lifeline systems, traffic, water areas and water environment, power equipment and power grids are the focus of public security, so we need to systematically study it from the perspective of whole life-cycle security. It includes the following research contents: Studying the operation, monitoring and early warning methods of lifeline system in megacities; Studying the multi-source data's mining and anomaly's identification method of "people-facilities-environment" in crowded places; Studying traffic system's risk characteristics, large-scale traffic evacuation's simulation method and strategy optimization; Studying the risk identification and quantitative analysis methods of rail, aviation and water transportation networks; Studying the method of whole process risk analysis and the whole lifecycle safe operation's theory in the field of dangerous chemicals' production, transportation, storage and use; Studying urban flood's simulation and disasters' rapid assessment based on Urban Internet of Things Rain and Flood Perception; Studying the evolution process and coping measures of water security in changeable environment; Studying the security characteristics and defense system of power grid's stable operation; Studying anti-terrorism and risks' perception, prevention and control methods based on multi-source heterogeneous data's integration; Studying the theory of resilient city under the new urban structure and productivity layout; Studying the emergency management process, mechanism and method under the blending of flat management, ubiquitous information and prominent individual role; Studying the quantitative management theory and method supported by big data and computing; Studying the theory and method of collaborative emergency response based on software services; Studying the methods of situation's deduction and decision-making's support by integration of "data-computing-deduction"; Studying the new emergency management theory and method from passive rescue to active prevention before disaster, and from single disaster to various disasters.

(7) **Principles and methods of active safety for important infrastructure and efficient and accurate emergency technology and equipment**

Important infrastructure has important basic functions for social and economic development and long-term, reliable, active and safe operation of infrastructure is an important foundation for ensuring public security and even national security. The development of new technologies such as artificial intelligence and additive manufacturing provides powerful technical support for the efficiency and accuracy of infrastructure and emergency rescue. Important infrastructure and emergency rescue technology and equipment need to develop to meet the needs of new situation. There are many aspects: Studying the mechanism and risk assessment methods of extreme meteorological conditions causing damage to infrastructure such as power grid, communication and transportation; Studying the robustness and vulnerability analysis theory of urban subway network under emergency, the principle of human–robot interactive sharing and proactive perception technology and the principle of robots' group intelligence under emergency; Studying the traffic capacity, emergency evacuation strategy and system loss and recovery model of underground space such as subway; Studying the mechanism of infrastructure security under multi-field network-physical joint attack; Studying the complex self-adaption's evolution mechanism, system security modeling and prevention and control strategy for important infrastructure; Studying the proactive security method for important infrastructure based on resilience principle; Studying the integration and fast information processing method of multi-sensor system based on unmanned platform and accurate acquisition and analysis method of on-site information based on multi-source sensors' information and data; Studying the dynamic deduction, prediction and comprehensive judgment methods of "data-model-case" coupling and "on-site and simulation" coupling; Studying the technical principle of highly integrated wearable micro sensors for the detection of hazardous substance and the technical principle of multi-functional emergency rescue equipment.

(8) **Basic scientific issues of decision-making system of national security management**

In recent years, with the change of international pattern and the continuous enhancement of China's comprehensive national strength, China's national security situation has shown an overall improvement momentum. However, under the new situation, the threats to China's national security are also characterized by complexity and pluralism: external security problems and internal stability problems are intertwined, traditional security and non-traditional security challenges overlap, and strategic decisions of national security coexist with decisions of major incidents. National security is significant and urgent, and requires research on the decision-making system of national security management. The decision-making system of national security management contains profound and complex issues of science and technology, policy choice and institutional mechanism. The research is open, and there is a huge space for innovation, which requires deep integration of multiple disciplines and provides bright future for innovative research of national security management. The research on the decision-making system for national security management is also an empirical study of the decision-making theory and method in the direction of national security's management science, and we should improve and enrich the management discipline and propel its development. The basic scientific issues of the decision-making system of national security management involve the following aspects: Management mechanism for major national security incidents, especially the issues of multi-department coordination, multi-level cooperation, cross-regional connection, government and society coordination, information sharing and so on;

The methods of national security big data's comprehensive integration and analysis to provide information support for national security incidents' management; The theory and method for national security risks' management and comprehensive judgment to study security risks' identification, trend judgment and threat prediction at home and abroad, and provide trend judgment for national security incidents' management; The theory and method of intelligent prevention and control in national security to study proactive intelligent prevention and control and crisis response methods of national security, and to establish an intelligent decision-making platform for national security incidents' prevention and control and crisis response; The design of the decision-making system of national security management to carry out forward-looking cross-cutting innovation and research in the field of national security, provide the basic theory for the decision-making system of national security and provide decision-making reference for macro-management of national security.

Bibliography

1. 陈安，陈宁，武艳南，等. 现代应急管理技术与系统 [M]. 北京:科学出版社, 2011. Chen A, Chen N, Wu YN et al (2011) Modern Emergency management technology and system. Science China Press, Beijing
2. 陈达强. 基于应急系统特性分析的应急物资分配优化决策模型研究 [M]. 杭州:浙江工商大学出版社, 2010. Chen DQ (2010) Decision-making model research on emergency resources distribution based on a feature analysis of emergency system. Zhejiang Gongshang University Press, Hangzhou
3. 陈玲丽. 个体主义—集体主义的结构与跨文化研究 [M]. 北京:中国社会科学出版社, 2013. Chen LL (2013) The Structure of Individualism-Collectivism and Cross-Cultural Comparison. China Social Sciences Press, Beijing
4. 陈炘钧,等. 传染病信息学:公共卫生症候群监测与生物防御 [M]. 宋宏彬,王立贵,曹志冬,等译. 北京:科学出版社, 2011. Chen H, Zeng D, Yan P (2011) Infectious disease informatics: syndromic surveillance for public health and bio-defense (Song HB, Wang LG, Cao ZD et al trans). Science China Press, Beijing
5. 陈雪龙. 非常规突发事件情景的构建与动态推演方法 [M]. 天津:天津大学出版社, 2013. Chen XL (2013) An Approach to scenario construction and dynamic scenario deduction of unconventional emergency. Tianjin University Press, Tianjin
6. 陈雪龙. 面向复杂决策问题的模型构造与管理方法 [M]. 天津:天津大学出版社, 2013. Chen XL (2013) Construction and management method of model orienting complex decision-making problem. Tianjin University Press, Tianjin
7. 迟菲,胡成,李凤,等. 密集人群流动规律与模拟技术 [M]. 北京:化学工业出版社, 2012. Chi F, Hu C, Li F et al (2012) Crowd flow pattern and technique of simulation. Chemical Industry Press, Beijing
8. 邓青,马晔风,刘艺,等. 基于BP神经网络的微博转发量的预测 [J]. 清华大学学报 (自然科学版), 2015, 55 (12):1342–1347. Deng Q, Ma YF, Liu Y et al (2015) Prediction of retweet counts by a back propapagtion neural network. J Tsinghua Univ (Sci Technol) 55(12):1342–1347
9. 杜治娟,王硕,王秋月,等. 社会媒体大数据分析研究综述 [J]. 计算机科学与探索, 2017, 11 (1):1–23. Du ZJ, Wang S, Wang QY et al (2017) Survey on social media big data analytics. J Front Comput Sci Technol 11(1):1–23
10. 多恩,哈勒维,艾夫斯. 数据集成原理 [M]. 孟小峰,马如霞,马友忠,等译.北京:机械工业出版社, 2014. Doan A, Halevy A, Ives Z (2014) Principles of data integration. Meng XF, Ma RX, Ma YZ et al trans. China Machine Press, Beijing
11. 樊博. G2G电子政务的协同效率改进研究 [M]. 北京:清华大学出版社, 2013. Fan B (2013) The improvement on collaborative efficiency of G2G e-government. Tsinghua University Press, Beijing

W. Fan (ed.), *Unconventional Emergency Management Research*, Reports of China's Basic Research, https://doi.org/10.1007/978-981-99-6798-8

12. 樊治平,姜艳萍,刘洋,等. 突发事件应急方案选择的决策方法研究 [M]. 北京:科学出版社, 2016. Fan ZP, Jiang YP, Liu Y et al (2016) Decision-making method research on selecting emergency response plans. Science China Press, Beijing
13. 范维澄,刘奕,翁文国,等. 公共安全科学导论 [M]. 北京:科学出版社, 2013. Fan WC, Liu Y, Weng WG et al (2013) An introduction to public safety sciences. Science China Press, Beijing
14. 范维澄,闪淳昌,等. 公共安全与应急管理 [M]. 北京:科学出版社, 2017. Fan WC, Shan CC et al (2017) Public safety and emergency management. Science China Press, Beijing
15. 方志耕,刘思峰,施红星,等. 灰色博弈理论及其经济应用研究 [M]. 北京:高等教育出版社, 2016. Fang ZG, Liu SF, Shi HX et al (2016) Grey game theory and its applications in economic decision-making. Higher Education Press, Beijing
16. 高德华,邓修权. 组织惯例的微观机理与演化复杂性——基于多主体仿真的研究 [M]. 北京:北京邮电大学出版社, 2016. Gao DH, Deng XQ (2016) Micromechanism and complexity of evolution of organizational rutines—a multi-agent simulation-based research. Beijing University of Posts and Telecommunications Press, Beijing
17. 郭旦怀. 基于空间场景相似性的地理空间分析 [M]. 北京:科学出版社, 2016. Guo DH (2016) Geographic spatial analysis based on similarity of spacial scenarios. Science China Press, Beijing
18. 国务院应急办赴英培训考察团. 英国应急管理的特点及启示 [J]. 中国应急管理, 2007 (7):54–58. Training and investigation mission to the UK, the State Council Emergency Management Office. Characteristics of British emergency management and their inspirations. China Emerg Manag 2007(7):54–58
19. 韩传峰. 灾难管理系统论:方法与应用 [M]. 北京:科学出版社, 2013. Simonovic SP (2013) Systems approach to management of disasters: methods and applications (Trans. by Han CF). Science China Press, Beijing
20. 韩传峰. 国家关键基础设施系统安全与保护 [M]. 上海:同济大学出版社, 2015. Han CF (2015) National critical infrastructure system safety and protection. Tongji University Press, Shanghai
21. 韩传峰. 国家安全管理 [M]. 上海:同济大学出版社, 2016. Han CF (2016) National safety management. Tongji University Press, Shanghai
22. 韩传峰,等. 关键基础设施系统保护建模与仿真 [M]. 北京:科学出版社, 2016. Han CF et al (2016) Modeling and simulation of critical infrastructure system protection. Science China Press, Beijing
23. 黄金才,刘忠,张维明,等. 指挥控制辅助决策方法 [M]. 北京:国防科技大学出版社, 2013. Huang JC, Liu Z, Zhang WM et al (2013) Auxilary decision-making methods for command and control. National University of Defense Technology Press, Beijing
24. 黄全义,钟少波,孙超,等. 灾害性气象事件影响预评估理论与方法 [M]. 北京:科学出版社, 2017. Huang QY, Zhong SB, Sun C et al (2017) Preassessment of influence of disastrous meteorological incidents: theory and mehods. Science China Press, Beijing
25. 黄炜,赵来军. 蓝藻水华相关因素识别、预测与治理 [M]. 上海:复旦大学出版社, 2015. Huang W, Zhao LJ (2015) Identification, prediction and control of cyanobacterial bloom. Fudan University Press, Shanghai
26. 贾传亮,佘廉,李慧嘉,等. 突发事件案例提示顿悟理论与方法 [M]. 北京:科学出版社, 2017. Jia CL, She L, Li HJ et al (2017) Hints and insights from emergency cases: theory and methods. Science China Press, Beijing
27. 贾庆山,杨玉,夏俐,等. 基于事件的优化方法简介及其在能源互联网中的应用 [J]. 控制理论与应用, 2018, 35 (1):32–40. Jia QS, Yang Y, Xia L et al (2018) A tutorial on event-based optimization with application in energy Internet. Control Theory & Appl 35(1):32–40
28. 姜振飞,刘紫涵. 小布什政府与奥巴马政府核战略比较研究 [M]. 北京:中国社会科学出版社, 2017. Jiang ZF, Liu ZH (2017) Comparison of nuclear strategies between the bush administration and the Obama administration. China Social Sciences Press, Beijing
29. 李和瀚,孟小峰,邹磊. 面向ScholarSpace知识库的关键词查询方法 [J]. 通信学报, 2015,36 (12):28–36. Li HH, Meng XF, Zou L (2015) Keyword search approach for knowledge base in ScholarSpace. J Commun 36(12):28–36

30. 李湖生. 应急准备体系规划建设理论与方法 [M]. 北京:科学出版社, 2016. Li HS (2016) Planning and construction of emergency preparedness system: theory and methods. Science China Press, Beijing
31. 李丽华,马亚萍,丁宁,等. 应急疏散中社会关系网络与“领导—追随”行为变化 [J]. 清华大学学报 (自然科学版), 2016,56 (3):334–340. Li LH, Ma YP, Ding N et al (2016) Changes in social relation networks and leader-follower behavior in emergency evacuations. J Tsinghua Univ (Sci Technol) 56(3):334–340
32. 李旺民. 基于移动设备的全站仪内外业一体化方法研究 [J]. 地理空间信息, 2018, 16 (1):71–73. Li WM (2018) Research on the integration method of total station indoor and field work based on mobile device. Geospatial Inf 16(1):71–73
33. 李向阳,等. 大规模灾害应对准备的容错规划方法 [M]. 北京:科学出版社, 2017. Li XY et al (2017) Fault-tolerant planning methods on large-scale disaster response and prepardness. Science China Press, Beijing
34. 李雪峰. 突发事件应对案例研究:方法与实践 [M]. 北京:国家行政学院出版社, 2017. Li XF (2017) Case studies on emergency response: methods and practice. National Academy of Governance Press, Beijing
35. 李勇建,乔晓娇,孙晓晨,等. 非常规突发事件运行机理与应急决策 [M]. 北京:科学出版社, 2016. Li YJ, Qiao XJ, Sun XC et al (2016) Running mechanism and decision-making for unconventional emergencies. Science China Press, Beijing
36. 廖东升. 基于多智能体的心理战效果评估方法 [M]. 北京:军事科学出版社, 2011. Liao DS (2011) Research on evaluation method and application of effects of psychology operation based on multi-agent. Military Science Publishing House, Beijing
37. 刘川意. 拨得云开见日出:解构一个典型的云计算系统 [M]. 北京:电子工业出版社, 2012. Liu CY (2012) Deconstruction of a typical cloud computing system. Publishing House of Electronic Industry, Beijing
38. 刘栋,李素,曹志冬. 基于改进人工蜂群算法的支持向量机时序预测 [J]. 传感器与微系统, 2018, 37 (5):51–54. Liu D, Li S, Cao Z D (2018) Time sequence prediction of SVM based on improved ABC algorithm. Transducer Microsyst Technol 37(5):51–54
39. 刘南,葛洪磊. 应急资源配置决策的理论、方法及应用 [M]. 北京:科学出版社, 2015. Liu N, Ge HL (2015) Emergency resources distribution decision-making: theory, methods and application. Science China Press, Beijing
40. 刘思峰,郭本海,方志耕,等. 系统评价:方法、模型、应用 [M]. 北京:科学出版社, 2016. Liu SF, Guo BH, Fang ZG et al (2016) System evaluation: methods, models and application. Science China Press, Beijing
41. 刘思峰,杨英杰,吴利丰,等. 灰色系统理论及其应用 [M]. 7版. 北京:科学出版社, 2014. Liu SF, Yang YJ, Wu LF et al (2014) Grey System: theory and application, 7th edn. Science China Press, Beijing
42. 刘铁民,等. 重大生产安全事故情景构建理论与方法:基于高含硫油气田井喷等重大事故应急准备研究 [M]. 北京:科学出版社, 2017. Liu TM et al (2017) Scenario Construction theory and methods of major production safety accidents: major accident preparedness research based on high sulfur-bearing oil and gas field blowout. Science China Press, Beijing
43. 刘霞,向良云. 公共危机治理 [M]. 上海:上海交通大学出版社, 2010. Liu X, Xiang LY (2010) Public crisis governance. Shanghai Jiao Tong University Press, Shanghai
44. 刘小月,赵秋红,鞠彦兵,等. 非常规突发事件应急管理系统评估方法 [M]. 北京:科学出版社, 2016. Liu XY, Zhao QH, Ju YB et al (2016) Unconventional emergency management system evaluation methods. Science China Press, Beijing
45. 刘宴兵,尚明生,肖云鹏. 网格高性能调度及资源管理技术 [M]. 北京:科学出版社, 2010. Liu YB, Shang MS, Xiao YP (2010) Grid high performance scheduling and resources management techniques. Science China Press, Beijing
46. 刘怡君,李倩倩,马宁,等. 社会舆情的网络分析方法与建模仿真 [M]. 北京:科学出版社, 2016. Liu YJ, Li QQ, Ma N et al (2016) Network analysis methods and modeling simulation of public opinions. Science China Press, Beijing
47. 马怀德. 非常规突发事件应急管理的法律问题研究 [M]. 北京:中国法制出版社, 2015. Ma HD (2015) Research on the legal issues of unconventional emergency management. China Legal Publishing House, Beijing

48. 马祖军,等. 非常规突发事件应急血液保障理论与方法 [M]. 北京:科学出版社, 2015. Ma ZJ et al (2015) Emergency blood supply security for unconventional emergencies: theory and methods. Science China Press, Beijing
49. 彭宗超,陶鹏,等. 中国应急管理研究报告 (2012—2013):应急准备文化建设 [M]. 北京:清华大学出版社, 2014. Peng ZC, Tao P et al (2014) China Emergency Management Research Report (2012–2013): Emergency preparedness culture construction. Tsinghua University Press, Beijing
50. 齐佳音,张一文. 突发性公共危机事件与网络舆情作用机制研究 [M]. 北京:科学出版社, 2016. Qi J Y, Zhang Y W. Public Crisis Emergencies and Network Public Opinion Working Mechanism. Science China Press, Beijing2016
51. 乔仁毅,龚维斌. 政府应急管理 [M]. 北京:国家行政学院出版社, 2014. Qiao RY, Gong WB (2014) Government emergency management. National Academy of Governance Press, Beijing
52. 秦绪伟,刘晓. 风险环境下的供应链可靠设计与应急计划 [M]. 北京:科学出版社, 2012. Qin XW, Liu X (2012) Reliable design and emergency planning of supply chains under risk conditions. Science China Press, Beijing
53. 邱晓刚,陈彬. 基于HLA的分布仿真环境设计 [M]. 北京:国防工业出版社, 2016. Qiu XG, Chen B (2016) HLA based distributed simulation environment design. National Defense Industry Press, Beijing
54. 邱晓刚,陈彬,张鹏. 面向应急管理的人工社会构建与计算实验 [M]. 北京:科学出版社, 2017. Qiu X G, Chen B, Zhang P (2017) Artificial society construction and computational experiment toward emergency management. Science China Press, Beijing
55. 裘江南,王雪华,等.突发事件应急知识管理的模型与方法 [M]. 北京:科学出版社, 2016. Qiu JN, Wang XH et al (2016) Emergency knowledge management models and methods. Science China Press, Beijing
56. 闪淳昌,等. 中国突发事件应急体系顶层设计 [M]. 北京:科学出版社, 2017. Shan CC et al (2017) China's top-level design of emergency response system. Science China Press, Beijing
57. 尚明生,陈端兵,高辉,等. 舆情信息分析与处理技术 [M]. 北京:科学出版社, 2015. Shang MS, Chen DB, Gao H et al (2015) Analysis and handling technique on public opinion information. Science China Press, Beijing
58. 佘廉,黄超. 突发事件案例生成理论与方法 [M]. 北京:科学出版社, 2017. She L, Huang C (2017) Generative theory on and methods of emergency cases. Science China Press, Beijing
59. 施俊琦,汪默. 管理心理学:21世纪新进展 [M]. 北京:北京大学出版社, 2013. Shi JQ, Wang M (2013) Management psychology: progress in the 21st century. Peking University Press, Beijing
60. 时勘. 救援人员应对非常规突发事件的抗逆力模型 [M]. 北京:科学出版社, 2017. Shi K (2017) Resilience models of rescuers in unconventional emergencies. Science China Press, Beijing
61. 宋劲松. 突发事件应急指挥 [M]. 北京:中国经济出版社, 2011. Song JS (2011) Emergency response command. Economic Press China, Beijing
62. 宋劲松. 应急管理社会动员 [M]. 北京:中国经济出版社, 2012. Song JS (2012) Emergency management social mobilization. Economic Press China, Beijing
63. 孙金华,宋卫国,等.突发事件中大规模人群的行为规律 [M]. 北京:科学出版社, 2016. Sun JH, Song WG et al (2016) Large-scale crowd behavior pattern under unconventional emergencies. Science China Press, Beijing
64. 滕五晓. 应急管理能力评估:基于案例分析的研究 [M]. 北京:社会科学文献出版社, 2014. Teng WX (2014) The evaluation of emergency management capacity based on case studies. Social Sciences Academic Press (China), Beijing
65. 王红卫,祁超. 基于层次任务网络规划的应急响应决策理论与方法 [M]. 北京:科学出版社, 2015. Wang HW, Qi C (2015) Theory and methods of emergency response decision-making based on HTN. Science China Press, Beijing
66. 王红卫,王剑,陈曦. 国民经济动员管理决策与仿真演练方法、技术及其应用 [M]. 北京:科学出版社,2013. Wang HW, Wang J, Chen X (2013) National economy mobilization management decision-making and simulation and training methods, techniques and applications. Science China Press, Beijing

67. 王慧敏,刘高峰,陶飞飞,等. 非常规突发水灾害应急合作管理与决策 [M]. 北京:科学出版社, 2016. Wang HM, Liu GF, Tao FF et al (2016) Unconventional flood emergency response cooperation management and decision-making. Science China Press, Beijing
68. 王雷,赵秋红,王欣. 应急管理技术与方法 [M]. 北京:北京航空航天大学出版社, 2016. Wang L, Zhao QH, Wang X (2016) Emergency management techniques and methods. Beihang University Press, Beijing
69. 王宁,仲秋雁. 非常规突发事件演化与推演方法 [M]. 北京:科学出版社, 2016. Wang N, Zhong QY (2016) Unconventional emergency evolution and deduction methods. Science China Press, Beijing
70. 王伟. 人格心理学 [M]. 2版.北京:人民卫生出版社, 2013. Wang W (2013) Personality psychology, 2nd edn. People's Medical Publishing House, Beijing
71. 王筱莉,赵来军,等. 基于群体差异的谣言传播规律与政府辟谣策略研究 [M]. 上海:复旦大学出版社, 2016. Wang XL, Zhao LJ et al (2016) Rumor spreading pattern and government rumor reputing strategies based on group difference. Fudan University Press, Shanghai
72. 王旭坪,阮俊虎,赵颖,等.非常规突发事件的应急资源协调优化及仿真研究 [M]. 北京:科学出版社, 2016. Wang XP, Ruan JH, Zhao Y et al (2016) Coordinated optimization of resources dispatch and simulation under unconventional emergencies. Science China Press, Beijing
73. 吴启迪,康琦,汪镭,等. 自然计算导论 [M].上海:上海科学技术出版社, 2011. Wu QD, Kang Q, Wang L et al (2011) Introduction to natural computing. Shanghai Scientific and Technical Publishers, Shanghai
74. 谢科范,陈刚,吴倩,等. 非常规突发事件的集群决策 [M]. 北京:知识产权出版社, 2013. Xie KF, Chen G, Wu Q et al (2013) Group decision-making for unconventional emergencies. Intellectual Property Publishing House, Beijing
75. 许伟. 基于网络大数据的社会经济监测预警研究 [M]. 北京:科学出版社, 2016. Xu W (2016) Social economy monitoring and early warning based on online big data. Science China Press, Beijing
76. 薛澜,刘冰. 应急管理体系新挑战及其顶层设计 [J]. 国家行政学院学报, 2013 (1):10–14. Xue L, Liu B (2013) New challenges and top-level design of emergency management. J Chin Acad Govern 2013(1):10–14
77. 薛澜,曾光,等. 防控"甲流":中国内地甲型H1N1流感应对评估 [M]. 北京:社会科学文献出版社, 2014. Xue L, Zeng G et al (2014) Control and prevention of influenza a: response and evaluation of influenza A H1N1 in the mainland of China. Social Sciences Academic Press (China), Beijing
78. 杨立中. 建筑内人员运动规律与疏散动力学 [M]. 北京:科学出版社, 2012. Yang LZ (2012) Pedestrian moving pattern and evacuation dynamics inside buildings. Science China Press, Beijing
79. 杨青,杨帆,刘星星,等. 基于免疫学的非常规突发事件识别和预控 [M]. 北京:科学出版社, 2015. Yang Q, Yang F, Liu XX et al (2015) Uncoventional crisis identification and pre-control based on immunology. Science China Press, Beijing
80. 尤天慧,张尧,樊治平,等. 信息不完全确定的多指标决策理论与方法 [M]. 北京:科学出版社, 2010. You TH, Zhang Y, Fan ZP et al (2010) Theories and methods for multiple attribute decision making with incomplete information. Science China Press, Beijing
81. 游志斌,杨永斌. 国外政府风险管理制度的顶层设计与启示 [J]. 行政管理改革, 2012 (5):76–79. You ZB, Yang YB (2012) Top-level design of government hazard management system and its inspirations in other countries. Adm Reform 2012(5): 76–79
82. 喻凯,熊祥瑞,高涛. 基于Hadoop的地图瓦片云存储系统的设计与实现 [J]. 测绘地理信息, 2017,42 (3):74–77. Yu K, Xiong XR, Gao T (2017) Design and implementation of cloud storage system for map tiles based on Hadoop. J Geomat 42(3):74–77
83. 袁宏永,黄全义,苏国锋,等. 应急平台体系关键技术研究的理论与实践 [M]. 北京:清华大学出版社, 2012. Yuang HY, Huang QY, Su GF et al (2012) Theory and practice of key technologies of emergency platform system. Tsinghua University Press, Beijing
84. 袁宏永,苏国锋,陈建国,等. 突发事件及其链式效应理论研究与应用 [M]. 北京:科学出版社, 2016. Yuang HY, Su GF, Chen JG et al (2016) Studies on and application of emergency and its chain effect theory. Science China Press, Beijing

85. 袁玉宇,刘川意,郭松柳. 云计算时代的数据中心 [M]. 北京:电子工业出版社, 2012. Yuan YY, Liu CY, Guo SL (2012) Data center in the cloud computing era. Publishing House of Electronic Industry, Beijing
86. 苑盛成,王刚桥,马晔风,等. 基于实时数据的应急交通疏散仿真方法研究 [J]. 系统工程理论与实践, 2015, 35 (10):2484–2489. Yuan SC, Wang GQ, Ma YF et al (2015) Study on emergency traffic evacuation with real-time data. Syst Eng Theory Pract 35(10):2484–2489
87. 曾大军,等. 传染病信息学与生物监测 [M]. 宋宏彬,王立贵,蒲卫,等译.北京:科学出版社, 2014. Zeng D et al (2014) Infectious disease informatics and biosurveillance (Song HB, Wang LG, Pu W et al trans). Science China Press, Beijing
88. 詹承豫. 从危机管理到风险治理:基于理论、制度及实践的分析 [M]. 北京:中国法制出版社, 2016. Zhan CY (2016) From crisis management to hazard control: an analysis based on theory, system and pracitice. China Legal Publishing House, Beijing
89. 詹承豫,王敬波,等. 学校突发事件应急管理理论与实务 [M]. 北京:中国法制出版社, 2014. Zhan CY, Wang JB et al (2014) School emergency management theory and practice. China Legal Publishing House, Beijing
90. 张辉,刘奕,刘艺. 突发事件应急决策支持的理论与方法 [M]. 北京:科学出版社, 2017. Zhang H, Liu Y, Liu Y (2017) Theory and methods of emergency response decision-making support. Science China Press, Beijing
91. 张鹭鹭,康鹏,顾洪. 地震应急医学救援"两期三段"研究:基于玉树地震医学救援实证分析 [M]. 北京:科学出版社, 2016. Zhang LL, Kang P, Gu H (2016) "Two periods and three stages" in earthquake emergency medical rescue: a reference to the case of Yushu earthquake. Science China Press, Beijing
92. 张鹭鹭,张义. 卫勤循证决策之卫勤决策支持平台 [M]. 上海:第二军医大学出版社, 2015. Zhang LL, Zhang Y (2015) Military health service evidence-informed decision-making: military health service decision-making support platform. Second Military Medical University Press, Shanghai
93. 张鹏翥. 电子政务 [M]. 北京:电子工业出版社, 2011. Zhang PZ (2011) E-Government. Publishing House of Electronic Industry, Beijing
94. 张郁. 无人机低空摄影的精度分析与研究 [J]. 测绘地理信息, 2018,43 (4):59–61. Zhang Y (2018) Precision study and analysis of unmanned aerial vehicle low altitude photography. J Geomat 43(4):59–61
95. 赵来军. 公共危机与社会治理 [M]. 北京:社会科学文献出版社, 2011. Zhao LJ (2011) Public crisis and social governance. Social Sciences Academic Press (China), Beijing
96. 赵来军. 城市危险化学品无缝隙化安全管理研究 [M]. 北京:科学出版社, 2011. Zhao LJ (2011) Seamless safety management of hazardous chemicals in China. Science China Press, Beijing
97. 赵来军,王珂,汪建. 城市应急避难场所规划建设理论与方法 [M]. 北京:科学出版社, 2014. Zhao LJ, Wang K, Wang J (2014) Theory and methods of urban emergency shelters planning and construction. Science China Press, Beijing
98. 赵来军,赵筱莉,王佳佳,等. 谣言传播规律与突发事件应对策略研究 [M]. 北京:科学出版社, 2016. Zhao LJ, Zhao XL, Wang JJ et al (2016) Rumor spreading pattern and emergency response strategies. Science China Press, Beijing
99. 赵秋红,郗蒙浩. 非常规突发事件应急管理体系的组织设计:理论、方法与应用 [M]. 北京:科学出版社, 2017. Zhao QH, Xi MH (2017) Organizational design of unconventional emergency management system: theory, methods and applications. Science China Press, Beijing
100. 郑涛. 生物安全学 [M]. 北京:科学出版社, 2014. Zheng T (2014) Biosecurity. Science China Press, Beijing
101. 郑涛. 生物事件应急管理 [M]. 北京:科学出版社, 2016. Zheng T (2016) Biological emergency management. Science China Press, Beijing
102. 钟开斌. 中外政府应急管理比较 [M]. 北京:国家行政学院出版社, 2012. Zhong KB (2012) Government emergency management comparison between china and other countries. National Academy of Governance Press, Beijing

103. 钟开斌. 应急决策:理论与案例 [M]. 北京:社会科学文献出版社, 2014. Zhong KB (2014) Emergency decision-making: theoretical framework and case studies. Social Sciences Academic Press (China), Beijing
104. Ai P, Feng P, Mu P et al (2012) Research on hydrological information organization based on virtualization. In: Proceedings of the 11th international symposium on distributed computing and application to business, engineering & science (DCABES 2012), pp 86–90
105. An J, Kang Q, Wang L et al (2012) Population-based dynamic scheduling optimisation for complex production process. Int J Comput Appl Technol 43(4):304–310
106. An J, Kang Q, Wang L et al (2013) Mussels wandering optimization: an ecologically inspired algorithm for global optimization. Cogn Comput 5(2):188–199
107. Arsuaga J, Diao Y, Klingbeil M et al (2014) Properties of topological networks of flexible polygonal chains. Mol Based Math Biol 2(1):98–106
108. Bai QY, Lin WP, Wang L (2016) Family incivility and counterproductive work behavior: a moderated mediation model of self-esteem and emotional regulation. J Vocat Behav 94:11–19
109. Bullnheimer B, Hartl RF, Strauss C (1999) Applying the ANT system to vehicle routing problems. In: Voβ S, Martello S, Osman IH et al (eds) Meta-heuristics. Springer, New York
110. Cai HL, Wang L, Liu L et al (2010) A temporal relation-based method for extracting occurrence time in Chinese web news of emergencies. J Converg Inf Technol 5(3):31–37
111. Cai LQ, Chen WZ, Shen YD et al (2012) Recognition of facial expressions of emotion in panic disorder. Psychopathology 45(5):294–299
112. Cao SC, Song WG, Liu XD et al (2014) Simulation of pedestrian evacuation in a room under fire emergency. Procedia Eng 71:403–409
113. Cao Y, Liu YS, Fan HR et al (2013) SysML-based uniform behavior modeling and automated mapping of design and simulation model for complex mechatronics. Comput Aided Des 45(3):764–776
114. Cao ZD, Zeng DJ, Wang QY et al (2010) An epidemiological analysis of the Beijing 2008 hand-foot-mouth epidemic. Chin Sci Bull 55(12):1142–1149
115. Cao ZD, Zeng DJ, Wang QY et al (2010) Epidemiological features and spatio-temporal evolution in the early phase of the Beijing H1N1 epidemic. Acta Geogr Sin 65(3):361–368
116. Cao ZD, Zeng DJ, Zheng XL et al (2010) Spatio-temporal evolution of Beijing 2003 SARS epidemic. Sci China Earth Sci 53(7):1017–1028
117. Carciofo R, Du F, Song N et al (2012) Age-related chronotype differences in Chinese, and reliability assessment of a reduced version of the Chinese morningness eveningness questionnaire. Sleep Biol Rhythms 10(4):310–318
118. Carciofo R, Du F, Song N et al (2014) Chronotype and time-of-day correlates of mind wandering and related phenomena. Biol Rhythm Res 45(1):37–49
119. Carciofo R, Du F, Song N et al (2014) Mind wandering, sleep quality, affect and chronotype: an exploratory study. PLoS ONE 9(3):e91285
120. Chai CL, Liu X, Zhang WJ et al (2011) Application of social network theory to prioritizing oil & gas industries protection in a networked critical infrastructure system. J Loss Prev Process Ind 24(5):688–694
121. Chai H, Chen WZ, Zhu J et al (2012) Processing of facial expressions of emotions in healthy volunteers: an exploration with event-related potentials and personality traits. Neurophysiol Clin 42(6):369–375
122. Chen A, Chen N, Li JM (2012) During-incident process assessment in emergency management: concept and strategy. Saf Sci 50(1):90–102
123. Chen B, Zhang LB, Guo G et al (2015) KD-ACP: a software framework for social computing in emergency management. Math Probl Eng 2015:1–27
124. Chen BK, Chen MS, Zhang ZL et al (2011) Speed of last vehicle feedback strategy in intelligent transportation systems. Int J Mod Phys C 22(7):745–753
125. Chen BK, Sun XY, Wei H et al (2011) Piecewise function feedback strategy in intelligent traffic systems with a speed limit bottleneck. Int J Mod Phys C 22(8):849–860
126. Chen C, Liu Z, Lin WH et al (2013) Distributed modeling in a mapreduce framework for data-driven traffic flow forecasting. IEEE Trans Intell Transp Syst 14(1):22–33

127. Chen DB, Gao H, Lü LY et al (2013) Identifying influential nodes in large-scale directed networks: The role of clustering. PLoS ONE 8(10):e77455
128. Chen DB, Lü LY, Shang MS et al (2012) Identifying influential nodes in complex networks. Phys A: Stat Mech Appl 391(4):1777–1787
129. Chen DB, Shang MS, Lü ZH et al (2012) Detecting overlapping communities of weighted networks via a local algorithm. Phys A: Stat Mech Appl 389(19):4177–4187
130. Chen HP, Li MN, Dai ZX et al (2016) Factors influencing the perception of medical staff and outpatients of dual practice in Shanghai, People's Republic of China. Patient Prefer Adherence 10:1667–1678
131. Chen HP, Li MN, Wang JR et al (2016) Factors influencing inpatients' satisfaction with hospitalization service in public hospitals in Shanghai, People's Republic of China. Patient Prefer Adherence 10:469–477
132. Chen JF, Jin QJ, Chao J (2012) Design of deep belief networks for short-term prediction of drought index using data in the Huaihe River Basin. Math Probl Eng 2012:1–16
133. Chen JF, Jin QJ, Wang HM et al (2012) A risk assessment model of flood based on information diffusion method and BP neural network. Prz Elektrotech 88:33–36
134. Chen JF, Zhang L, Wang HM et al (2012) Losses assessment for region flood disasters based on entropy weight TOPSIS model. Adv Inf Sci Serv Sci 4(17):192–200
135. Chen JG, Su GF, Yuan HY (2011) Integration of the post-earthquake casualty estimation model and GIS platform. In: Proceedings of the international conference on information systems for crisis response and management (ISCRAM 2011), pp 523–527
136. Chen JG, Su GF, Zhao QL (2012) Study of emergency management information system for hydropower project. In: Proceedings of the 3rd world congress on software engineering (WCSE 2012), pp 198–201
137. Chen JG, Zhang XL, Su GF et al (2012) Study on fire-rescue command system and key technology of ultra high-rise building. In: 5th international congress on image and signal processing (CISP 2012), pp 1906–1910
138. Chen L, Song WG (2013) New method to calculate exit flow based on RFID technology. Appl Mech Mater 444–445:1625–1629
139. Chen L, Song WG, Huo FZ (2014) Study of personnel monitoring and counting in single channel based on RFID technology. Procedia Eng 71:529–536
140. Chen ME, Jin XM, Shen D (2011) Short text classification improved by learning multi-granularity topics. In: Proceedings of the 22nd international joint conference on artificial intelligence (IJCAI 2011), pp 1776–1781
141. Chen ML, Ma QG, Li ML et al (2010) Cognitive and emotional conflicts of counter-conformity choice in purchasing books online: an event-related potentials study. Biol Psychol 85(3):437–445
142. Chen ML, Ma QG, Li ML et al (2010) The neural and psychological basis of herding in purchasing books online: an event-related potential study. Cyberpsychol Behav Soc Netw 13(3):321–328
143. Chen NH, Cai P, Zhou TG et al (2016) Perceptual learning modifies the functional specializations of visual cortical areas. Proc Natl Acad Sci 113(20):5724–5729
144. Chen NS, Chen A (2012) Emergency evaluation model and method: a perspective overview. J Saf Crisis Manag 2(1):11–23
145. Chen NS, Chen A, Ribeiro B (2013) Influence of class distribution on cost-sensitive learning: a case study of bankruptcy analysis. Intell Data Anal 17(3):423–437
146. Chen NS, Ribeiro B, Chen A (2011) Using non-negative matrix factorization for bankruptcy analysis. INFOCOMP J Comput Sci 10(4):57–64
147. Chen NS, Ribeiro B, Chen A (2015) Comparative study of classifier ensembles for cost-sensitive credit risk assessment. Intell Data Anal 19(1):127–144
148. Chen P, Yuan HY, Li DS (2013) Space-time analysis of burglary in Beijing. Secur J 26(1):1–15
149. Chen T, Huang J (2012) Exploratory research on the system of China relief reserve. Syst Eng Procedia 5:99–106

150. Chen X, Peng L, Zhang MH et al (2016) A public traffic demand forecast method based on computational experiments. IEEE Trans Intell Transp Syst 18(4):984–995
151. Chen X, Xiong X, Zhang MH et al (2016) Public authority control strategy for opinion evolution in social networks. Chaos: Interdiscipl J Nonlinear Sci 26(8):083105
152. Chen YH, He KJ, Yu LA (2015) The information content of OVX for crude oil returns analysis and risk measurement: evidence from the Kalman filter model. Ann Data Sci 2(4):471–487
153. Chen Y, Li KW, Xu HY et al (2009) A DEA-TOPSIS method for multiple criteria decision analysis in emergency management. J Syst Sci Syst Eng 18(4):489–507
154. Chen Y, Zhai PM (2013) Persistent extreme precipitation events in China during 1951–2010. Climate Res 57(2):143–155
155. Chen YZ, Zhao QH, Wang L et al (2016) The regional cooperation-based warehouse location problem for relief supplies. Comput Ind Eng 102:259–267
156. Chen ZK, Yang SQ, Tan S et al (2014) The data allocation strategy based on load in NoSQL database. Appl Mech Mater 513–517:1464–1469
157. Chen ZY, Fan ZP, Sun MH (2012) A hierarchical multiple kernel support vector machine for customer churn prediction using longitudinal behavioral data. Eur J Oper Res 223(2):461–472
158. Chen ZK, Yang SQ, Tan S et al (2015) A new fragment re-allocation strategy for NoSQL database systems. Front Comput Sci 9(1):111–127
159. Chen ZY, Zhang HT (2013) A remark on collective circular motion of heterogeneous multi-agents. Automatica 49(5):1236–1241
160. Cheng GQ, Huang JC, Liu Z et al (2011) Image quality assessment using natural image statistics in gradient domain. AEU—Int J Electron Commun 65(5):392–397
161. Cheng Q, Liu Z, Huang JC et al (2016) Community detection in hypernetwork via density-ordered tree partition. Appl Math Comput 276(3):384–393
162. Cheng Q, Lu X, Liu Z et al (2015) Mining research trends with anomaly detection models: the case of social computing research. Scientometrics 103(2):453–469
163. Cheng Q, Lu X, Liu Z et al (2016) Spatial clustering with density-ordered tree. Phys A: Stat Mech Appl 460:188–200
164. Cheng Q, Lu X, Wu JT et al (2016) Analysis of heterogeneous dengue transmission in Guangdong in 2014 with multivariate time series model. Sci Rep 6:33755
165. Cheng W, Jin XM, Sun JT et al (2014) Searching dimension incomplete databases. IEEE Trans Knowl Data Eng 26(3):725–738
166. Cheng W, Jin XM, Sun JT (2009) Probabilistic similarity query on dimension incomplete data. In: 9th IEEE international conference on data mining (ICDM 2009), pp 81–90
167. Chi F, Chen A (2011) Emergency loan of bank after disaster in China. J Saf Crisis Manag 1(2):1–10
168. Chu X, Zhong QY (2015) Post-earthquake allocation approach of medical rescue teams. Nat Hazards 79(3):1809–1824
169. Chu X, Zhong QY, Khokhar SG (2015) Triage scheduling optimization for mass casualty and disaster response. Asia-Pac J Oper Res 32(6):1550041
170. Chuang YT, Xie XF, Liu C (2016) Interdependent orientations increase pro-environmental preferences when facing self-interest conflicts: the mediating role of self-control. J Environ Psychol 46:96–105
171. Cui AX, Wang W, Tang M et al (2014) Efficient allocation of heterogeneous response times in information spreading process. Chaos: Interdisciplinary J Nonlinear Sci 24(3):033113
172. Dai J, Ding ZM, Xu JJ (2016) Context-based moving object trajectory uncertainty reduction and ranking in road network. J Comput Sci Technol 31(1):167–184
173. Dai WH, Hu HZ, Wu TN et al (2014) Information spread of emergency events: path searching on social networks. Sci World J 179620
174. Dai WH, Su LH, Liu XY (2010) Dynamic monitoring and emergency response system for environment protection. Adv Mater Res 113–116:571–574
175. Dai WH, Wan XQ, Liu XY (2011) Emergency event: internet spread, psychological impacts and emergency management. J Comput 6(8):1748–1755

176. Dai WH, Wang CS, Li GX (2011) Research on 3G-based mobile self-rescue system for medical emergency. J Netw 6(8):1214–1221
177. Dai YR, Wang J (2014) Application of integrated concept in industries for maximizing energy saving potential. Appl Mech Mater 496–500:2598–2602
178. Dang YG, Liu SF, Wang YH (2010) Optimization of regional industrial structures and applications. CRC Press, New York
179. Deng N (2013) Study on the statistical method of direct economic losses from work safety accidents. In: Huang CF, Kahrama C (eds) Intelligent systems and decision making for risk analysis and crisis response. CRC Press, London
180. Deng O, Su GF, Huang QY et al (2013) Forest fire risk mapping based on spatial logistic model of northeastern China forest zone. In: Geo-informatics in resource management and sustainable ecosystem. Springer, Berlin, Heidelberg, pp 181–192
181. Deng O, Su GF, Huang QY et al (2015) Development and analysis of 11-year MODIS forest fire datasets across the Heilongjiang Province of northeastern China. In: Proceedings of the 2015 international symposium on computers and informatics (ISCI 2015), pp 2062–2069
182. Deng Q, Liu Y, Li LH et al (2016) Emergency online attention and psychological distance under risk. In: Proceedings of the 2nd ACM SIGSPATIAL international workshop on the use of GIS in emergency management, pp 1–6
183. Deng Q, Liu Y, Zhang H et al (2016) A new crowdsourcing model to assess disaster using microblog data in typhoon Haiyan. Nat Hazards 84(2):1241–1256
184. Deng QY, Tang BH, Xue C et al (2016) Comparison of the ability to predict mortality between the injury severity score and the new injury severity score: a meta-analysis. Int J Environ Res Public Health 13(8):825
185. Deng XL, Yu YL, Guo DH et al (2018) Efficient CPS model based online opinion governance modeling and evaluation for emergency accidents. GeoInformatica 22(2):479–502
186. Deng XX, Liu Z, Xie JX et al (2014) Web system upgrading with transaction failure and strategic customers. IEEE Trans Syst Man Cybern: Syst 44(2):209–219
187. Diao Y, Ernst C, Montemayor A et al (2014) The knot spectrum of confined random equilateral polygons. Comput Math Biophys 2(1):19–33
188. Diao Y, Ernst C, Ziegler U (2014) Random walks and polygons in tight confinement. J Phys: Conf Ser 544(1):012017
189. Diao Y, Rodriguez V, Klingbeil M et al (2015) Orientation of DNA minicircles balances density and topological complexity in kinetoplast DNA. PLoS ONE 10(6):e0130998
190. Ding H, Liu J, Zhao B et al (2012) Application of feature matching in the vehicle violation detection in urban traffic. Int J Digit Content Technol Appl 6(11):131–138
191. Ding H, Zhang WS (2012) Multi-target tracking with occlusions via skeleton points assignment. Neurocomputing 83(15):165–175
192. Ding N, Chen T, Zhang H (2017) Experimental study of elevator loading and unloading time during evacuation in high-rise buildings. Fire Technol 53(1):29–42
193. Ding N, Chen T, Zhang H (2017) Simulation of high-rise building evacuation considering fatigue factor based on cellular automata: a case study in China. Build Simul 10(3):407–418
194. Ding N, Zhang H, Chen T et al (2015) Stair evacuation simulation based on cellular automata considering evacuees' walk preferences. Chin Phys B 24(6):068801
195. Ding Q, Tong QH, Tong QL et al (2016) A new definite linear algorithm for the structural equation model with application in fire risk evaluation. Commun Stat—Theory Methods 45(23):7075–7085
196. Ding YC, Yang LZ, Rao P (2013) Investigating the merging behavior at the floor-stair interface of high-rise building based on computer simulations. Procedia Eng 62:463–469
197. Ding YC, Yang LZ (2014) Occupant evacuation process study of public buildings based on computer modeling and simulation. J Appl Fire Sci 23(3):365–380
198. Ding ZJ, Jiang R, Wang BH (2011) Traffic flow in the Biham-Middleton-Levine model with random update rule. Phys Rev E 83(4):047101
199. Ding ZM, Chen ZK, Yang Q (2014) IoT-SVKSearch: a real-time multimodal search engine mechanism for the internet of things. Int J Commun Syst 27(6):871–897

200. Ding ZM, Xu JJ, Yang Q (2013) SeaCloudDM: a database cluster framework for managing and querying massive heterogeneous sensor sampling data. J Supercomput 66(3):1260–1284
201. Ding ZM, Yang B, Chi YY et al (2016) Enabling smart transportation systems: a parallel spatio-temporal database approach. IEEE Trans Comput 65(5):1377–1391
202. Ding ZM, Yang B, Güting RH et al (2015) Network-matched trajectory-based moving- object database: models and applications. IEEE Trans Intell Transp Syst 16(4):1918–1928
203. Ding ZY, Jia Y, Zhou B et al (2013) Mining topical influencers based on the multi-relational network in micro-blogging sites. China Commun 10(1):93–104
204. Ding ZY, Zhang JF, Jia Y et al (2013) Detecting spammers in microblogs. J Internet Technol 14(2):289–296
205. Dong CF, Ma X, Wang BH (2010) Advanced information feedback strategy in intelligent two-route traffic flow systems. Sci China Inf Sci 53(11):2265–2271
206. Dong XS, Zou QB, Guan Y (2012) Set-similarity joins based semi-supervised sentiment analysis. In: International conference on neural information processing (ICONIP 2012), pp 176–183
207. Dong XX, Wang YL, Zhu JP et al (2014) Influence of evacuees number and elevator load on mixed evacuation process. In: Hydraulic engineering II—proceedings of the 2nd SREE conference on hydraulic engineering (CHE 2013), pp 211–216
208. Du F, Yang JY, Yin Y et al (2013) On the automaticity of contingent capture: disruption caused by the attentional blink. Psychon Bull Rev 20(5):944–950
209. Du J, Wong SC, Shu CW et al (2013) Revisiting Jiang's dynamic continuum model for urban cities. Transp Res Part B: Methodol 56:96–119
210. Duan HX, Yuan YR, Yang C et al (2015) Anticipatory processes under academic stress: an ERP study. Brain Cogn 94:60–67
211. Duan HX, Yuan YR, Zhang L et al (2013) Chronic stress exposure decreases the cortisol awakening response in healthy young men. Stress 16(6):630–637
212. Duan W, Cao ZD, Wang YZ et al (2013) An ACP approach to public health emergency management: using a campus outbreak of H1N1 influenza as a case study. IEEE Trans Syst Man Cybern: Syst 43(5):1028–1041
213. Duan W, Fan ZC, Zhang P et al (2015) Mathematical and computational approaches to epidemic modeling: a comprehensive review. Front Comput Sci 9(5):806–826
214. Duan W, Qiu XG (2012) Fostering artificial societies using social learning and social control in parallel emergency management systems. Front Comput Sci 6(5):604–610
215. Duan W, Qiu XG, Cao ZD et al (2013) Heterogeneous and stochastic agent-based models for analyzing infectious diseases' super spreaders. IEEE Intell Syst 28(4):18–25
216. Duan W, Quax R, Lees M et al (2014) Topology dependent epidemic spreading velocity in weighted networks. J Stat Mech: Theory Exp 12:P12020
217. Dynich A, Wang YZ (2016) On the e-commercialization of science: a step toward the future? Electron Commer Res Appl 20:183–188
218. Ergu D, Kou G, Peng Y et al (2016) Estimating the missing values for the incomplete decision matrix and consistency optimization in emergency management. Appl Math Model 40(1):254–267
219. Ergu D, Kou G (2013) Data inconsistency and incompleteness processing model in decision matrix. Stud Inf Control 22(4):359–366
220. Ergu D, Kou G, Peng Y et al (2014) Simulation experiments for improving the consistency ratio of reciprocal matrices. Int J Comput Commun Control 9(4):408–418
221. Ergu D, Peng Y (2014) A framework for SaaS software packages evaluation and selection with virtual team and BOCR of analytic network process. J Supercomput 67(1):219–238
222. Fan B (2013) Achieving horizontal integration of municipal e-government in China: assessment of managerial mechanisms. Inf Dev 29(4):333–348
223. Fan B (2013) The impact of information technology capability, information sharing and government process redesign on the operational performance of emergency incident management systems. Inf Res 18(4):591

224. Fan B (2014) Hybrid spatial data mining methods for site selection of emergency response centers. Nat Hazards 70(1):643–656
225. Fan B, Chen L, Chong YT et al (2016) Mining for spatio-temporal distribution rules of illegal dumping from large dataset. ACM SIGMIS Datab: DATAB Adv Inf Syst 47(3):41–53
226. Fan B, Jiang H (2016) Legislative policies for the dominant tasks of e-government push in the Chinese context. Inf Dev 32(4):953–968
227. Fan B, Li JH (2011) Emergency resource planning by using spatial data association rule mining and linear programming method. In: 4th international joint conference on computational sciences and optimization (CSO 2011), pp 613–617
228. Fan B, Luo JF (2014) Benchmarking scale of e-government stage in Chinese municipalities from government chief information officers' perspective. Inf Syst e-Bus Manag 12(2):259–284
229. Fan B, Luo JF (2013) Spatially enabled emergency event analysis using a multi-level association rule mining method. Nat Hazards 67(2):239–260
230. Fang BX, Jia Y, Han Y et al (2014) A survey of social network and information dissemination analysis. Chin Sci Bull 59(32):4163–4172
231. Fang ZG, Liu SF, Shi HX et al (2016) Grey game theory and its applications in economic decision-making. CRC Press, New York
232. Fang ZG, Wei HW, Yang BH et al (2010) The design of the driving-coupling algorithm in unconventional incidents based on the GERTS network. In: Liu SF, Forrest JY-L (eds) Advances in grey systems research. Springer, Berlin, Heidelberg, pp 3–11
233. Fang ZM, Song WG, Li ZJ et al (2012) Experimental study on evacuation process in a stairwell of a high-rise building. Build Environ 47:316–321
234. Fang ZM, Song WG, Liu X et al (2012) A continuous distance model (CDM) for the single-file pedestrian movement considering step frequency and length. Phys A: Stat Mech Appl 391(1–2):307–316
235. Fang ZM, Song WG, Zhang J et al (2012) A multi-grid model for evacuation coupling with the effects of fire products. Fire Technol 48(1):91–104
236. Fu LB, Song WG, Lü W et al (2014) Simulation of emotional contagion using modified SIR model: a cellular automaton approach. Phys A: Stat Mech Appl 405:380–391
237. Fu M, Weng WG, Yuan HY (2014) Effects of multiple air gaps on the thermal performance of firefighter protective clothing under low-level heat exposure. Text Res J 84(9):968–978
238. Fu M, Weng WG, Yuan HY (2014) Numerical simulation of the effects of blood perfusion, water diffusion, and vaporization on the skin temperature and burn injuries. Numer Heat Transf Part A: Appl 65(12):1187–1203
239. Fu M, Weng WG, Yuan HY (2014) Thermal insulations of multilayer clothing systems measured by a bench scale test in low level heat exposures. Int J Cloth Sci Technol 26(5):412–423
240. Fu TJ, Abbasi A, Zeng D et al (2012) Sentimental spidering: leveraging opinion information in focused crawlers. ACM Trans Inf Syst (TOIS) 30(4):24
241. Fu TR, Tan P, Zhong MH (2012) Experimental research on the influence of surface conditions on the total hemispherical emissivity of iron-based alloys. Exp Thermal Fluid Sci 40:159–167
242. Fu TR, Zhao H, Zeng J et al (2010) Improvements to three-color optical CCD-based pyrometer system. Appl Opt 49(31):5997–6005
243. Fu ZJ, Yang LZ, Chen YQ et al (2013) The effect of individual tendency on crowd evacuation efficiency under inhomogeneous exit attraction using a static field modified FFCA model. Phys A: Stat Mech Appl 392(23):6090–6099
244. Fu ZJ, Yang LZ, Rao P et al (2013) Interactions of pedestrians interlaced in T-shaped structure using a modified multi-field cellular automaton. Int J Mod Phys C 24(4):1350024
245. Fu ZJ, Zhou XD, Zhu KJ et al (2015) A floor field cellular automaton for crowd evacuation considering different walking abilities. Phys A: Stat Mech Appl 420:294–303
246. Gan YQ, Wen Y, Wang JY et al (2014) Is expressive suppression an effective coping strategy? A study of Chinese rescue medical staff following an earthquake. Asian J Soc Psychol 17(4):264–276

247. Gan YQ, Xie XF, Wang T et al (2013) Thriving in the shadow of the 2008 Sichuan earthquake: two studies on resilience in adolescents. J Health Psychol 18(9):1232–1241
248. Gao D, Deng X, Bai B (2014) The emergence of organizational routines from habitual behaviours of multiple actors: an agent-based simulation study. J Simul 8(3):215–230
249. Gao F, Fei FX, Lee XJ et al (2014) Inversion mechanism with functional extrema model for identification incommensurate and hyper fractional chaos via differential evolution. Expert Syst Appl 41(4):1915–1927
250. Gao F, Lee T, Cao WJ et al (2016) Self-evolution of hyper fractional order chaos driven by a novel approach through genetic programming. Expert Syst Appl 52:1–15
251. Gao F, Lee XJ, Fei FX et al (2013) Parameter identification for Van Der Pol-Duffing oscillator by a novel artificial bee colony algorithm with differential evolution operators. Appl Math Comput 222:132–144
252. Gao F, Lee XJ, Fei FX et al (2014) Identification time-delayed fractional order chaos with functional extrema model via differential evolution. Expert Syst Appl 41(4):1601–1608
253. Gao F, Lee XJ, Tong HQ et al (2013) Identification of unknown parameters and orders via Cuckoo search oriented statistically by differential evolution for noncommensurate fractional-order chaotic systems. Abstr Appl Anal 2013:1–19
254. Gao HM, Cao HB, Zhou Y et al (2014) Taking pleasure at another's misfortune: the implicit Schadenfreude of disaster spectators. Psychol Rep 114(2):439–460
255. Gao HM, Zhang Y, Wang F et al (2014) Regret causes ego-depletion and finding benefits in the regrettable events alleviates ego-depletion. J Gen Psychol 141(3):169–206
256. Gao TT, Rong LL (2014) Study on management of the life cycle of emergency plan system based on effectiveness. Int J Knowl Syst Sci 5(1):49–64
257. Gao Y, Chen T, Luh PB et al (2016) Modified social force model based on predictive collision avoidance considering degree of competitiveness. Fire Technol 53(1):331–351
258. Ge Y, Qu WN, Jiang CH et al (2014) The effect of stress and personality on dangerous driving behavior among Chinese drivers. Accid Anal Prev 73:34–40
259. Ge Y, Wu JH, Sun XH et al (2011) Enhanced mismatch negativity in adolescents with posttraumatic stress disorder (PTSD). Int J Psychophysiol 79(2):231–235
260. Ge YZ, Liu L, Qiu XG et al (2013) A framework of multilayer social networks for communication behavior with agent-based modeling. Simulation 89(7):810–828
261. Ge YZ, Meng RQ, Cao ZD et al (2014) Virtual city: an individual-based digital environment for human mobility and interactive behavior. Simulation 90(8):917–935
262. Ge YZ, Song ZC, Qiu XG et al (2013) Modular and hierarchical structure of social contact networks. Phys A: Stat Mech Appl 392(19):4619–4628
263. Gong K, Tang M, Yang H et al Variability of contact process in complex networks. Chaos: An Interdiscip J Nonlinear Sci 21(4):043130 (2011)
264. Gualdi S, Medo M, Zhang YC (2011) Influence, originality and similarity in directed acyclic graphs. EPL (Europhys Lett) 96(1):18004
265. Gualdi S, Yeung CH, Zhang YC (2011) Tracing the evolution of physics on the backbone of citation networks. Phys Rev E 84(4):046104
266. Guan NY, Tao DC, Luo ZG et al (2011) Manifold regularized discriminative nonnegative matrix factorization with fast gradient descent. IEEE Trans Image Process 20(7):2030–2048
267. Guan NY, Tao DC, Luo ZG et al (2011) Non-negative patch alignment framework. IEEE Trans Neural Netw 22(8):1218–1230
268. Guan NY, Tao DC, Luo ZG et al (2012) NeNMF: an optimal gradient method for nonnegative matrix factorization. IEEE Trans Signal Process 60(6):2882–2898
269. Guan NY, Tao DC, Luo ZG et al (2012) Online nonnegative matrix factorization with robust stochastic approximation. IEEE Trans Neural Netw Learn Syst 23(7):1087–1099
270. Guan YY, Cai SM, Shang MS (2014) Recommendation algorithm based on item quality and user rating preferences. Front Comput Sci 8(2):289–297
271. Guo DK, Liu YH, Jin H et al (2012) Theory and network applications of balanced Kautz tree structures. ACM Trans Internet Technol 12(1):3

272. Guo DH, Zhou H, Zou Y et al (2013) Geographical analysis of the distribution and spread of human rabies in China from 2005 to 2011. PLoS ONE 8(8):e72352
273. Guo G, Qiu XG (2014) Accuracy of lattice Monte Carlo simulation of biased diffusion models. Phys Rev E 90(4):043305
274. Guo G, Qiu XG (2014) First passage time distributions of anomalous biased diffusion with double absorbing barriers. Phys A: Stat Mech Appl 411:80–86
275. Guo SN, Meng XF (2017) Density peaks clustering with differential privacy. In: 8th biennial conference on innovative data systems research
276. Guo XJ, Liu SF, Wu LF et al (2015) A multi-variable grey model with a self-memory component and its application on engineering prediction. Eng Appl Artif Intell 42:82–93
277. Guo XT, Yuan JQ, Cao XF et al (2012) Understanding the acceptance of mobile health services: a service participants analysis. In: International conference on management science and engineering (ICMSE 2012), pp 1868–1873
278. Guo XY, Zhao N, Wang S et al (2010) Player-spectator discrepancies on risk preference during decision making. J Gen Psychol: Exp Psychol Comput Psychol 137(2):210–224
279. Guo XY, Zheng L, Zhu L et al (2011) Acquisition of conscious and unconscious knowledge of semantic prosody. Conscious Cogn 20(2):417–425
280. Guo XY, Zhu L, Zheng L et al (2013) Effects of task orientation on subsequent source memory as revealed by functional MRI. Neural Regen Res 8(26):2424–2431
281. Han SH, Northoff G, Vogeley K et al (2013) A cultural neuroscience approach to the biosocial nature of the human brain. Annu Rev Psychol 64:335–359
282. Han Y, Davidson RA, Su G et al (2011) An efficient simulation-based seismic hazard analysis method. In: Proceedings of the 3rd international conference on risk analysis and crisis response, pp 250–255
283. Han Y, Fang BX, Jia Y (2014) Predicting the topic influence trends in social media with multiple models. Neurocomputing 144:463–470
284. Han ZY, Weng WG, Huang QY et al (2015) Aerodynamic characteristics of human movement behaviours in full-scale environment: comparison of limbs pendulum and body motion. Indoor Built Environ 24(1):87–100
285. Han ZY, Weng WG, Huang QY (2013) Characterizations of particle size distribution of the droplets exhaled by sneeze. J R Soc Interface 10(88):20130560
286. Han Z, Lu X, Hörhager EI et al (2017) The effects of trust in government on earthquake survivors' risk perception and preparedness in China. Nat Hazards 86(1):437–452
287. Han ZY, Weng WG, Huang QY (2014) Numerical and experimental investigation on the dynamic airflow of human movement in a full-scale cabin. HVAC&R Res 20(4):444–457
288. Han ZY, Weng WG, Zhao QL et al (2013) Investigation on an integrated evacuation route planning method based on real-time data acquisition for high-rise building fire. IEEE Trans Intell Transp Syst 14(2):782–795
289. He H, Mei S, Zhu YF (2014) Timed influence nets based evaluation of the effectiveness of public health emergency scenario. Adv Mater Res 971–973:2442–2447
290. He KJ, Wang LJ, Zou YC et al (2014) Exchange rate forecasting using entropy optimized multivariate wavelet denoising model. Math Probl Eng 2014:389598
291. He KJ, Wang LJ, Zou YC et al (2014) Value at risk estimation with entropy-based wavelet analysis in exchange markets. Phys A: Stat Mech Appl 408:62–71
292. He KJ, Yu LA, Tang L (2015) Electricity price forecasting with a BED (Bivariate EMD Denoising) methodology. Energy 91:601–609
293. He L, Jia Y, Ding ZY et al (2014) Hierarchical classification with a topic taxonomy via LDA. Int J Mach Learn Cybern 5(4):491–497
294. He L, Jia Y, Han WH et al (2014) Mining user interest in microblogs with a user-topic model. China Commun 11(8):131–144
295. He SX, Liu XF, Wang CM (2016) A nonlinear Lagrange algorithm for minimax problems with general constraints. Numer Funct Anal Optim 37(6):680–698
296. He SX, Wei M, Tong HQ (2015) A smooth penalty-based sample average approximation method for stochastic complementarity problems. J Comput Appl Math 287:20–31

297. He SX, Wu LX (2014) Convergence analysis of a nonlinear Lagrange algorithm for general nonlinear constrained optimization problems. Appl Math—J Chin Univ 29(3):352–366
298. He SX, Zhang LW, Zhang J (2015) The rate of convergence of an NLM based on F-B NCP for constrained optimization problems without strict complementarity. Asia-Pac J Oper Res 32(3):1–27
299. He W, Zhang YC, Zhu JP et al (2011) Could sex difference in color preference and its personality correlates fit into social theories? Let Chinese university students tell you. Personal Individ Differ 51(2):154–159
300. He YY, Liu Z, Shi JM et al (2015) K-shortest-path-based evacuation routing with police resource allocation in city transportation networks. PLoS ONE 10(7):e0131962
301. He Z, Fan B, Cheng TCE et al (2016) A mean-shift algorithm for large-scale planar maximal covering location problems. Eur J Oper Res 250(1):65–76
302. He ZY, Liu SJ, Li M et al (2013) Learning entity representation for entity disambiguation. In: Proceedings of the 51st annual meeting of the association for computational linguistics. Short papers, vol 2, pp 30–34
303. He ZY, Wang HF (2012) A comparison and improvement of online learning algorithms for sequence labeling. In: Proceedings of COLING 2012, pp 1147–1162
304. Hou L, Lao YT, Wang YH et al (2013) Modeling freeway incident response time: a mechanism-based approach. Transp Res Part C: Emerg Technol 28:87–100
305. Hou L, Liu JG, Pan X et al (2014) A social force evacuation model with the leadership effect. Phys A: Stat Mech Appl 400:93–99
306. Hu CL, Liu X, Hua YK (2016) A bi-objective robust model for emergency resource allocation under uncertainty. Int J Prod Res 54(24):7421–7438
307. Hu SL, Han CF, Meng LP (2015) A scenario planning approach for propositioning rescue centers for urban waterlog disasters. Comput Ind Eng 87:425–435
308. Hu SL, Han CF, Meng LP (2016) Stochastic optimization for investment in facilities in emergency prevention. Transp Res Part E: Logist Transp Rev 89:14–31
309. Hu TY, Li JY, Jia HY et al (2016) Helping others, warming yourself: altruistic behaviors increase warmth feelings of the ambient environment. Front Psychol 7:1349
310. Hu XH, Mu T, Dai WH et al (2012) Analysis of browsing behaviors with ant colony clustering algorithm. J Comput 7(12):3096–3102
311. Hu XH, Zhou X, Dai WH et al (2012) Tourism crisis management system based on ecological mechanism. J Softw 7(12):2808–2815
312. Hu YQ, Gan YQ, Liu Y (2012) How Chinese people infer helpers' ambiguous intentions: helper effort and interpersonal relationships. Int J Psychol 47(5):393–404
313. Huang C, Huang QY, Zhong SB et al (2013) Case reuse based on fuzzy reasoning—adaption for emergency management. Appl Mech Mater 333–335:1324–1327
314. Huang D, Zhang YB, Zhang K (2014) The effects of the relationships between object features on multiple-identity tracking. Exp Psychol 61(5):340–346
315. Huang H, Kato S, Hu R (2012) Optimum design for indoor humidity by coupling genetic algorithm with transient simulation based on contribution ratio of indoor humidity and climate analysis. Energy Build 47:208–216
316. Huang J, Xue F, Song XF (2013) Simulation analysis on policy interaction effects between emission trading and renewable energy subsidy. J Mod Power Syst Clean Energy 1(2):195–201
317. Huang JC, Cheng GQ, Liu Z et al (2012) Synthetic aperture radar image compression using tree-structured edge-directed orthogonal wavelet packet transform. AEU—Int J Electron Commun 66(3):195–203
318. Huang JY, Yao J, Sun J et al (2012) A new coupled map car-following model under inter-vehicle communication. In: 12th international conference on control, automation, robotics and vision (ICARCV 2012), pp 430–435
319. Huang QL, Jia QS, Guan XH (2018) Robust scheduling of EV charging load with uncertain wind power integration. IEEE Trans Smart Grid 9(2):1043–1054
320. Huang QW, Jia Q, Li Q et al (2012) Research on capability of continuous reconnaissance from UAV formation in condition of multi-target. J Syst Simul 7:31

321. Huang YQ, Choe Y, Lee S et al (2018) Drinking tea improves the performance of divergent creativity. Food Qual Prefer 66:29–35
322. Huang YH, Zhang WS, Wang J (2012) Multiple instance learning with correlated features. International Conference on Information Science and Technology (ICIST 2012), pp 441–446
323. Huang ZB, Liu D, Lü D et al (2014) Signature-based data association for multiple target tracking. In: International conference on software engineering and information technology (SEIT 2014)
324. Ji J, Fan CG, Zhong W et al (2012) Experimental investigation on influence of different transverse fire locations on maximum smoke temperature under the tunnel ceiling. Int J Heat Mass Transf 55(17–18):4817–4826
325. Ji J, Gao ZH, Fan CG et al (2012) A study of the effect of plug-holing and boundary layer separation on natural ventilation with vertical shaft in urban road tunnel fires. Int J Heat Mass Transf 55(21–22):6032–6041
326. Jia HY, Lu JY, Xie XF et al (2015) When your strength threatens me: supervisors show less social comparison bias than subordinates. J Occup Organ Psychol 89(3):568–587
327. Jia QS (2010) A structural property of optimal policies for multi-component maintenance problems. IEEE Trans Autom Sci Eng 7(3):677–680
328. Jia QS (2011) An adaptive sampling algorithm for simulation-based optimization with descriptive complexity preference. IEEE Trans Autom Sci Eng 8(4):720–731
329. Jia QS (2011) On solving event-based optimization with average reward over infinite stages. IEEE Trans Autom Control 56(12):2912–2917
330. Jia QS (2012) Efficient computing budget allocation for simulation-based policy improvement. IEEE Trans Autom Sci Eng 9(2):342–352
331. Jia QS (2011) On solving optimal policies for finite-stage event-based optimization. IEEE Trans Autom Control 56(9):2195–2200
332. Jia QS (2011) On state aggregation to approximate complex value functions in large-scale Markov decision processes. IEEE Trans Autom Control 56(2):333–344
333. Jia QS (2018) On supply demand coordination in vehicle-to-grid—a brief literature review. In: 33rd youth academic annual conference of chinese association of automation (YAC 2018), pp 1083–1088
334. Jia QS, Shi L, Mo YL et al (2012) On optimal partial broadcasting of wireless sensor networks for Kalman filtering. IEEE Trans Autom Control 57(3):715–721
335. Jia QS, Wu JJ, Wu ZJ et al (2018) Event-based HVAC control—a complexity-based approach. IEEE Trans Autom Sci Eng 15(4):1909–1919
336. Jia QS, Zhang YM, Zhao QC (2018) Controlling the internet of things: from energy saving to fast evacuation in smart buildings. In: Intelligent building control systems. Springer, Cham, Switzerland, pp 293–310
337. Jia QS, Zhao QC (2009) Strategy optimization for controlled Markov process with descriptive complexity constraint. Sci China Ser F: Inf Sci 52(11):1993
338. Jian LR, Liu SF, Lin Y (2011) Hybrid rough sets and applications in uncertain decision-making. CRC Press, New York
339. Jian M, Song WG, Lo SM (2011) Flow dilemma in the game of bottleneck conflicts. In: International conference on remote sensing, environment and transportation engineering (RSETE 2011), pp 369–372
340. Jiang CS, Zheng S, Yuan F et al (2012) Experimental assessment on the moving capabilities of mobility-impaired disabled. Saf Sci 50(4):974–985
341. Jiang FY, Song LT, Fan YD et al (2013) Applying fuzzy weighted average approach for the selection of the emergency supplies storage location. J Comput Inf Syst 9(10):4101–4110
342. Jiang FY, Song LT, Fan YD et al (2013) Disaster rescue simulation based on complex adaptive theory. Indones J Electr Eng Comput Sci 11(5):2402–2408
343. Jiang Y, Shu CW, Zhang MP (2013) An alternative formulation of finite difference weighted ENO schemes with Lax-Wendroff time discretization for conservation laws. SIAM J Sci Comput 35(2):A1137–A1160

344. Jiang ZY, Jia QS, Guan XH (2017) Optimization of sensor location for improving wind power prediction accuracy. In: 13th IEEE conference on automation science and engineering (CASE 2017), pp 1220–1225
345. Jing JB, Luo XF, Xuan JY et al (2014) Cognition-based semantic annotation for web images. In: 4th international conference on big data and cloud computing (BdCloud 2014), pp 540–546
346. Kang P, Lü YP, Hao L et al (2015) Psychological consequences and quality of life among medical rescuers who responded to the 2010 Yushu earthquake: a neglected problem. Psychiatry Res 230(2):517–523
347. Kang P, Tang BH, Liu Y et al (2015) Medical efforts and injury patterns of military hospital patients following the 2013 Lushan earthquake in China: a retrospective study. Int J Environ Res Public Health 12(9):10723–10738
348. Kang P, Zhang LL, Liang WN et al (2012) Medical evacuation management and clinical characteristics of 3,255 inpatients after the 2010 Yushu earthquake in China. J Trauma Acute Care Surg 72(6):1626–1633
349. Kang Q, An J, Wang L et al (2012) Unification and diversity of computation models for generalized swarm Intelligence. Int J Artif Intell Tools 21(3):1–18
350. Kang Q, Lan T, Yan Y et al (2011) Swarm-based optimal power flow considering generator fault in distribution systems. In: IEEE international conference on systems, man, and cybernetics (ICSMC 2011), pp 786–790
351. Kang Q, Lan T, Yan Y et al (2012) Group search optimizer based optimal location and capacity of distributed generations. Neurocomputing 78(1):55–63
352. Kang Q, Zhou MC, An J et al (2013) Swarm intelligence approaches to optimal power flow problem with distributed generator failures in power networks. IEEE Trans Autom Sci Eng 10(2):343–353
353. Ke H, Ma JJ (2014) Modeling project time-cost trade-off in fuzzy random environment. Appl Soft Comput 19:80–85
354. Kong JJ, Han CF (2013) Interaction of multi-interorganizational relationships of emergency organizations of 2008 Wenchuan Earthquake in China. In: IEEE international conference on systems, man, and cybernetics, pp 1276–1281
355. Kou G, Ergu D, Shang J (2014) Enhancing data consistency in decision matrix: adapting hadamard model to mitigate judgment contradiction. Eur J Oper Res 236(1):261–271
356. Lai MH, Li YJ, Liu YL (2010) Determining the optimal scale width for a rating scale using an integrated discrimination function. Measurement 43(10):1458–1471
357. Lam L, Bellavia DC, Han XP et al (2010) Bilinear effect in complex systems. EPL (Europhys Lett) 91(6):68004
358. Lan T, Kang Q, An J et al (2013) Sitting and sizing of aggregator controlled park for plug-in hybrid electric vehicle based on particle swarm optimization. Neural Comput Appl 22(2):249–257
359. Li B, Zheng CH, Huang DS et al (2010) Gene expression data classification using locally linear discriminant embedding. Comput Biol Med 40(10):802–810
360. Li CM, Wang FZ, Wei XD et al (2011) Solution method of optimal scheme set for water resources scheduling group decision making based on multi-agent computation. Intell Autom Soft Comput 17(7):871–883
361. Li DD, Yang YC, Dai WH (2014) Cost-sensitive learning for emotion robust speaker recognition. Sci World J 2014:628516
362. Li FY, Guo XY, Zhu L et al (2013) Implicit learning of mappings between forms and metaphorical meanings. Conscious Cogn 22(1):174–183
363. Li GH, Lu S, Chen XD et al (2014) Study on correlation factors that influence terrorist attack fatalities using global terrorism database. Procedia Eng 84:698–707
364. Li GH, Lu S, Mei P et al (2014) Influences of time, location, and cause factors on the probability of fire loss in China: a correspondence analysis. Fire Technol 50(5):1181–1200
365. Li GH, Lu S, Zhang HP et al (2013) Correspondence analysis on exploring the association between fire causes and influence factors. Procedia Eng 62:581–591

366. Li H, Li J, Chang PC et al (2013) Parametric prediction on default risk of Chinese listed tourism companies by using random oversampling, isomap, and locally linear embeddings on imbalanced samples. Int J Hosp Manag 35:141–151
367. Li HJ, Daniels JJ (2015) Social significance of community structure: statistical view. Phys Rev E 91(1):012801
368. Li J, Du WH, Yang FM et al (2014) Evolutionary game analysis of remanufacturing closed-loop supply chain with asymmetric information. Sustainability 6(9):6312–6324
369. Li J, Du WH, Yang FM et al (2014) The carbon subsidy analysis in remanufacturing closed-loop supply chain. Sustainability 6(6):3861–3877
370. Li J, Liu XF, Wu J et al (2014) Coordination of supply chain with a dominant retailer under demand disruptions. Math Probl Eng 2014:854681
371. Li JH, Zhou YC, Shang W et al (2013) A cloud computation architecture for unconventional emergency management. In: International conference on web-age information management (WAIM 2013). Springer, Berlin, Heidelberg, pp 187–198
372. Li JH, Jia QS, Guan XH et al (2011) Tracking a moving object via a sensor network with partial information broadcasting scheme. Inf Sci 181(20):4733–4753
373. Li LH, Jin XM, Long MS (2012) Topic correlation analysis for cross-domain text classification. In: Proceedings of the 26th AAAI conference on artificial intelligence (AAAI 2012), pp 998–1004
374. Li LH, Jin XM, Pan SJL et al (2012) Multi-domain active learning for text classification. In: Proceedings of the 18th ACM SIGKDD international conference on knowledge discovery and data mining, pp 1086–1094
375. Li L, Luo XF (2014) Automatic student grouping method for foreign language learning. In: 10th international conference on semantics, knowledge and grids (SKG 2014), pp 63–66
376. Li M, Liu L, Li CB (2011) An approach to expert recommendation based on fuzzy linguistic method and fuzzy text classification in knowledge management systems. Expert Syst Appl 38(7):8586–8596
377. Li M, Liu RR, Jia CX et al (2013) Critical effects of overlapping of connectivity and dependence links on percolation of networks. New J Phys 15(9):093013
378. Li N, Ma N, Liu Y et al (2013) Resting-state functional connectivity predicts impulsivity in economic decision-making. J Neurosci 33(11):4886–4895
379. Li WM (2018) Research on the integration method of total station indoor and field work based on mobile device. Geospatial Inf
380. Li X, Guo SN, Yu LA (2015) Skewness of fuzzy numbers and its applications in portfolio selection. IEEE Trans Fuzzy Syst 23(6):2135–2143
381. Li X, He KJ, Lai KK et al (2014) Forecasting crude oil price with multiscale denoising ensemble model. Math Probl Eng 2014:716571
382. Li X, Ma J, Shang W et al (2014) How does public attention influence natural gas price: new evidence with Google search data. Int J Knowl Syst Sci 5(2):65–80
383. Li X, Shang W, Wang SY et al (2015) A MIDAS modelling framework for Chinese inflation index forecast incorporating Google search data. Electron Commer Res Appl 14(2):112–125
384. Li X, Su GF, Zhong SB et al (2014) Study on scene-driven emergency drill method. In: Practical applications of intelligent systems. Springer, Berlin, Heidelberg, pp 1089–1097
385. Li XB, Lei YL, Vangheluwe H et al (2013) A multi-paradigm decision modeling framework for combat system effectiveness measurement based on domain-specific modeling. J Zhejiang University Sci C 14(5):311–331
386. Li XB, Lei YL, Vangheluwe H et al (2014) Domain-specific decision modelling and statistical analysis for combat system effectiveness simulation. J Stat Comput Simul 84(6):1261–1279
387. Li XC, Mao WJ, Zeng D (2012) Forecasting complex group behavior via multiple plan recognition. Front Comput Sci 6(1):102–110
388. Li Y, Shan SQ, Liu L (2012) A clustering coefficient preference model on scale-free networks. J Converg Inf Technol 7(4):231–237
389. Li Y, Wang YZ, Cui Q (2016) Energy efficiency measures for airlines: an application of virtual frontier dynamic range adjusted measure. J Renew Sustain Energy 8(1):207–232

390. Li Y, Wang YZ, Cui Q (2016) Has airline efficiency affected by the inclusion of aviation into European Union emission trading scheme? Evidences from 22 airlines during 2008–2012. Energy 96:8–22
391. Li YX, Xu JH (2011) Legislation concerning the protection of the right to online privacy in China: a comparative study with EU. In: 2nd worldwide cybersecurity summit (WCS 2011), pp 1–5
392. Li Z, Xu W, Zhang LK et al (2014) An ontology-based Web mining method for unemployment rate prediction. Decis Support Syst 66:114–122
393. Liang J, Ma HY, Bos K et al (2016) Ease-of-retrieval effects on procedural justice judgements under conditions of informational and personal uncertainty. Asian J Soc Psychol 19(4):336–346
394. Liao ZM, Yang S, Liang JN (2010) Detection of abnormal crowd distribution. In: Proceedings of the 2010 IEEE/ACM int'l conference on green computing and communications and int'l conference on cyber, physical and social computing, pp 600–604
395. Lin WP, Ma JJ, Wang L et al (2015) A double-edged sword: the moderating role of conscientiousness in the relationships between work stressors, psychological strain, and job performance. J Organ Behav 36(1):94–111
396. Lin WP, Wang L, Bamberger PA et al (2016) Leading future orientations for current effectiveness: the role of engagement and supervisor coaching in linking future work self salience to job performance. J Vocat Behav 92:145–156
397. Lin YT, Yang HX, Rong ZH et al (2010) Effects of heterogeneous influence of individuals on the global consensus. Int J Mod Phys C 21(8):1011–1019
398. Lin ZJ, Ding GG, Wang JM (2011) Image annotation based on recommendation model. In: Proceedings of the 34th international ACM SIGIR conference on research and development in information retrieval, pp 1097–1098
399. Linghu B, Chen F, Guo XT et al (2013) A conceptual model for flood risk assessment based on agent-based modeling. In: International conference on IEEE computer sciences and applications (CSA 2013), pp 369–373
400. Liu C, Xie JR, Chen HS et al (2015) Interplay between the local information based behavioral responses and the epidemic spreading in complex networks. Chaos: Interdiscip J Nonlinear Sci 25(10):103111
401. Liu C, Yeung CH, Zhang ZK (2011) Self-organization in social tagging systems. Phys Rev E 83(6):066104
402. Liu C, Yu S, Qian J et al (2017) A scenario representation model for emergency decision support. Procedia Comput Sci 107:301–305
403. Liu D, Liu X, Wang WJ et al (2014) Semi-supervised community detection based on discrete potential theory. Phys A: Stat Mech Appl 416:173–182
404. Liu D, Zhong SB, Huang QY (2015) Study on risk assessment framework for snowmelt flood and hydro-network extraction from watersheds. In: Bian FL, Xie YC (eds) Geo-informatics in resource management and sustainable ecosystem. Springer, Berlin, Heidelberg, pp 638–651
405. Liu DY, Huang ZP, Wu JK (2013) Model-driven multi-target tracking in crowd scenes. In: 16th international conference on information fusion, vol 1, pp 1495–1501
406. Liu G, Sha YZ (2011) Building inventory management model for relief supplies based on OGC and SOA. In: 4th international symposium on computational intelligence and design, pp 368–371
407. Liu GF, Wang HM, Tong JP (2011) Scenario construction of flood emergency management in river basin based on scene perception. In: Wu DDD, Zhou Y (eds) Modeling risk management for resources and environment in China. Springer, Berlin, Heidelberg, pp 119–127
408. Liu J, Li B, Zhang WS (2012) Feature extraction using maximum variance sparse mapping. Neural Comput Appl 21(8):1827–1833
409. Liu J, Li J, Sun XP et al (2014) An Embedded Co-AdaBoost based construction of software document relation coupled resource spaces for cyber-physical society. Futur Gener Comput Syst 32:198–210

410. Liu J, Zhou J, Wang JF et al (2012) Irregular community discovery for cloud service improvement. J Supercomput 61(2):317–336
411. Liu J, Zhu JM, Huang J (2012) Multi-stage multi-objective engineering evaluation method for the ability of the emergency resources reserve system. Syst Eng Procedia 5:43–48
412. Liu JF, Fan YD, Shi PJ (2011) Response to a high-altitude earthquake: the Yushu earthquake example. Int J Disaster Risk Sci 2(1):43–53
413. Liu JF, Zhou YS (2012) Predicting earthquakes: the Mw9.0 Tohoku earthquake and historical earthquakes in northeastern Japan. Int J Disaster Risk Sci 3(3):155–162
414. Liu KF, Liu X (2011) Emergency policy for perishable products supply chain when nodes fail. Adv Mater Res 317–319:2097–2102
415. Liu KJ, Wang W (2011) Reliability of statistical tests for detecting changes in extreme precipitation. In: Ren LL, Wang W, Yuan F (eds) Hydrological cycle and water resources sustainability in changing environments. IAHS Publication, Wallingford, UK, pp 645–650
416. Liu KP, Fang BX, Zhang WZ (2011) Exploring social relations for personalized tag recommendation in social tagging systems. IEICE Trans Inf Syst 94(3):542–551
417. Liu MW, Feng MY, Wong CY (2014) Flexible service policies for a Markov inventory system with two demand classes. Int J Prod Econ 151:180–185
418. Liu N, Ye Y (2014) Humanitarian logistics planning for natural disaster response with Bayesian information updates. J Ind Manag Optim (JIMO) 10(3):665–689
419. Liu RR, Wang WX, Lai YC et al (2011) Optimal convergence in naming game with geography-based negotiation on small-world networks. Phys Lett A 375(3):363–367
420. Liu SF, Forrest JYL, Yang YJ (2013) Advances in grey systems research. J Grey Syst 25(2):1–18
421. Liu SF, Forrest JYL, Yang YJ (2015) Grey system: thinking, methods, and models with applications. In: Zhou MC, Li HX, Weijnen M (eds) Contemporary issues in systems science and engineering. Wiley, Hoboken, NJ
422. Liu SF, Lin Y (2011) Grey systems: theory and applications. Springer, Berlin
423. Liu SF, Sheng KQ, Forrest J et al (2012) On uncertain systems and uncertain models. Kybernetes 41(5–6):548–558
424. Liu SF, Tao LY, Xie NM et al (2015) On the new model system and framework of grey system theory. In: 2015 IEEE international conference on grey systems and intelligent services (GSIS 2015), pp 1–11
425. Liu SF, Xie NM, Yuan CQ et al (2012) Systems evaluation: methods, models, and applications. CRC Press, New York
426. Liu SF, Xu B, Forrest J et al (2013) On uniform effect measure functions and a weighted multi-attribute grey target decision model. J Grey Syst 25(1):1–11
427. Liu SF, Yang YJ, Forrest J (2016) Grey data analysis. Springer, Singapore
428. Liu SF, Zeng B, Liu JF et al (2015) Four basic models of GM (1,1) and their suitable sequences. Grey Syst: Theory Appl 5(2):141–156
429. Liu T, Bu RG, Huang J (2012) A cost-efficiency equilibrium problem of regional single emergency resource guarantee with multi-objective programming. Syst Eng Procedia 5:29–36
430. Liu X, Ghorpade A, Tu YL et al (2012) A novel approach to probability distribution aggregation. Inf Sci 188:269–275
431. Liu X, Ghorpade A (2012) An optimization approach to distribution aggregation for operations management. Inf Sci 46(20):5781–5795
432. Liu X, Li W, Tu YL et al (2011) An expert system for an emergency response management in networked safe service systems. Expert Syst Appl 38(9):11928–11938
433. Liu X, Liu Y, Zhang LL et al (2013) Mass aeromedical evacuation of patients in an emergency: experience following the 2010 Yushu earthquake. J Emerg Med 45(6):865–871
434. Liu X, Zhang LL, Liu Y et al (2012) Critical care aeromedical evacuation staff in Batang airport after the Yushu earthquake at high altitude. Aviat Space Environ Med 83(4):436–440
435. Liu XC, Wang C, Zhou BB et al (2013) Priority-based consolidation of parallel workloads in the cloud. IEEE Trans Parallel Distrib Syst 24(9):1874–1883

436. Liu XJ, Yuan YY, Fang BX et al (2014) A strategy optimization approach for mission deployment in distributed systems. Math Probl Eng 404681
437. Liu Y, Fan ZP, Chen FD (2011) A risk decision method for emergency response based on cumulative prospect theory. In: 4th international joint conference on computational sciences and optimization (CSO 2011), pp 618–622
438. Liu Y, Fan ZP, Yuan Y et al (2014) A FTA-based method for risk decision making in emergency response. Comput Oper Res 42:49–57
439. Liu Y, Fan ZP, Zhang Y (2011) A method for stochastic multiple criteria decision making based on dominance degrees. Inf Sci 181(19):4139–4153
440. Liu Y, Fan Z, Zhang Y (2014) Risk decision analysis in emergency response: a method based on cumulative prospect theory. Comput Oper Res 42:75–82
441. Liu Y, Luo XF, Xuan JY (2015) Online hot event discovery based on association link network. Concurr Comput: Pract Exp 27(15):4001–4014
442. Liu Y, Tang M, Zhou T et al (2015) Core-like groups result in invalidation of identifying super-spreader by k-shell decomposition. Sci Rep 5:9602
443. Liu Y, Yuan Y, Fan ZP (2011) A FTA-based method for risk decision making in emergency response. In: 4th international joint conference on computational sciences and optimization (CSO 2011), pp 608–612
444. Liu YJ, Guo B (2014) A lexicographic approach to postdisaster relief logistics planning considering fill rates and costs under uncertainty. Math Probl Eng 939853
445. Liu YJ, Li QQ, Tang XY et al (2014) Superedge prediction: what opinions will be mined based on an opinion supernetwork model? Decis Support Syst 64:118–129
446. Liu YJ, Zhang T, Lei HT et al (2013) Stochastic programming approach for earthquake disaster relief mobilization with multiple objectives. J Syst Eng Electron 24(4):642–654
447. Liu YL, Shen YF, Han CF (2011) Development of an interpretive structural model of factors in emergency materials support. In: International conference on information systems for crisis response and management, pp 503–514
448. Liu YH (2013) Uncertain random variables: a mixture of uncertainty and randomness. Soft Comput 17(4):625–634
449. Liu Z, Yang DS, Wen D et al (2011) Cyber-physical-social systems for command and control. IEEE Intell Syst 26(4):92–96
450. Liu Z, Zhang QM, Lü L et al (2011) Link prediction in complex networks: a local naïve Bayes model. EPL (Europhys Lett) 96(4):48007
451. Liu ZM, Liu L (2011) A novel Chinese web news source extraction algorithm. J Converg Inf Technol 6(8):99–106
452. Liu ZM, Liu L, Li H (2012) Determinants of information retweeting in microblogging. Internet Res 22(4):443–466
453. Long M, Cheng W, Jin X et al (2010) Transfer learning via cluster correspondence inference. In: 10th international conference on data mining (ICDM 2010), pp 917–922
454. Lu CX, Yan XW (2015) The break-even distance of road and inland waterway freight transportation systems. Marit Econ Logist 17(2):246–263
455. Lu HM, Zeng D, Chen H (2010) Prospective infectious disease outbreak detection using Markov switching models. IEEE Trans Knowl Data Eng 22(4):565–577
456. Lu JF, Shu CW, Zhang MP (2013) Stability analysis and a priori error estimate of explicit Runge-Kutta discontinuous Galerkin methods for correlated random walk with density-dependent turning rates. Sci China Math 56(12):2645–2676
457. Lu JY, Jia HY, Xie XF et al (2016) Missing the best opportunity; who can seize the next one? Agents show less inaction inertia than personal decision makers. J Econ Psychol 54:100–112
458. Lu JY, Xie XF, Wang M et al (2015) Double reference points: The effects of social and financial reference points on decisions under risk. J Behav Decis Mak 28(5):451–463
459. Lu L, Peng C, Zhu JP et al (2014) Correlation between fire attendance time and burned area based on fire statistical data of Japan and China. Fire Technol 50(4):851–872
460. Lü LY, Chen DB, Zhou T (2011) The small world yields the most effective information spreading. New J Phys 13(12):123005

461. Lü LY, Liu WP (2011) Information filtering via preferential diffusion. Phys Rev E 83(6):066119
462. Lü LY, Zhang YC, Yeung CH et al (2011) Leaders in social networks, the delicious case. PLoS ONE 6(6):e21202
463. Lü LY, Zhou T (2011) Link prediction in complex networks: a survey. Phys A: Stat Mech Appl 390(6):1150–1170
464. Lu QH, Liu N (2013) Pricing games of mixed conventional and e-commerce distribution channels. Comput Ind Eng 64(1):122–132
465. Lu S, Liang CJ, Song WG et al (2013) Frequency-size distribution and time-scaling property of high-casualty fires in China: analysis and comparison. Saf Sci 51(1):209–216
466. Lu S, Mei P, Li GH et al (2013) Time-scaling properties of high-casualty fires in China. Procedia Eng 62:602–608
467. Lu S, Mei P, Wang J et al (2012) Fatality and influence factors in high-casualty fires: a correspondence analysis. Saf Sci 50(4):1019–1033
468. Lu S, Wang J, Yang H et al (2011) Bayesian network model for fast disaster assessment in unconventional emergencies management. In: International conference on information systems for crisis response and management (ISCRAM 2011), pp 375–381
469. Lu S, Wu D, Lu SC et al (2011) A Bayesian network model for the Asian Games fire risk assessment. In: International conference on information systems for crisis response and management (ISCRAM 2011), pp 350–355
470. Lu X, Brelsford C (2014) Network structure and community evolution on Twitter: human behavior change in response to the 2011 Japanese earthquake and Tsunami. Sci Rep 4:6773
471. Lu XL, Xue L (2016) Managing the unexpected: sense-making in the Chinese Emergency management system. Public Adm 94(2):414–429
472. Lu XL (2017) Managing uncertainty in crisis: exploring the impact of institutionalization on organizational sensemaking. Springer, Singapore
473. Luo C, Zheng XL, Zeng DJ et al (2014) Portrayal of electronic cigarettes on YouTube. BMC Public Health 14(1):1028
474. Luo SL, Du YY, Liu P et al (2015) A study on coevolutionary dynamics of knowledge diffusion and social network structure. Expert Syst Appl 42(7):3619–3633
475. Luo SY, Yu D, Han SH (2016) Genetic and neural correlates of romantic relationship satisfaction. Soc Cogn Affect Neurosci 11(2):337–348
476. Luo XF, Liu HM, Xuan JY (2014) Website interaction network. J Organ Comput Electron Commer 24(2–3):215–235
477. Luo XF, Wei X, Zhang J (2010) Guided game-based learning using fuzzy cognitive maps. IEEE Trans Learn Technol 3(4):344–357
478. Luo XF, Xu Z, Yu J et al (2011) Building association link network for semantic link on web resources. IEEE Trans Autom Sci Eng 8(3):482–494
479. Luo XF, Xuan JY, Liu HM (2014) Web event state prediction model: combining prior knowledge with real time data. J Web Eng 13(5–6):483–506
480. Luo XF, Zhang J, Li Q et al (2014) Measuring algebraic complexity of text understanding based on human concept learning. IEEE Trans Hum-Mach Syst 44(5):638–649
481. Luo XF, Zhang J, Ye FY et al (2014) Power series representation model of text knowledge based on human concept learning. IEEE Trans Syst Man Cybern: Syst 44(1):86–102
482. Lü W, Fang ZM, Wei XG et al (2013) Experiment and modelling for pedestrian following behavior using velocity-headway relation. Procedia Eng 62:525–531
483. Lü W, Song WG, Fang ZM et al (2013) Modelling of lane-changing behaviour integrating with merging effect before a city road bottleneck. Phys A: Stat Mech Appl 392(20):5143–5153
484. Lü W, Song WG, Fang ZM (2011) Three-lane changing behaviour simulation using a modified optimal velocity model. Phys A: Stat Mech Appl 390(12):2303–2314
485. Lü W, Song WG, Liu XD et al (2013) A microscopic lane changing process model for multilane traffic. Phys A: Stat Mech Appl 392(5):1142–1152
486. Lü W, Song WG, Ma J et al (2013) A two-dimensional optimal velocity model for unidirectional pedestrian flow based on pedestrian's visual hindrance field. IEEE Trans Intell Transp Syst 14(4):1753–1763

487. Lü YP, Tang BH, Liu X et al (2015) A comparative study of scientific publications in health care sciences and services from China, Japan, and India (2007–2014). Int J Environ Res Public Health 13(1):79
488. Lü YP, Xue C, Ge Y et al (2016) Analysis of factors influencing inpatient and outpatient satisfaction with the Chinese military health service. PLoS ONE 11(3):e0151234
489. Ma J, Lo SM, Song WG et al (2013) Modeling pedestrian space in complex building for efficient pedestrian traffic simulation. Autom Constr 30:25–36
490. Ma J, Lo SM, Xu X et al (2011) Dynamic features of pedestrian-vehicle counter flow conflicts. In: 3rd international conference on transportation engineering (ICTE), pp 697–702
491. Ma J, Song WG (2013) Automatic clustering method of abnormal crowd flow pattern detection. Procedia Eng 62:509–518
492. Ma J, Song WG, Tian W et al (2012) Experimental study on an ultra high-rise building evacuation in China. Saf Sci 50(8):1665–1674
493. Ma N, Li N, He XS et al (2012) Rejection of unfair offers can be driven by negative emotions, evidence from modified ultimatum games with anonymity. PLoS ONE 7(6):e39619
494. Ma N, Liu YJ (2014) SuperedgeRank algorithm and its application in identifying opinion leader of online public opinion supernetwork. Expert Syst Appl 41(4):1357–1368
495. Ma QC, Luo XF, Luo Y (2014) Information entropy based the stability measure of user behaviour network in microblog. In: 10th international conference on semantics, knowledge and grids (SKG), pp 67–74
496. Ma QG, Bian J, Ji WJ et al (2011) Research on warnings with new thought of neuro-IE. Procedia Eng 26:1633–1638
497. Ma QG, Feng YD, Xu Q et al (2012) Brain potentials associated with the outcome processing in framing effects. Neurosci Lett 528(2):110–113
498. Ma QG, Jin J, Wang L (2010) The neural process of hazard perception and evaluation for warning signal words: evidence from event-related potentials. Neurosci Lett 483(3):206–210
499. Ma QG, Shen Q, Xu Q et al (2011) Empathic responses to others' gains and losses: an electrophysiological investigation. Neuroimage 54(3):2472–2480
500. Ma YF, Wu N, Zhang H et al (2014) Thermal annealing system and process design to improve quality of large size glasses. Int J Heat Mass Transf 72:411–422
501. Ma YF, Zhang H (2017) Enhancing knowledge management and decision-making capability of China's emergency operations center using big data. Intell Autom Soft Comput 24(1):1–8
502. Ma YH, Zhao QH, Xi MH (2016) Decision-makings in safety investment: an opportunity cost perspective. Saf Sci 83:31–39
503. Ma YN, Li SY, Wang CB et al (2016) Distinct oxytocin effects on belief updating in response to desirable and undesirable feedback. Proc Natl Acad Sci 113(33):9256–9261
504. Ma YP, Li LH, Zhang H et al (2017) Experimental study on small group behavior and crowd dynamics in a tall office building evacuation. Phys A: Stat Mech Appl 473:488–500
505. Mai X, Lü W, Wei XG et al (2013) Analyzing the characteristics of unidirectional bicycle movement around a track based on digital image processing. Procedia Eng 62:519–524
506. Mao XJ, Dong MG, Liu L et al (2014) An integrated approach to developing self-adaptive software. Int J Inf Sci Eng 30(4):1071–1085
507. Mao ZL, Mu HL, Xiao HH et al (2013) Reliability analysis of occupant safety evacuation in public assembly occupancies. Procedia Eng 62:493–500
508. Mao ZL, Xiao HH, Mu H et al (2014) Uncertainty analysis of parameters on occupant evacuation in building fires. J Appl Fire Sci 24(1):67–80
509. Mei P, Li GH, Lu S et al (2013) Influence of government on the power-law distribution of city fires in Jiangxi, China. Procedia Eng 62:1057–1062
510. Mei P, Li GH, Lu S et al (2014) An analytic network process approach for rapid loss assessment of high casualty fires in China. Fire Technol 50(5):1163–1179
511. Mei P, Lu S, Li GH et al (2014) Evaluation of fire protection performance of eight countries based on fire statistics: an application of data envelopment analysis. Fire Technol 50(2):349–361

512. Mei S, Chen B, Zhu YF et al (2015) Simulating city-level airborne infectious diseases. Comput Environ Urban Syst 51:97–105
513. Mei S, Quax R, van de Vijver D et al (2011) Increasing risk behaviour can outweigh the benefits of antiretroviral drug treatment on the HIV incidence among men-having-sex-with-men in Amsterdam. BMC Infect Dis 11(1):118
514. Mei S, Zarrabi N, Lees M et al (2015) Complex agent networks: an emerging approach for modeling complex systems. Appl Soft Comput 37:311–321
515. Mei S, Zhu YF, Qiu XG et al (2014) Individual decision making can drive epidemics: a fuzzy cognitive map study. IEEE Trans Fuzzy Syst 22(2):264–273
516. Meng LP, Kang Q, Han CF et al (2013) A game theoretical model for location of terror response facilities under capacitated resources. Sci World J 742845
517. Meng LP, Kang Q, Han CF et al (2016) A multi-agent model for simulation of public crisis information dissemination. Int J Wirel Mobile Comput 11(1):33–41
518. Meng LP, Ma Q, Han CF et al (2013) Determining the number of facilities for large-scale emergency. Int J Comput Sci Math 4(3):242–251
519. Meng RQ, Ge YZ, Cao ZD et al (2015) A framework for generating geospatial social computing environments. IEEE Intell Syst 30(1):44–52
520. Meng XF, Ding ZM, Xu JJ (2014) Moving objects management: models, techniques and applications, 2nd edn. Tsinghua University Press & Springer, Beijing & New York
521. Meng XF, Wei FR, Liu XH et al (2012) Cross-lingual mixture model for sentiment classification. In: Proceedings of the 50th annual meeting of the association for computational linguistics. Long papers, vol 1. Association for Computational Linguistics, Jeju Island, ROK, pp 572–581
522. Meng XF, Wei FR, Liu XH et al (2012) Entity-centric topic-oriented opinion summarization in twitter. In: Proceedings of the 18th ACM SIGKDD international conference on knowledge discovery and data mining, pp 379–387
523. Meng XF, Wei FR, Xu G et al (2012) Lost in translations? Building sentiment lexicons using context based machine translation. In: Proceedings of COLING 2012: posters, pp 829–838
524. Meng Z, Dong H, Li JH et al (2015) DarwinTree: a molecular data analysis and application environment for phylogenetic study. Data Sci J 14:10
525. Miao QL, Li QD, Zeng D (2010) Fine-grained opinion mining by integrating multiple review sources. J Am Soc Inform Sci Technol 61(11):2288–2299
526. Mu H, Wang JW, Mao ZL et al (2013) Pre-evacuation human reactions in fires: an attribution analysis considering psychological process. Procedia Eng 52:290–296
527. Mu HL, Lo SM, Song WG et al (2016) An experimental and numerical study of imbalanced door choice during an announced evacuation drill. Fire Technol 52(3):801–815
528. Mu HL, Lo SM, Song WG et al (2015) Impact of wedge-shaped design for building bottlenecks on evacuation time for efficiency optimization. Simulation 91(11):1014–1021
529. Mu N, Song WG, Qi XX et al (2014) Simulation of evacuation in a twin bore tunnel: analysis of evacuation time and egress selection. Procedia Eng 71:333–342
530. Ni SJ, Weng WG, Zhang H (2011) Modeling the effects of social impact on epidemic spreading in complex networks. Phys A: Stat Mech Appl 390(23–24):4528–4534
531. Ni ZJ, Wang YZ, Yin ZG (2016) Relative risk model for assessing domino effect in chemical process industry. Saf Sci 87:156–166
532. Ouyang M, Dueñas-Osorio L (2012) Time-dependent resilience assessment and improvement of urban infrastructure systems. Chaos:Interdiscip J Nonlinear Sci 22(3):033122
533. Pan G, Xu YQ, Wu ZH et al (2011) Task shadow: toward seamless task migration across smart environments. IEEE Intell Syst 26(3):50–57
534. Pan W, Yu LA, Wang SY et al (2014) A fuzzy multi-objective model for provider selection in data communication services with different QoS levels. Int J Prod Econ 147:689–696
535. Peng M, Peng YJ, Chen H (2014) Post-seismic supply chain risk management: a system dynamics disruption analysis approach for inventory and logistics planning. Comput Oper Res 42:14–24

536. Peng ZC, Ma B, Liu TX (2017) Chinese cooperative-harmonious democracy. Routledge, New York
537. Ping J, Jiang JB, Yu ZY (2011) Emergency management organizational structure optimization based on information transmission reliability. J Converg Inf Technol 6(10):235–242
538. Pu Q, Wang J, Wu SH et al (2013) Secure verifier-based three-party password-authenticated key exchange. Peer-to-peer Netw Appl 6(1):15–25
539. Pu Q, Wang J, Wu SH (2013) Scalable and efficient mobile authentication scheme preserving user privacy. Int J Ad Hoc Ubiquitous Comput 12(2):65–74
540. Pu Q, Wang J, Wu SH (2013) Secure SIP authentication scheme supporting lawful interception. Secur Commun Netw 6(3):340–350
541. Qi JY, Qu QX, Tan Y (2012) Topic evolution prediction of user generated contents considering enterprise generated contents. In: Proceedings of the first ACM international workshop on hot topics on interdisciplinary social networks research, pp 73–76
542. Qi JY, Zhou YP, Chen WJ et al (2012) Are customer satisfaction and customer loyalty drivers of customer lifetime value in mobile data services: a comparative cross-country study. Inf Technol Manag 13(4):281–296
543. Qian X, Miao QL, Zhai PM et al (2014) Cold-wet spells in mainland of China during 1951–2011. Nat Hazards 74(2):931–946
544. Qin CS, Fan B (2016) Factors that influence information sharing, collaboration, and coordination across administrative agencies at a Chinese university. Inf Syst e-Bus Manag 14(3):637–664
545. Qin XW, Liu X, Tang LX (2013) A two-stage stochastic mixed-integer program for the capacitated logistics fortification planning under accidental disruptions. Comput Ind Eng 65(4):614–623
546. Qiu CY, Wang CL, Fang BX et al (2014) A multiobjective particle swarm optimization-based partial classification for accident severity analysis. Appl Artif Intell 28(6):555–576
547. Qiu JN, Wang ZQ, Ye X et al (2014) Modeling method of cascading crisis events based on merging Bayesian network. Decis Support Syst 62(1246):94–105
548. Qiu XY, Zhao LJ, Wang JJ et al (2016) Effects of time-dependent diffusion behaviors on the rumor spreading in social networks. Phys Lett A 380(24):2054–2063
549. Qu T, Guo DK, Shen YL et al (2016) Minimizing traffic migration during network update in IaaS datacenters. IEEE Trans Serv Comput 12(4):577–589
550. Qu WN, Ge Y, Jiang CH et al (2014) The Dula dangerous driving index in China: an investigation of reliability and validity. Accid Anal Prev 64:62–68
551. Qu WN, Ge Y, Xiong YX et al (2015) Dangerous driving in a Chinese sample: associations with morningness-eveningness preference and personality. PLoS ONE 10(1):e0116717
552. Qu Y, Shang W, Wang SY (2013) Webpage mining for inflation emergency early warning. In: Web-age information management. Springer, Berlin, Heidelberg, pp 211–222
553. Qu YP, Yu H (2015) Robust planning models for uncertain emergency evacuation problem. Int J Ind Syst Eng 20(4):415–436
554. Rao P, Yang LZ, Zhu KJ et al (2012) Analysis of characteristics of pedestrian flow on staircases during fire. J Appl Fire Sci 21(2):123–142
555. Ren XD, Zhu JM, Huang J (2012) Multi-period dynamic model for emergency resource dispatching problem in uncertain traffic network. Syst Eng Procedia 5:37–42
556. Ruan JH, Wang XP, Shi Y et al (2013) Scenario-based path selection in uncertain emergency transportation networks. Int J Innov Comput Inf Control 9(8):3293–3305
557. Ruan JH, Wang XP, Shi Y (2013) Developing fast predictors for large-scale time series using fuzzy granular support vector machines. Appl Soft Comput 13(9):3981–4000
558. Sha Y, Huang J (2012) The multi-period location-allocation problem of engineering emergency blood supply systems. Syst Eng Procedia 5:21–28
559. Sha YZ, Liu Y, Yan J (2012) The attitude analysis of Internet users in the event involving PM2.5 governance. In: 5th international conference on business intelligence and financial engineering (BIFE 2012), pp 443–445

560. Shan SQ, Xin TL, Wang L et al (2013) Identifying influential factors of knowledge sharing in emergency events: a virtual community perspective. Syst Res Behav Sci 30(3):367–382
561. Shang MS, Chen DB, Zhou T (2010) Detecting overlapping communities based on community cores in complex networks. Chin Phys Lett 27(5):058901
562. Shang RX, Zhang PH, Zhong MH (2013) Investigation and analysis on evacuation behavior of large scale population in campus. Procedia Eng 52:302–308
563. Shang Z, Tao TX, Wang L (2016) Don't always prefer my chosen objects: low level of trait autonomy and autonomy deprivation decreases mere choice effect. Front Psychol 7:524
564. Shang Z, Wang L, Wu H (2017) The interactive influence of perceived ownership and perceived choosership of stocks on brain response to stock outcomes. Front Psychol 8:8
565. Shen Y, Liu SF, Fang ZG et al (2011) Modeling and simulation of stranded passengers' transferring decision-making on the basis of herd behaviors. In: IEEE international conference on grey systems and intelligent services (GSIS 2011), pp 795–802
566. Shen YF, Liu YL, Han CF (2011) Research on relationship between public security and economic growth in China. In: International conference on information systems for crisis response and management (ISCRAM 2011), pp 204–208
567. Shen Z, Zhao QC, Jia QS (2010) Quantifying heuristics in the ordinal optimization framework. Discr Event Dyn Syst 20(4):441–471
568. Shi CL, Zhong MH, Nong XZ et al (2012) Modeling and safety strategy of passenger evacuation in a metro station in China. Saf Sci 50(5):1319–1332
569. Shi DM, Wang BH (2013) Evacuation of pedestrians from a single room by using snowdrift game theories. Phys Rev E 87(2):022802
570. Shi HB, Li H, Li W et al (2012) A synthetic evaluation model for consumer products quality safety based on the VIKOR method. Int J Inf 15(10):4037–4047
571. Shi L, Jia QS, Mo YL et al (2011) Sensor scheduling over a packet-delaying network. Automatica 47(5):1089–1092
572. Shi WX, Ji J, Sun JH et al (2014) Experimental study on influence of stack effect on fire in the compartment adjacent to stairwell of high rise building. J Civ Eng Manag 20(1):121–131
573. Shi Y, Wan Y, Wu KF et al (2017) Non-negativity and locality constrained Laplacian sparse coding for image classification. Expert Syst Appl 72:121–129
574. Shu PP, Wang W, Tang M et al (2015) Numerical identification of epidemic thresholds for susceptible-infected-recovered model on finite-size networks. Chaos: Interdiscip J Nonlinear Sci 25(6):063104
575. Slam N, Wang WJ, Xue GX et al (2015) A framework with reasoning capabilities for crisis response decision-support systems. Eng Appl Artif Intell 46(B):346–353
576. Song C, Wang YL, Dong XX et al (2014) The impacts of uncertain locations of extra evacuees in source nodes on evacuation process. In: Hydraulic engineering II—proceedings of the 2nd SREE conference on hydraulic engineering (CHE 2013), pp 179–184
577. Song WG, Lü W, Fang ZM (2013) Experiment and modeling of microscopic movement characteristic of pedestrians. Procedia Eng 62:56–70
578. Song ZC, Ge YZ, Luo L et al (2015) An effective immunization strategy for airborne epidemics in modular and hierarchical social contact network. Phys A: Stat Mech Appl 439:142–149
579. Su GF, Chen JG, Liu FZ et al (2014) Research on key technologies for jail incident prevention and response system. In: Practical applications of intelligent systems. Springer, Berlin, Heidelberg, pp 765–774
580. Su ZX, Chen MY, Xia GP et al (2011) An interactive method for dynamic intuitionistic fuzzy multi-attribute group decision making. Expert Syst Appl 38(12):15286–15295
581. Su ZX, Xia GP, Chen MY (2011) Some induced intuitionistic fuzzy aggregation operators applied to multi-attribute group decision making. Int J Gen Syst 40(8):805–835
582. Su ZX, Xia GP, Chen MY et al (2012) Induced generalized intuitionistic fuzzy OWA operator for multi-attribute group decision making. Expert Syst Appl 39(2):1902–1910
583. Sun A, Zeng DJ, Chen H (2010) Burst detection from multiple data streams: a network-based approach. IEEE Trans Syst Man Cybern—Part C: Appl Rev 40(3):258–267

584. Sun C, Deng Y, Tang BT et al (2015) Urban traffic operation pattern and spatiotemporal mode based on big data (taking Beijing urban area as an example). In: Geo-informatics in resource management and sustainable ecosystem. Springer, Berlin, Heidelberg, pp 32–47
585. Sun C, Zhang FS, Zhong SB et al (2015) Expression and deduction of emergency scenario based on scenario element model. In: Proceedings of the ISCRAM 2015 conference
586. Sun JS, Xu W, Ma J et al (2015) Leverage RAF to find domain experts on research social network services: a big data analytics methodology with MapReduce framework. Int J Prod Econ 165:185–193
587. Sun QY, Li XY, Yu F (2016) Designing the emergency continuity plan for a megacity government: a conceptual framework for coping with natural catastrophes. Int J Crit Infrastruct Prot 13:28–35
588. Sutin AR, Stephan Y, Wang L et al (2015) Personality traits and body mass index in Asian populations. J Res Pers 58:137–142
589. Tan S, Jia Y (2014) NaEPASC: a novel and efficient public auditing scheme for cloud data. J Zhejiang Univ Sci C 15(9):794–804
590. Tan S, Tan L, Li XL et al (2014) An efficient method for checking the integrity of data in the cloud. China Commun 11(9):68–81
591. Tang BH, Ge Y, Liu ZP et al (2015) Health-related quality of life for medical rescuers one month after Ludian earthquake. Health Qual Life Outcomes 13(1):88
592. Tang BH, Ge Y, Xue C et al (2015) Health status and risk factors among adolescent survivors one month after the 2014 Ludian earthquake. Int J Environ Res Public Health 12(6):6367–6377
593. Tang BH, Kang P, Liu X et al (2014) Post-traumatic psychological changes among survivors of the Lushan earthquake living in the most affected areas. Psychiatry Res 220(1–2):384–390
594. Tang BH, Liu X, Liu Y et al (2014) A meta-analysis of risk factors for depression in adults and children after natural disasters. BMC Public Health 14(1):623
595. Tang B, Zhang LL (2013) Ya'an earthquake. The Lancet 381(9882):1984–1985
596. Tang L, Dai W, Yu LA et al (2015) A novel CEEMD-based EELM ensemble learning paradigm for crude oil price forecasting. Int J Inf Technol Decis Mak 14(1):141–169
597. Tang L, Lü HL, Yang FM et al (2015) Complexity testing techniques for time series data: a comprehensive literature review. Chaos Solitons Fractals 81:117–135
598. Tang L, Shi JR, Yu LA et al (2017) Economic and environmental influences of coal resource tax in China: a dynamic computable general equilibrium approach. Resour Conserv Recycl 117:34–44
599. Tang L, Wang ZS, Li XX et al (2015) A novel hybrid FA-based LSSVR learning paradigm for hydropower consumption forecasting. J Syst Sci Complex 28(5):1080–1101
600. Tang L, Wu JQ, Yu LA et al (2015) Carbon emissions trading scheme exploration in China: a multi-agent-based model. Energy Policy 81:152–169
601. Tang L, Yu LA, He KJ (2014) A novel data-characteristic-driven modeling methodology for nuclear energy consumption forecasting. Appl Energy 128:1–14
602. Tang L, Yu LA, Liu FT et al (2013) An integrated data characteristic testing scheme for complex time series data exploration. Int J Inf Technol Decis Mak 12(3):491–521
603. Tang L, Yu LA, Wang S et al (2012) A novel hybrid ensemble learning paradigm for nuclear energy consumption forecasting. Appl Energy 93:432–443
604. Tang M, Zhou T (2011) Efficient routing strategies in scale-free networks with limited bandwidth. Phys Rev E 84(2):026116
605. Tang P, Wang HW, Qi C et al (2011) A new decision model for handling asynchronous incident objectives with priorities in emergency response. J Comput Inf Syst 7(14):5251–5258
606. Tang P, Wang HW, Qi C et al (2012) Anytime heuristic search in temporal HTN planning for developing incident action plans. AI Commun 25(4):321–342
607. Tao FF, Wang HM, Kang JL et al (2012) Study on knowledge base system of extreme flood magnitude based on data driven reasoning. Int Rev Comput Softw 7(5):2226–2230
608. Tao FF, Wang HM, Liu GF (2012) Application study on visual discussion technique in group decision making of sudden disasters. J Digit Inf Manag 10(6):373–378

609. Teng MM, Han CF, Lin XQ (2016) A novel model for evaluating environmental governance performance based on trends analysis. Int J Wirel Mob Comput 11(1):75–82
610. Tian RY, Liu YJ (2014) Isolation, insertion, and reconstruction: Three strategies to intervene in rumor spread based on supernetwork model. Decis Support Syst 67:121–130
611. Tian RY, Ma N, Li QQ et al (2013) The evolution and application of network analysis methods. In: IEEE international conference on systems, man, and cybernetics, pp 2197–2201
612. Tian W, Song WG, Lü W et al (2011) Experiment and analysis on microscopic characteristics of pedestrian movement in building bottleneck. Sci China Technol Sci 54(7):1730–1736
613. Tian W, Song WG, Ma J et al (2012) Experimental study of pedestrian behaviors in a corridor based on digital image processing. Fire Saf J 47:8–15
614. Tong JP, Ma JF, Liu GF (2011) Evolutionary game analysis of emergency management cooperation for water disaster. In: 8th international conference on service systems and service management (ICSSSM 2011), pp 1–4
615. Tong JJ, Wang L, Peng KP (2015) From person-environment misfit to job burnout: theoretical extensions. J Manag Psychol 30(2):169–182
616. Tong QL, Zou XC, Ding Q et al (2013) Latent regression with constrained parameters to determine the weight coefficients in summary index model. Commun Stat—Simul Comput 42(7):1628–1642
617. Wan Y, Chen XL, Shi Y (2017) Adaptive cost dynamic time warping distance in time series analysis for classification. J Comput Appl Math 319:514–520
618. Wang CM, Li T (2011) Study on the release mechanism of national economy mobilization resource based on military and civilian integration: taking tent for example. Acta Armamentarii 32(SUPPL. 1):112–117
619. Wang C, Wang YQ (2011) Discovering consumer's purchasing behavior based on efficient association rules. In: 8th international conference on fuzzy systems and knowledge discovery (FSKD 2011), vol 2, pp 937–941
620. Wang CL, Zhong SB, Zhang QY et al (2015) Urban disaster comprehensive risk assessment research based on GIS: a case study of Changsha City, Hunan Province, China. In: Bian FL, Xie YC (eds) Geo-informatics in resource management and sustainable ecosystem. Springer, Berlin, Heidelberg, pp 95–106
621. Wang D, Qi C, Wang HW (2014) Improving emergency response collaboration and resource allocation by task network mapping and analysis. Saf Sci 70:9–18
622. Wang DJ, Liu F, Wang JJ et al (2016) Integrated rescheduling and preventive maintenance for arrival of new jobs through evolutionary multi-objective optimization. Soft Comput 20(4):1635–1652
623. Wang DJ, Liu F, Wang YZ et al (2015) A knowledge-based evolutionary proactive scheduling approach in the presence of machine breakdown and deterioration effect. Knowl-Based Syst 90:70–80
624. Wang DJ, Liu F, Yin YQ et al (2015) Prioritized surgery scheduling in face of surgeon tiredness and fixed off-duty period. J Comb Optim 30(4):967–981
625. Wang DY, Pan LW, Lu L et al (2013) Emergency management business process reengineering and integrated emergency response system structure design for a city in China. Procedia Eng 52:371–376
626. Wang DY, Weng T, Zhu JP et al (2013) Sequential decision analysis of fire emergency and rescue on urban successional building fires. Procedia Eng 62:1087–1095
627. Wang FH, Chen WZ, Huang JY et al (2011) Preliminary study of relationships between hypnotic susceptibility and personality disorder functioning styles in healthy volunteers and personality disorder patients. BMC Psychiatry 11(1):121
628. Wang GH, Liu YJ, Li JM et al (2015) Superedge coupling algorithm and its application in coupling mechanism analysis of online public opinion supernetwork. Expert Syst Appl 42(5):2808–2823
629. Wang HB, Fan TH, Shi AY et al (2010) Fuzzy integral based information fusion for water quality monitoring using remote sensing data. Int J Commun Netw Syst Sci 3(9):737–744

630. Wang HX, Li KF, Xie XF (2017) Individual differences in pain sensitivity predict the experience of unfairness. J Health Psychol 24(7):953–963
631. Wang HW, Liu Y, Yin P (2012) Study on user preferences modelling based on web mining. Int J Inf Technol Manag 11(4):307–322
632. Wang J, Rong LL (2013) Similarity index based on the information of neighbor nodes for link prediction of complex network. Mod Phys Lett B 27(6):1350039
633. Wang J, Yang HX, Zhu JM (2012) A two-stage stochastic programming model for emergency resources storage region division. Syst Eng Procedia 5:125–130
634. Wang J, Zhu JM, Yang HX et al (2013) Reliable path selection problem in uncertain traffic network after natural disaster. Math Probl Eng 1–5.
635. Wang JJ, Cao ZD, Zeng DJ et al (2014) Epidemiological analysis, detection, and comparison of space-time patterns of Beijing hand-foot-mouth disease (2008–2012). PLoS ONE 9(3):e92745
636. Wang JH (2011) Dynamics of urban fire correlations with detrended fluctuation analysis. J Risk Anal Crisis Response 1(2):126–132
637. Wang JH, Lo SM, Sun JH et al (2012) Qualitative simulation of the panic spread in large-scale evacuation. Simulation 88(12):1465–1474
638. Wang JH, Lo SM, Wang QS et al (2013) Risk of large-scale evacuation based on the effectiveness of rescue strategies under different crowd densities. Risk Anal 33(8):1553–1563
639. Wang JH, Sun JH, Lo SM (2015) Randomness in the evacuation route selection of large-scale crowds under emergencies. Appl Math Model 39(18):5693–5706
640. Wang JH, Xie S, Sun JH (2011) Self-organized criticality judgment and extreme statistics analysis of major urban fires. Chin Sci Bull 56(6):567–572
641. Wang JJ, Zhao LJ, Huang RB (2014) 2SI2R rumor spreading model in homogeneous networks. Phys A: Stat Mech Appl 413:153–161
642. Wang JJ, Zhao LJ, Huang RB (2014) SIRaRu rumor spreading model in complex networks. Phys A: Stat Mech Appl 398:43–55
643. Wang JZ, Kitayama S, Han SH (2011) Sex difference in the processing of task-relevant and task-irrelevant social information: an event-related potential study of familiar face recognition. Brain Res 1408:41–51
644. Wang KM, Ma ZJ (2015) Age-based policy for blood transshipment during blood shortage. Transp Res Part E: Logist Transp Rev 80:166–183
645. Wang L, Ma QG, Song ZF et al (2011) N400 and the activation of prejudice against rural migrant workers in China. Brain Res 1375:103–110
646. Wang LG, Xu YY, Wang Y et al (2010) The epidemiological investigation and intelligent analytical system for foodborne disease. Food Control 21(11):1466–1471
647. Wang M, Burlacu G, Truxillo DM et al (2015) Age differences in feedback reactions: the roles of employee feedback orientation on social awareness and utility. J Appl Psychol 100(4):1296–1308
648. Wang Q, Zhao LJ, Huang RB et al (2015) Interaction of media and disease dynamics and its impact on emerging infection management. Discr Contin Dyn Syst B 20(1):215–230
649. Wang S, Shang W (2014) Forecasting direction of China security index 300 movement with least squares support vector machine. Procedia Comput Sci 31:869–874
650. Wang S, Yu LA, Tang L et al (2011) A novel seasonal decomposition based least squares support vector regression ensemble learning approach for hydropower consumption forecasting in China. Energy 36(11):6542–6554
651. Wang SD, He YP, Zou JJ et al (2014) A flame detection synthesis algorithm. Fire Technol 50(4):959–975
652. Wang SD, He YP, Zou JJ et al (2014) Early smoke detection in video using swaying and diffusion feature. J Intell Fuzzy Syst: Appl Eng Technol 26(1):267–275
653. Wang SD, Rong JZ, Zhou DC et al (2013) A swaying object detection algorithm. Proc SPIE—Int Soc Opt Eng 8878(4):32
654. Wang T, Liu Z, Xiu BX et al (2014) Characterizing the evolution of social computing research. IEEE Intell Syst 29(5):48–56

655. Wang T, Zhang QP, Liu Z et al (2012) On social computing research collaboration patterns: a social network perspective. Front Comput Sci 6(1):122–130
656. Wang W, Tang M, Zhang HF et al (2014) Epidemic spreading on complex networks with general degree and weight distributions. Phys Rev E 90(4):042803
657. Wang W, Wang XG, Zhou X (2011) Impacts of Californian dams on flow regime and maximum/minimum flow probability distribution. Hydrol Res 42(4):275–289
658. Wang W, Zhang ZZ, Wang XG et al (2012) Evaluation of using the modified water deficit index derived from MODIS vegetation index and land surface temperature products for monitoring drought. In: IEEE international geoscience and remote sensing symposium (IGARSS 2012), pp 5951–5954
659. Wang WJ, Jiao PF, He DX et al (2016) Autonomous overlapping community detection in temporal networks. Knowl-Based Syst 110:121–134
660. Wang WJ, Liu D, Liu X et al (2013) Fuzzy overlapping community detection based on local random walk and multidimensional scaling. Phys A: Stat Mech Appl 392(24):6578–6586
661. Wang WJ, Pan L, Yuan N et al (2015) A comparative analysis of intra-city human mobility by taxi. Phys A: Stat Mech Appl 420:134–147
662. Wang WJ, Yuan N, Pan L et al (2015) Temporal patterns of emergency calls of a metropolitan city in China. Phys A: Stat Mech Appl 436:846–855
663. Wang X, Jia Y, Chen RH et al (2013) Improving text categorization with semantic knowledge in Wikipedia. IEICE Trans Inf Syst 96(12):2786–2794
664. Wang X, Jin XM, Chen ME et al (2012) Topic mining over asynchronous text sequences. IEEE Trans Knowl Data Eng 24(1):156–169
665. Wang XL, Wu XL (2012) A novel knowledge representation method based on ontology for natural disaster decision-making. In: IEEE international conference on computer science and automation engineering (CSAE 2012), vol 3, pp 241–245
666. Wang XL, Zeng DJ, Seale H et al (2010) Comparing early outbreak detection algorithms based on their optimized parameter values. J Biomed Inform 43(1):97–103
667. Wang XZ, Wang JY, Zhang H et al (2016) Sentiment analysis of name entity for text. In: The 28th international conference on software engineering and knowledge engineering, pp 37–41
668. Wang XZ, Zhang HM, Zhao JH et al (2015) An interactive web-based analysis framework for remote sensing cloud computing. ISPRS Ann Photogram Remote Sens Spat Inf Sci 43–50
669. Wang XY, Huang YJ, Ma QG et al (2012) Event-related potential P2 correlates of implicit aesthetic experience. NeuroReport 23(14):862–866
670. Wang XZ, Liu Y, Zhang H et al (2017) Public health emergency management and multi-source data technology in China. Intell Autom Soft Comput 24(1):1–8
671. Wang XZ, Luo XF, Liu HM (2015) Measuring the veracity of web event via uncertainty. J Syst Softw 102:226–236
672. Wang XZ, Luo XF, Zhang H et al (2016) Outbreak power measurement for evolution course of web events. Int J Web Eng Technol 15(3–4):226–248
673. Wang XY, Ma QG, Wang CC (2012) N400 as an index of uncontrolled categorization processing in brand extension. Neurosci Lett 525(1):76–81
674. Wang XP, Ruan JH, Shi Y (2012) A recovery model for combinational disruptions in logistics delivery: considering the real-world participators. Int J Prod Econ 140(1):508–520
675. Wang XZ, Zhang H (2017) Propagation of social emotion in cyber space based on cognitive social psychology. IEEE Access 5:1005–1012
676. Wang XZ, Zhang H, Liu Y (2018) Sentence vector model based on implicit word vector expression. IEEE Access 6:17455–17463
677. Wang X Z, Zhang H, Wang JY et al (2016) Sentiment analysis of name entity for text. SEKE 37–41
678. Wang XZ, Zhang H, Xu Z (2016) Public sentiments analysis based on fuzzy logic for text. Int J Softw Eng Knowl Eng 26(09n10):1341–1360
679. Wang XZ, Zhang H, Yuan SC et al (2016) Sentiment processing of social media information from both wireless and wired network. EURASIP J Wirel Commun Netw 1:164

680. Wang Y, Han CF, Liu L (2013) Network analysis of the inter-organizational collaboration in emergency management system of China. IEEE Int Conf Syst Man Cybern 1271–1275
681. Wang Y, Wu CG, Ji ZH et al (2011) Non-parametric change-point method for differential gene expression detection. PLoS ONE 6(5):e20060
682. Wang Y, Yao F, Liu K (2013) The safety efficiency of an overhead lines undergrounding of cities cabling project. In: Huang CF, Kahrama C (eds) Intelligent systems and decision making for risk analysis and crisis response. CRC Press, London, pp 427–433
683. Wang YH, Hu XH, Dai WH et al (2014) Vocal emotion of humanoid robots: a study from brain mechanism. Sci World J 216341
684. Wang YW, Zhou YC, Liu Y et al (2013) A grid-based clustering algorithm for wild bird distribution. Front Comput Sci 7(4):475–485
685. Wang Z, Chen YF, Tjosvold D et al (2010) Cooperative goals and team agreeableness composition for controversy in China. Asia Pac J Manag 27(1):139–153
686. Wang Z, Liang W, Hu X (2014) A metaheuristic based on a pool of routes for the vehicle routing problem with multiple trips and time windows. J Oper Res Soc 65(1):37–48
687. Wang Z, Wang HW, Qi C et al (2013) A resource enhanced HTN planning approach for emergency decision-making. Appl Intell 38(2):226–238
688. Wei D, Zhou T, Cimini G et al (2011) Effective mechanism for social recommendation of news. Phys A: Stat Mech Appl 390(11):2117–2126
689. Wei JC, Bu B, Liang L (2012) Estimating the diffusion models of crisis information in micro blog. J Informet 6(4):600–610
690. Wei JC, Guo XM, Marinova D et al (2014) Industrial SO_2 pollution and agricultural losses in China: evidence from heavy air polluters. J Clean Prod 64:404–413
691. Wei JC, Jia RY, Marinova D et al (2012) Modeling pollution control and performance in China's provinces. J Environ Manag 113:263–270
692. Wei JC, Wang F, Zhao DT (2012) A risk perception model: simulating public response to news reports in China. Inf Res 17(2):411–436
693. Wei JC, Wang H, Fan J et al (2013) Corporate accidents, media coverage, and stock market responses: empirical study of the Chinese listed firms. Chin Manag Stud 7(4):617–630
694. Wei JC, Zhao DT, Marinova D (2013) Disaster relief drivers: China and U.S. in comparative perspective. China: Int J 11(2):93–116
695. Wei JC, Zhou L, Wei YF et al (2014) Collective behavior in mass incidents: a study of contemporary China. J Contemp China 23(88):715–735
696. Wei MY, Li X, Yu LA (2015) Time-dependent fuzzy random location-scheduling programming for hazardous materials transportation. Transp Res Part C: Emerg Technol 57:146–165
697. Wei WQ, Lu JG, Galinsky AD et al (2017) Regional ambient temperature is associated with human personality. Nat Hum Behav 1(12):890–895
698. Wei WQ, Ma JJ, Wang L (2015) The, "warm" side of coldness: cold promotes interpersonal warmth in negative contexts. Br J Soc Psychol 54(4):712–727
699. Wei WQ, Wang L, Shang Z et al (2015) Non-sympathetic FRN responses to drops in others' stocks. Soc Neurosci 10(6):616–623
700. Wei X, Luo XF, Li Q (2013) Improving the compression efficiency for news web service using semantic relations among webpages. Int J Cogn Inf Nat Intell (IJCINI 2013) 7(2):49–64
701. Wei XG, Lü W, Song WG (2013) Rescue route reselection model and algorithm for the unexpected accident. Procedia Eng 62:532–537
702. Wei XG, Mai X, Lü W et al (2014) Microscopic character and movement consistency of pedestrian group: an experimental study in campus. Fire Saf Sci 11:1103–1114
703. Wei XG, Song WG, Lü W (2013) Study on exit flow rate of an one-exit room. Appl Mech Mater 444–445:1569–1573
704. Wei XG, Song WG, Lü W et al (2014) Defining static floor field of evacuation model in large exit scenario. Simul Model Pract Theory 40:122–131
705. Wei ZW, Wang BH, Han XP (2013) Renormalization and small-world model of fractal quantum repeater networks. Sci Rep 3:1222

706. Wu JH, Ge Y, Shi ZB et al (2010) Response inhibition in adolescent earthquake survivors with and without posttraumatic stress disorder: a combined behavioral and ERP study. Neurosci Lett 486(3):117–121
707. Wu JH, Yuan YR, Cao CQ et al (2015) The relationship between response inhibition and posttraumatic stress symptom clusters in adolescent earthquake survivors: an event-related potential study. Sci Rep 5:8844
708. Wu JH, Yuan YR, Duan HX et al (2014) Long-term academic stress increases the late component of error processing: an ERP study. Biol Psychol 99:77–82
709. Wu LF, Liu SF, Fang ZG et al (2015) Properties of the GM (1,1) with fractional order accumulation. Appl Math Comput 252:287–293
710. Wu LF, Liu SF, Yang YJ (2016) A gray model with a time varying weighted generating operator. IEEE Trans Syst Man Cybern: Syst 46(3):427–433
711. Wu LF, Liu SF, Yang YJ (2015) A model to determine OWA weights and its application in energy technology evaluation. Int J Intell Syst 30(7):798–806
712. Wu LF, Liu SF, Yang YJ (2016) Grey double exponential smoothing model and its application on pig price forecasting in China. Appl Soft Comput 39:117–123
713. Wu LF, Liu SF, Yang YJ et al (2016) Multi-variable weakening buffer operator and its application. Inf Sci 339:98–107
714. Wu LF, Liu SF, Yao LJ et al (2015) Using fractional order accumulation to reduce errors from inverse accumulated generating operator of grey model. Soft Comput 19(2):483–488
715. Wu N, Yang R, Zhang H (2015) An advanced fire estimation model for decentralized building control. Build Simul 8(5):579–591
716. Wu YY, Xue YS, Wang HH et al (2012) Extension of power system early-warning defense schemes by integrating typhoon information. In: International conference on sustainable power generation and supply (SUPERGEN 2012), pp 1–7
717. Wu ZL, Liu X (2011) Emergency strategy of civic critical infrastructure to the urgent events. Adv Mater Res 317–319:2091–2096
718. Xiao T, Li T (2011) Research on the civil-military integration system of military oil storage and supply. Acta Armamentarii 32(SUPPL. 1):134–138
719. Xiao YP, Wang B, Wu B et al (2012) A hybrid human dynamics model on analyzing hotspots in social networks. Discr Dyn Nat Soc 678286
720. Xiao YP, Wang B, Wu B et al (2014) Experimental study of choice behavior during building evacuation considering long-range auditory information and particular groups. J Appl Fire Sci 24(2):335–357
721. Xie JR, Jiang R, Ding ZJ et al (2013) Dynamical traffic light strategy in the Biham-Middleton-Levine model. Phys Rev E 87(2):022812
722. Xie KF, Chen G, Wu Q et al (2012) Research on the group decision-making about emergency event based on network technology. Inf Technol Manag 12(2):137–147
723. Xie KF, Liu J, Chen G et al (2012) Group decision-making in an unconventional emergency situation using agile Delphi approach. Inf Technol Manag 13(4):351–361
724. Xie KF, Liu J, Chen Y et al (2014) Escape behavior in factory workshop fire emergencies: a multi-agent simulation. Inf Technol Manag 15(2):141–149
725. Xie KL, Zhang ZL, Zhang ZQ et al (2016) Effects of managerial response on consumer eWOM and hotel performance: evidence from TripAdvisor. Int J Contemp Hosp Manag 28(9):2013–2034
726. Xie NM, Liu SF (2015) Interval grey number sequence prediction by using non-homogenous exponential discrete grey forecasting model. J Syst Eng Electron 26(1):96–102
727. Xie WW, Ho B, Meier S et al (2017) Rank reversal aversion inhibits redistribution across societies. Nat Hum Behav 1(8):0142
728. Xie XF, Liu HM, Gan YQ (2011) Belief in a just world when encountering the 5/12 Wenchuan earthquake. Environ Behav 43(4):566–586
729. Xie XF, Stone E, Zheng R et al (2011) The "Typhoon Eye Effect": determinants of distress during the SARS epidemic. J Risk Res 14(9):1091–1107

730. Xie XF, Wang M, Zhang RG et al (2011) The role of emotions in risk communication. Risk Anal: Int J 31(3):450–465
731. Xie Y, Qu YY, Li CH et al (2012) Online multiple instance gradient feature selection for robust visual tracking. Pattern Recogn Lett 33(9):1075–1082
732. Xie Y, Zhang WS, Qu YY et al (2014) Discriminative subspace learning with sparse representation view-based model for robust visual tracking. Pattern Recogn 47(3):1383–1394
733. Xu B, Liu L, You WJ (2011) Importance of tie strengths in the prisoner's dilemma game on social networks. Phys Lett A 375(24):2269–2273
734. Xu EHW, Wang W, Xu C et al (2015) Suppressed epidemics in multirelational networks. Phys Rev E 92(2):022812
735. Xu JH, Peng ZC (2015) People at risk of influenza pandemics: the evolution of perception and behavior. PLoS ONE 10(12):e0144868
736. Xu KQ, Guo XT, Li JX et al (2012) Discovering target groups in social networking sites: an effective method for maximizing joint influential power. Electron Commer Res Appl 11(4):318–334
737. Xu LY, Wang HM, Chen JF (2011) Application of extreme value analysis to extreme drought disaster area in China. In: Wu DS, Zhou Y (eds) Modeling risk management for resources and environment in China. Springer, Berlin, Heidelberg, pp 349–357
738. Xu Q, Shen Q, Chen PS et al (2011) How an uncertain cue modulates subsequent monetary outcome evaluation: an ERP study. Neurosci Lett 505(2):200–204
739. Xu Q, Zu ZH, Xu ZJ et al (2013) Space P-based empirical research on public transport complex networks in 330 cities of China. J Transp Syst Eng Inf Technol 13(1):193–198
740. Xu X, Zhong MH, Chu BQ et al (2012) Study of optimized evacuation route of commercial street based on microscopic model. In: TRB 91st annual meeting, pp 1–26
741. Xu X, Zhong MH, Shi CL (2013) An optimization algorithm for exit choice and its application in theater evacuation. J Northeast Univ (Nat Sci Edn) 34(1):44–52
742. Xu YJ, Da QL, Wang HM (2011) A note on group decision-making procedure based on incomplete reciprocal relations. Soft Comput 15(7):1289–1300
743. Xu YJ, Wang HM (2011) Approaches based on 2-tuple linguistic power aggregation operators for multiple attribute group decision making under linguistic environment. Appl Soft Comput 11(5):3988–3997
744. Xu YJ, Wang HM (2011) Distance measure for linguistic decision making. Syst Eng Procedia 1:450–456
745. Xu YJ, Wang HM, Palacios-Marqués D (2013) An interactive approach based on alternative achievement scale and alternative comprehensive scale for multiple attribute decision making under linguistic environment. Int J Comput Intell Syst 6(1):87–95
746. Xu ZY, Li T (2011) A comparative study on pheromone updating strategies of MP^2AS for VRP. In: International conference on system science, engineering design and manufacturing informatization (ICSSEM 2011), vol 2, pp 142–145
747. Xu Z, Liu YH, Mei L et al (2015) Crowd sensing based burst computing of events using social media. Int J Distrib Sens Netw 1–8
748. Xu Z, Luo XF, Mei L et al (2014) Measuring the semantic discrimination capability of association relations. Concurr Comput: Pract Exp 26(2):380–395
749. Xu Z, Luo XF, Yu J et al (2011) Measuring semantic similarity between words by removing noise and redundancy in web snippets. Concurr Comput: Pract Exp 23(18):2496–2510
750. Xu Z, Zhang H, Hu CP et al (2016) Building knowledge base of urban emergency events based on crowdsourcing of social media. Concurr Comput: Pract Exp 28(15):4038–4052
751. Xu Z, Zhang H, Sugumaran V et al (2016) Participatory sensing-based semantic and spatial analysis of urban emergency events using mobile social media. EURASIP J Wirel Commun Netw 1:44
752. Xu ZJ, Zu ZH, Zheng T et al (2014) Comparative analysis of the effectiveness of three immunization strategies in controlling disease outbreaks in realistic social networks. PLoS ONE 9(5):e95911

753. Xu ZC, Yuan YF, Ji SB (2009) A decision analysis framework for emergency notification: the case of the Sichuan earthquake. Int J Emerg Manag 6(2):227–243
754. Xu ZY, Xu WS, Shao WH et al (2014) Real-time pricing control on generation-side: optimal demand-tracking model and information fusion estimation solver. IEEE Trans Power Syst 29(4):1522–1535
755. Xue C, Ge Y, Tang BH et al (2015) A meta-analysis of risk factors for combat-related PTSD among military personnel and veterans. PLoS ONE 10(3):e0120270
756. Xue L, Zeng G (2019) A comprehensive evaluation on emergency response in China: the case of pandemic influenza (H1N1) 2009. Springer, Singapore
757. Yan KS, Rong LL, Yu K (2016) Discriminating complex networks through supervised NDR and Bayesian classifier. Int J Mod Phys C 27(5):1650051
758. Yan SL, Liu SF, Liu JF et al (2015) Dynamic grey target decision making method with grey numbers based on existing state and future development trend of alternatives. J Intell Fuzzy Syst 28(5):2159–2168
759. Yan XY, Han XP, Wang BH et al (2013) Diversity of individual mobility patterns and emergence of aggregated scaling laws. Sci Rep 3:2678
760. Yan XY, Han XP, Zhou T et al (2011) Exact solution of the gyration radius of an individual's trajectory for a simplified human regular mobility model. Chin Phys Lett 28(12):120506
761. Yang BH, Fang SL, Zhao JS (2012) The study on the coupling model of unconventional incidents. Kybernetes 41(5–6):664–673
762. Yang C, Guo Q, Meng XF et al (2017) Revisiting performance in big data systems: An resource decoupling approach. In: Proceedings of the 2017 symposium on cloud computing 639
763. Yang F, Yang Q, Huang W (2012) Multi-agents model of prevention and control emergencies with isolation measure. J Converg Inf Technol 7(22):492–499
764. Yang F, Yang Q, Liu XX et al (2015) SIS evolutionary game model and multi-agent simulation of an infectious disease emergency. Technol Health Care 23(s2):S603–S613
765. Yang F, Yang Q (2012) Multi-agents model of active defense mode with immunization vaccination in unconventional emergencies management. Int J Adv Comput Technol 4(19):607–616
766. Yang FM, Li M, Huang AQ et al (2014) Forecasting time series with genetic programming based on least square method. J Syst Sci Complex 27(1):117–129
767. Yang FM, Wang YK, Pei J et al (2015) Electronic markets selection in supply chain with uncertain demand and uncertain price. Math Probl Eng 939284
768. Yang H, Tang M, Gross T (2015) Large epidemic thresholds emerge in heterogeneous networks of heterogeneous nodes. Sci Rep 5:13122
769. Yang H, Tang M, Zhang HF (2012) Efficient community-based control strategies in adaptive networks. New J Phys 14(12):123017
770. Yang HX, Wang BH (2011) Universal role of migration in the evolution of cooperation. Chin Sci Bull 56(34):3693–3696
771. Yang HX, Wang WX, Xie YB et al (2011) Transportation dynamics on networks of mobile agents. Phys Rev E 83(1):016102
772. Yang HX, Tang M, Lai YC (2015) Traffic-driven epidemic spreading in correlated networks. Phys Rev E 91(6):062817
773. Yang LL, Su GF, Yuan HY (2012) Design principles of integrated information platform for emergency responses: the case of 2008 Beijing Olympic Games. Inf Syst Res 23(3-part-1):761–786
774. Yang LZ, Liu SB, Li J et al (2009) Information-based evacuation experiment and its cellular automaton simulation. Int J Mod Phys C 20(10):1583–1596
775. Yang LZ, Liu SB, Rao P et al (2012) Subconscious environmental information perceiving behavior and its attenuation in information-based evacuation experiment. Int J Mod Phys C 23(7):1250049
776. Yang LZ, Rao P, Zhu KJ et al (2012) Observation study of pedestrian flow on staircases with different dimensions under normal and emergency conditions. Saf Sci 50(5):1173–1179

777. Yang Q, Ma H, Yu Y (2012) Multi-agent risk identifier model of emergency management system engineering based on immunology. Syst Eng Procedia 4:385–392
778. Yang Q, Wu XC, Zhou XY et al (2013) Diverging effects of clean versus dirty money on attitudes, values, and interpersonal behavior. J Pers Soc Psychol 104(3):473–489
779. Yang Q, Yang F (2011) Multi-agents simulation on unconventional emergencies evolution mechanism in public health. Adv Intell Soft Comput (AINSC) 129(2):509–514
780. Yang Q, Yang F, Yu YL (2013) Multi-agents computational experiment on twice-linked model of unconventional emergencies evolution mechanism. J Comput 8(4):1035–1042
781. Yang RR, Li XY, Zhang T (2014) Analysis of linkage effects among industry sectors in China's stock market before and after the financial crisis. Phys A: Stat Mech Appl 411(10):12–20
782. Yang S (2013) On feature selection for traffic congestion prediction. Transp Res Part C: Emerg Technol 26:160–169
783. Yang S, Liu WH (2011) Anomaly detection on collective moving patterns: a hidden Markov model based solution. In: IEEE international conferences on internet of things, and cyber, physical and social computing, pp 291–296
784. Yang Y, Jia QS, Deconinck G et al (2018) Distributed coordination of EV charging with renewable energy in a microgrid of buildings. IEEE Trans Smart Grid 9(6):6253–6264
785. Yang Y, Jia QS, Guan XH et al (2019) Decentralized EV-based charging optimization with building integrated wind energy. IEEE Trans Autom Sci Eng 16(3):1002–1017
786. Yang YH, Hao CH (2011) Product selection for promotion planning. Knowl Inf Syst 29(1):223–236
787. Yang YJ, Liu SF, John R (2014) Uncertainty representation of grey numbers and grey sets. IEEE Trans Cybern 44(9):1508–1517
788. Yang ZM, Zhou T (2012) Epidemic spreading in weighted networks: an edge-based mean-field solution. Phys Rev E 85(5-part 2):056106
789. Yao J, Huang JY, Chen GR et al (2013) A new coupled-map car-following model based on a transportation supernetwork framework. Chin Phys B 22(6):145–153
790. Yao J, Lin F, Liu B (2011) H_∞ control for stochastic stability and disturbance attenuation in a class of networked hybrid systems. IET Control Theory Appl 5(15):1698–1708
791. Yao J, Su YQ, Guo YF et al (2013) Exponential synchronisation of hybrid impulsive and switching dynamical networks with time delays. IET Control Theory Appl 7(4):508–514
792. Yao W, Huang H, Shen SF et al (2013) Fire risk mapping based assessment method applied in performance based design. Fire Saf J 56:81–89
793. Yao XB, Han CF, Wan SL (2013) Incomplete rationality vs. incomplete strategy in minority game model. Adv Mater Res 629:735–740
794. Ye W, Fang ZG, Xu RT et al (2015) A quality detection GFTA-GERT model for trunk airplanes based on multi-core distributed coordination network. J Grey Syst 27(3):178–192
795. Ye X, Mackie P, Zhu LM (2011) A deadlock detection method for inter-organizational business process based on role network model. Adv Inf Sci Serv Sci 3(10):497–504
796. Yin SJ, Jing RT (2014) A schematic view of crisis threat assessment. J Contingencies Crisis Manag 22(2):97–107
797. Yu H, Liu HM (2013) Robust multiple objective game theory. J Optim Theory Appl 159(1):272–280
798. Yu H, Zhai J (2014) The distribution-free newsvendor problem with balking and penalties for balking and stockout. J Syst Sci Syst Eng 23(2):153–175
799. Yu K, Lin WP, Wang L et al (2016) The role of affective commitment and future work self salience in the abusive supervision-job performance relationship. J Occup Organ Psychol 89(1):28–45
800. Yu L, Yao X (2013) A total least squares proximal support vector classifier for credit risk evaluation. Soft Comput 17(4):643–650
801. Yu LA (2012) An evolutionary programming based asymmetric weighted least squares support vector machine ensemble learning methodology for software repository mining. Inf Sci 191:31–46

802. Yu LA (2014) Credit risk evaluation with a least squares fuzzy support vector machines classifier. Discr Dyn Nat Soc 564213
803. Yu LA, Dai W, Tang L (2016) A novel decomposition ensemble model with extended extreme learning machine for crude oil price forecasting. Eng Appl Artif Intell 47:110–121
804. Yu LA, Dai W, Tang L et al (2016) A hybrid grid-GA-based LSSVR learning paradigm for crude oil price forecasting. Neural Comput Appl 27(8):2193–2215
805. Yu LA, Lai KK (2011) A distance-based group decision-making methodology for multi-person multi-criteria emergency decision support. Decis Support Syst 51(2):307–315
806. Yu LA, Li JJ, Tang L (2015) Dynamic volatility spillover effect analysis between carbon market and crude oil market: a DCC-ICSS approach. Int J Glob Energy Issues 38(4–6):242–256
807. Yu LA, Li JJ, Tang L et al (2015) Linear and nonlinear Granger causality investigation between carbon market and crude oil market: a multi-scale approach. Energy Econ 51:300–311
808. Yu LA, Wang SY, Lai KK (2010) Developing an SVM-based ensemble learning system for customer risk identification collaborating with customer relationship management. Front Comput Sci China 4(2):196–203
809. Yu LA, Wang SY, Lai KK et al (2010) A multiscale neural network learning paradigm for financial crisis forecasting. Neurocomputing 73(4–6):716–725
810. Yu LA, Wang SY, Wen FH et al (2012) Genetic algorithm-based multi-criteria project portfolio selection. Ann Oper Res 197(1):71–86
811. Yu LA, Wang ZS, Tang L (2015) A decomposition-ensemble model with data-characteristic-driven reconstruction for crude oil price forecasting. Appl Energy 156:251–267
812. Yu LA, Yao X, Wang SY et al (2011) Credit risk evaluation using a weighted least squares SVM classifier with design of experiment for parameter selection. Expert Syst Appl 38(12):15392–15399
813. Yu LA, Zhao Y, Tang L (2014) A compressed sensing based AI learning paradigm for crude oil price forecasting. Energy Econ 46:236–245
814. Yu LA, Zheng J, Shen WC et al (2012) BC-PDM: data mining, social network analysis and text mining system based on cloud computing. In: Proceedings of the 18th ACM SIGKDD international conference on knowledge discovery and data mining, pp 1496–1499
815. Yu MZ, Tian X, Yu LA (2014) Pricing scheme of ocean carrier for inbound container storage for assistance of container supply chain finance. Discr Dyn Nat Soc 216057
816. Yu WY, Li MN, Xue C et al (2016) Determinants and influencing mechanism of outpatient satisfaction: a survey on tertiary hospitals in the People's Republic of China. Patient Prefer Adherence 10:601–612
817. Yuan CQ, Liu SF, Fang ZG (2016) Comparison of China's primary energy consumption forecasting by using ARIMA (the autoregressive integrated moving average) model and GM (1,1) model. Energy 100:384–390
818. Yuan SC, Chun SA, Spinelli B et al (2017) Traffic evacuation simulation based on multi-level driving decision model. Transp Res Part C: Emerg Technol 78:129–149
819. Yuan WJ, Wang L (2016) Optimism and attributional style impact on the relationship between general insecurity and mental health. Personal Individ Differ 101:312–317
820. Zang LJ, Cao C, Cao YN et al (2013) A survey of commonsense knowledge acquisition. J Comput Sci Technol 28(4):689–719
821. Zang LJ, Wang WM, Chen B et al (2015) The double-level default description logic D3L. In: Zhang SM, Wirsing M, Zhang ZL (eds) Knowledge science, engineering and management. Springer, Cham, pp 141–146
822. Zang LJ, Wang WM, Wang Y et al (2015) A Chinese framework of semantic taxonomy and description: preliminary experimental evaluation using web information extraction. In: Zhang SM, Wirsing M, Zhang ZL (eds) Knowledge science, engineering and management. Springer, Cham, pp 275–286
823. Zeng B, Liu SF, Xie NM (2010) Prediction model of interval grey number based on DGM (1,1). J Syst Eng Electron 21(4):598–603

824. Zeng DJ, Chen H, Lusch R et al (2010) Social media analytics and intelligence. IEEE Intell Syst 25(6):13–16
825. Zeng DJ, Wei DH, Chau M et al (2011) Domain-specific Chinese word segmentation using suffix tree and mutual information. Inf Syst Front 13(1):115–125
826. Zeng WS, Zeng A, Liu H et al (2014) Similarity from multi-dimensional scaling: solving the accuracy and diversity dilemma in information filtering. PLoS ONE 9(10):e111005
827. Zeng WS, Zeng A, Liu H et al (2014) Uncovering the information core in recommender systems. Sci Rep 4:6140
828. Zhan J, Sha YZ, Yan JS (2012) Design and implementation of logistics vehicle monitoring system based on the SaaS model. In: 5th international conference on business intelligence and financial engineering, pp 524–526
829. Zhan SL, Liu N (2016) Determining the optimal decision time of relief allocation in response to disaster via relief demand updates. Int J Syst Sci 47(3):509–520
830. Zhan SL, Liu N, Ye Y (2014) Coordinating efficiency and equity in disaster relief logistics via information updates. Int J Syst Sci 45(8):1607–1621
831. Zhang BS, Li XY, Wang SY (2015) A novel case adaptation method based on an improved integrated genetic algorithm for power grid wind disaster emergencies. Expert Syst Appl 42(21):7812–7824
832. Zhang FS, Yang ZT, Zhong SB et al (2016) Exploring mean annual precipitation values (2003–2012) in a specific area (36 °N–43 °N, 113 °E–120 °E) using meteorological, elevational, and the nearest distance to coastline variables. Adv Meteorol 2107908
833. Zhang FS, Zhong SB, Huang QY et al (2014) Pre-evaluation of contingency plans for meteorological disasters based on LINMAP method. In: Wen ZK, Li TR (eds) Foundations of intelligent systems. Springer, Berlin, Heidelberg
834. Zhang FS, Zhong SB, Sun C et al (2014) Research scheme on pre-assessment theory and method for influences of disastrous meteorological events. Int J Adv Manag Sci 3(4):117–122
835. Zhang FS, Zhong SB, Sun C et al (2015) Ontology-based modeling and reasoning framework for disastrous meteorological events. J Comput Inf Syst 11(11):3867–3873
836. Zhang FS, Zhong SB, Yang ZT et al (2015) Spatial estimation of mean annual precipitation (1951–2012) in Mainland of China based on collaborative Kriging interpolation. In: Bian FL, Xie YC (eds) Geo-informatics in resource management and sustainable ecosystem. Springer, Berlin, Heidelberg, pp 663–672
837. Zhang FS, Zhong SB, Yang ZT et al (2016) Spatial estimation of losses attributable to meteorological disasters in a specific area (105.0 °E–115.0 °E, 25 °N–35 °N) using Bayesian maximum entropy and partial least squares regression. Adv Meteorol 1547526
838. Zhang FS, Zhong SB, Yao SM et al (2016) Ontology-based representation of meteorological disaster system and its application in emergency management: illustration with a simulation case study of comprehensive risk assessment. Kybernetes 45(5):798–814
839. Zhang HF, Yang ZM, Wu ZX et al (2013) Braess's paradox in epidemic game: better condition results in less payoff. Sci Rep 3:3292
840. Zhang HF, Zhang J, Li P et al (2011) Risk estimation of infectious diseases determines the effectiveness of the control strategy. Phys D: Nonlinear Phenomena 240(11):943–948
841. Zhang HF, Zhang J, Zhou CS et al (2010) Hub nodes inhibit the outbreak of epidemic under voluntary vaccination. New J Phys 12(2):023015
842. Zhang HT, Wang N, Chen MZQ et al (2010) Spatially quantifying the leadership effectiveness in collective behavior. New J Phys 12(12):123025
843. Zhang JH, Li T, Wang C (2011) Research on agile mobilization mode from the perspective of civil-military integration. Acta Armamentarii 32(SUPPL. 1):15–19
844. Zhang JH, Li J, Liu ZP (2012) Multiple-resource and multiple-depot emergency response problem considering secondary disasters. Expert Syst Appl 39(12):11066–11071
845. Zhang J, Li Q, Luo XF et al (2013) A multi-level text representation model within background knowledge based on human cognitive process. 12th IEEE international conference on cognitive informatics and cognitive computing (ICCI*CC 2013), pp 324–331

846. Zhang J, Shen SF, Yang R (2010) Asymmetric information in combating terrorism: is the threat just a bluff? Tsinghua Sci Technol 15(5):604–612
847. Zhang J, Shen SF, Yang R (2010) The impacts of adaptive attacking and defending strategies on mitigation of intentional threats. Kybernetes 39(5):825–837
848. Zhang JH, Wang C (2013) System dynamics simulation of the civil-military integrated innovation system of military enterprises. J Beijing Inst Technol (English Edn) 1:106–113
849. Zhang JH, Zhang QQ (2013) Research on optimization models about production flexibility of China refined oil supply chain under normal conditions. J Converg Inf Technol 8(10):887
850. Zhang K, Ye W, Zhao LP (2012) The absolute degree of grey incidence for grey sequence base on standard grey interval number operation. Kybernetes 41(7–8):934–944
851. Zhang L K, Li L, He Z et al (2013) Improving Chinese word segmentation on micro-blog using rich punctuations. In: Proceedings of the 51st annual meeting of the association for computational linguistics. Short papers, vol 2, pp 177–182
852. Zhang LL, Liu X, Li YP et al (2012) Emergency medical rescue efforts after a major earthquake: lessons from the 2008 Wenchuan earthquake. The Lancet 379(9818):853–861
853. Zhang LL (2016) Modeling the injury flow following the major earthquake in China. Springer, Dordrecht
854. Zhang LM, Jia Y, Zhu X et al (2014) User-level sentiment evolution analysis in microblog. China Commun 11(12):152–163
855. Zhang M, Huang J, Zhu JM (2011) Facility location problem considering facility failure scenario. Syst Eng Procedia 87–94
856. Zhang M, Huang J, Zhu JM (2012) Reliable facility location problem considering facility failure scenarios. Kybernetes 41(10):1440–1461
857. Zhang P, Chen B, Ma L et al (2015) The large scale machine learning in an artificial society: prediction of the Ebola outbreak in Beijing. Comput Intell Neurosci 531650
858. Zhang P, Deng Q, Liu XD et al (2017) Emergency-oriented spatial temporal trajectory pattern recognition by intelligent sensor devices. IEEE Access 5:3687–3697
859. Zhang P, Yang R, Liu XD et al (2016) A GIS-based urban vulnerability and emergency response research after an earthquake disaster. In: Proceedings of the 2nd ACM SIGSPATIAL international workshop on the use of GIS in emergency management (EM-GIS 2016), pp 1–5
860. Zhang SX, Luo XF, Xuan JY et al (2014) Discovering small-world in association link networks for association learning. World Wide Web 17(2):229–254
861. Zhang WD, Zu ZH, Xu Q et al (2014) Optimized strategy for the control and prevention of newly emerging influenza revealed by the spread dynamics model. PLoS ONE 9(1):e84694
862. Zhang WS, Ding GG, Chen L et al (2013) Generating virtual ratings from Chinese reviews to augment online recommendations. ACM Trans Intell Syst Technol (TIST 2013) 4(1):9
863. Zhang X, Guan NY, Jia ZQ et al (2015) Semi-supervised projective non-negative matrix factorization for cancer classification. PLoS ONE 10(9):e0138814
864. Zhang X, Guan NY, Tao DC et al (2014) Online multi-modal robust non-negative dictionary learning for visual tracking. PLoS ONE 10(5):e0124685
865. Zhang XL, Su GF, Chen JG et al (2015) Iterative ensemble Kalman filter for atmospheric dispersion in nuclear accidents: An application to Kincaid tracer experiment. J Hazard Mater 297:329–339
866. Zhang XL, Su GF, Yuan HY et al (2014) Modified ensemble Kalman filter for nuclear accident atmospheric dispersion: prediction improved and source estimated. J Hazard Mater 280:143–155
867. Zhang XL, Weng WG, Yuan HY et al (2013) Empirical study of a unidirectional dense crowd during a real mass event. Phys A: Stat Mech Appl 392(12):2781–2791
868. Zhang XM, Jiang K, Wang HL et al (2012) An improved bean optimization algorithm for solving TSP. In: Tan Y, Shi Y H, Ji Z (eds) Advances in Swarm intelligence. Springer, Berlin, Heidelberg, pp 261–267
869. Zhang Y (2016) Functional diversity and group creativity: the role of group longevity. J Appl Behav Sci 52(1):97–123

870. Zhang Y, Huai MY (2016) Diverse work groups and employee performance: the role of communication ties. Small Group Res 47(1):28–57
871. Zhang Y, Waldman DA, Han YL et al (2015) Paradoxical leader behaviors in people management: antecedents and consequences. Acad Manag J 58(2):538–566
872. Zhang YH, Zhang WS, Xie Y (2013) Improved heuristic equivalent search algorithm based on maximal information coefficient for Bayesian network structure learning. Neurocomputing 117:186–195
873. Zhang YM, Jia QS (2017) Operational optimization for microgrid of buildings with distributed solar power and battery. Asian J Control 19(3):996–1008
874. Zhang YW, Qi JY, Fang BX et al (2011) The indicator system based on BP neural network model for net-mediated public opinion on unexpected emergency. China Commun 8(2):42–51
875. Zhang Z, Li X (2013) The optimal manufacturer's reserve investment and government's subsidy policy in emergency preparedness. J Inequalities Appl 2013(1):62
876. Zhang Z, Wang M, Shi JQ (2012) Leader-follower congruence in proactive personality and work outcomes: the mediating role of leader-member exchange. Acad Manag J 55(1):111–130
877. Zhang ZK, Zhou T, Zhang YC (2011) Tag-aware recommender systems: a state-of-the-art survey. J Comput Sci Technol 26(5):767
878. Zhang ZL, Li XY, Li HY (2015) A quantitative approach for assessing the critical nodal and linear elements of a railway infrastructure. Int J Crit Infrastruct Prot 8:3–15
879. Zhao B, Zhang WS, Liu J et al (2011) High-order MRF prior based Bayesian deblurring. Int J Digit Content Technol Appl 5(12):383–392
880. Zhao DT, Wang F, Wei JC et al (2013) Public reaction to information release for crisis discourse by organization: integration of online comments. Int J Inf Manag 33(3):485–495
881. Zhao HZ, Yan N, Zhang RX et al (2013) Optimization of emergency management and rescue command decision mode in coal mine. Disaster Adv 6:149–155
882. Zhao LJ, Cheng JJ, Qian Y et al (2012) USEIRS model for the contagion of individual aggressive behavior under emergencies. Simulation 88(12):1456–1464
883. Zhao LJ, Cui HX, Qiu XY et al (2013) SIR rumor spreading model in the new media age. Phys A: Stat Mech Appl 392(4):995–1003
884. Zhao LJ, Huang W (2014) Models for identifying significant environmental factors associated with cyanobacterial bloom occurrence and for predicting cyanobacterial blooms. J Great Lakes Res 40(2):265–273
885. Zhao LJ, Huang W, Gao HO et al (2014) A cooperative approach to reduce water pollution abatement cost in an interjurisdictional lake basin. J Am Water Resour Assoc 50(3):777–790
886. Zhao LJ, Li CM, Huang RB et al (2013) Harmonizing model with transfer tax on water pollution across regional boundaries in a China's lake basin. Eur J Oper Res 225(2):377–382
887. Zhao LJ, Qian Y, Huang RB et al (2012) Model of transfer tax on transboundary water pollution in China's river basin. Oper Res Lett 40(3):218–222
888. Zhao LJ, Qiu XY, Wang XL et al (2013) Rumor spreading model considering forgetting and remembering mechanisms in inhomogeneous networks. Phys A: Stat Mech Appl 392(4):987–994
889. Zhao LJ, Wang JJ, Chen YC et al (2012) SIHR rumor spreading model in social networks. Phys A: Stat Mech Appl 391(7):2444–2453
890. Zhao LJ, Wang JJ, Huang RB (2015) Immunization against the spread of rumors in homogenous networks. PLoS ONE 10(5):e0124978
891. Zhao LJ, Wang JJ, Huang RB et al (2014) Sentiment contagion in complex networks. Phys A: Stat Mech Appl 394:17–23
892. Zhao LJ, Wang Q, Cheng JJ et al (2011) Rumor spreading model with consideration of forgetting mechanism: a case of online blogging live journal. Phys A: Stat Mech Appl 390(13):2619–2625
893. Zhao LJ, Wang Q, Cheng JJ et al (2012) The impact of authorities' media and rumor dissemination on the evolution of emergency. Phys A: Stat Mech Appl 391(15):3978–3987
894. Zhao LJ, Wang XL, Qian Y (2012) Analysis of factors that influence hazardous material transportation accidents based on Bayesian networks: a case study in China. Saf Sci 50(4):1049–1055

895. Zhao LJ, Wang XL, Qiu XY et al (2013) A model for the spread of rumors in Barrat-Barthelemy-Vespignani (BBV) networks. Phys A: Stat Mech Appl 392(21):5542–5551
896. Zhao LJ, Wang XL, Wang JJ et al (2014) Rumor-propagation model with consideration of refutation mechanism in homogeneous social networks. Dyn Nat Soc 659273
897. Zhao LJ, Xie WL, Gao HO et al (2013) A rumor spreading model with variable forgetting rate. Phys A: Stat Mech Appl 392(23):6146–6154
898. Zhao QL, Su GF, Chen JG et al (2011) Crisis management for oil spill at sea in Hainan province. In: Proceedings of international conference on information systems for crisis response and management (ISCRAM 2011), pp 66–69
899. Zhao QS, Han LD, Luo NX (2018) A proposed semi-quantitative framework for comprehensive risk assessment of urban hazard installations considering rescue accessibility and evacuation vulnerability. Saf Sci 110:192–203
900. Zhao YJ, Chen X, Jia QS et al (2010) Long-term scheduling for cascaded hydro energy systems with annual water consumption and release constraints. IEEE Trans Autom Sci Eng 7(4):969–976
901. Zhao ZD, Cai SM, Huang JM et al (2012) Scaling behavior of online human activity. Europhys Lett 100(4):48004
902. Zhao ZD, Liu Y, Tang M (2012) Epidemic variability in hierarchical geographical networks with human activity patterns. Chaos: Interdiscip J Nonlinear Sci 22(2):023150
903. Zhao ZD, Xia H, Shang MS et al (2011) Empirical analysis on the human dynamics of a large-scale short message communication system. Chin Phys Lett 28(6):068901
904. Zheng LL, Chai H, Chen WZ et al (2011) Recognition of facial emotion and perceived parental bonding styles in healthy volunteers and personality disorder patients. Psychiatry Clin Neurosci 65(7):648–654
905. Zheng XL, Zhong Y, Zeng D et al (2012) Social influence and spread dynamics in social networks. Front Comput Sci 6(5):611–620
906. Zhong L, Luo Q, Wen D et al (2013) A task assignment algorithm for multiple aerial vehicles to attack targets with dynamic values. IEEE Trans Intell Transp Syst 14(1):236–248
907. Zhong MH, Shi CL, He L et al (2011) Full-scale experimental research on fire fume refluence of sloped long and large curved tunnel. Sci China Technol Sci 54(1):89–94
908. Zhong SB, Liu Y, Wang F et al (2013) Study on the application of GIS in comprehensive risk assessment of hazardous chemical plants. In: Bian FL, Xie YC, Cui XH et al (eds) Geo-informatics in resource management and sustainable ecosystem. Springer, Berlin, Heidelberg, pp 160–170
909. Zhong SB, Su GF, Wang F et al (2013) A preliminary research on incident chain modeling and analysis. In: Bian FL, Xie YC, Cui XH et al (eds) Geo-informatics in resource management and sustainable ecosystem. Springer, Berlin, Heidelberg, pp 171–180
910. Zhou B, Qin SJ, Han XP et al (2014) A model of two-way selection system for human behavior. PLoS ONE 9(1):e81424
911. Zhou L, Wei JC, Zhao DT (2014) Detecting the impacts of socioeconomic factors on regional severity of work-related casualties in China. Hum Ecol Risk Assess Int J 20(6):1469–1490
912. Zhou T, Medo M, Cimini G et al (2011) Emergence of scale-free leadership structure in social recommender systems. PLoS ONE 6(7):e20648
913. Zhou X, Chen B, Liu L et al (2015) An opinion interactive model based on individual persuasiveness. Comput Intell Neurosci 345160
914. Zhou XY, He LN, Yang Q et al (2012) Control deprivation and styles of thinking. J Pers Soc Psychol 102(3):460–478
915. Zhou XY, Wildschut T, Sedikides C et al (2012) Heartwarming memories: nostalgia maintains physiological comfort. Emotion 12(4):678–684
916. Zhou YC, Tang MJ, Pan WK et al (2014) Bird flu outbreak prediction via satellite tracking. IEEE Intell Syst 29(4):10–17
917. Zhu DQ, Xie XF, Gan YQ (2011) Information source and valence: How information credibility influences earthquake risk perception. J Environ Psychol 31(2):129–136

918. Zhu DQ, Xie XF, Xie JQ (2012) When do people feel more risk? The effect of ambiguity tolerance and message source on purchasing intention of earthquake insurance. J Risk Res 15(8):951–965
919. Zhu JJ, Hipel KW (2012) Multiple stages grey target decision making method with incomplete weight based on multi-granularity linguistic label. Inf Sci 212:15–32
920. Zhu JM, Gao MG, Huang J (2009) A robust approach to vehicle routing for medical supplies in large-scale emergencies. In: International symposium on engineering management (ISEM 2009). Scientific Research Publishing, Beijing, pp 535–538
921. Zhu JM (2012) Supply allocation and vehicle routing problem with multiple depots in large-scale emergencies. In: Eksioglu B (ed) Emergency management. InTech, Rijeka, pp 75–90
922. Zhu JM, Huang J, Liu DG et al (2012) Connected distribution center location problem under traffic network in emergency management. In: 5th international joint conference on computational sciences and optimization (CSO 2012), pp 160–163
923. Zhu JP, Jia JR, Wu J et al (2013) A modified model of direct estimation method for fine fuel moisture content prediction by considering crown density. Procedia Eng 62:1015–1022
924. Zhu JP, Wang Y, Dong XX et al (2013) Real-time evacuation path: unexpectedly increased numbers of evacuees in source nodes. Adv Mater Res 779–780:6
925. Zhu L, Guo XY, Li JQ et al (2012) Hippocampal activity is associated with self-descriptiveness effect in memory, whereas self-reference effect in memory depends on medial prefrontal activity. Hippocampus 22(7):1540–1552
926. Zhu L, Guo XY, Zheng L et al (2010) Graded contribution of hippocampus to multifeature binding across temporal delay. NeuroReport 21(13):902–906
927. Zhu Q, He KJ, Zou YC et al (2014) Day-ahead crude oil price forecasting using a novel morphological component analysis based model. Sci World J 341734
928. Zhu T, Wang B, Wu B et al (2011) Role defining using behavior-based clustering in telecommunication network. Expert Syst Appl 38(4):3902–3908
929. Zhu T, Wang B, Wu B et al (2012) Topic correlation and individual influence analysis in online forums. Expert Syst Appl 39(4):4222–4232
930. Zhu T, Wu B, Wang B et al (2010) Community structure and role analysis in biological networks. J Biomol Struct Dyn 27(5):573–579
931. Zhu TG, Zhou YC, Zhang Y et al (2013) An efficient strategy of building distributed index based on Lucene. In: Gao YJ, Shim K, Ding ZM et al (eds) Web-age information management. Springer, Berlin, Heidelberg, pp 40–45
932. Zhuang Q, Wang D, Fan Y et al (2012) Evolution of cooperation in a heterogeneous population with influential individuals. Phys A: Stat Mech Appl 391(4):1735–1741
933. Zou YC, Yu LA, He KJ (2015) Estimating portfolio value at risk in the electricity markets using an entropy optimized BEMD approach. Entropy 17(7):4519–4532
934. Zou YC, Yu LA, He KJ (2015) Wavelet entropy based analysis and forecasting of crude oil price dynamics. Entropy 17(10):7167–7184

Index

W. Fan (ed.), *Unconventional Emergency Management Research*, Reports of China's Basic Research, https://doi.org/10.1007/978-981-99-6798-8

MIX
Papier aus verantwortungsvollen Quellen
Paper from responsible sources
FSC® C105338
FSC
www.fsc.org

If you have any concerns about our products,
you can contact us on
ProductSafety@springernature.com

In case Publisher is established outside the EU,
the EU authorized representative is:
Springer Nature Customer Service Center GmbH
Europaplatz 3, 69115 Heidelberg, Germany

Printed by Libri Plureos GmbH
in Hamburg, Germany